advancing learning, changing lives

# Edexcel GCSE
# Geography B
## Evolving Planet

### Student Book

**David Flint • Lindsay Frost • Simon Oakes • Andy Palmer**
**Michael Witherick • Sue Warn • Nigel Yates**

Published by Pearson Education Limited, Edinburgh Gate, Harlow, Essex, CM20 2JE.

www.pearsonschoolsandfecolleges.co.uk

Copies of official specifications for all Edexcel qualifications may be found on the Edexcel website: www.edexcel.com

Edited by Graham Bradbury
Designed by HL Studios, Witney, Oxon
Illustrated by HL Studios, Witney, Oxon
Printed and bound by

The rights of David Flint, Lindsay Frost, Simon Oakes, Andy Palmer, Michael Witherick, Phil Wood and Nigel Yates to be identified as authors of this work have been asserted by them in accordance with the Copyright, Designs and Patents Act 1988.

First published 2013

16 15 14 13

10 9 8 7 6 5 4 3 2

British Library Cataloguing in Publication Data

A catalogue record for this book is available from the British Library

ISBN 9781446905814

Acknowledgements
The publisher would like to thank the following for their kind permission to reproduce their photographs:

(Key: b–bottom; c–centre; l–left; r–right; t–top)

**Alamy Images:** A.P.S. (UK) 222b, AfriPics.com 93, Allstar Picture Library 252, Bailey-Cooper Photography 251, Brett Baunton 122b, blinkwinkel 122t, brinkstock 235, Christina Bollen 232t, 232b, Thomas Cockrem 7b, 147, Brandon Cole Marine Photography 114, Colin Underhill 224, 225, Corbis Nomad 272, Craig Holmes Premium 234, David Bagnall 215, David Ball 229b, David Burton 216, dbimages 255, deco 279, dmark 90r, Chad Ehlers 5t, Chad Ehlers 5t, Julio Etchart 97, Greg Balfour Evans 261t, Frans Lemmens 120, Gary Roebuck 238, geogphotos 79b, Chris Gomersall 53, Melvin Green 84, Images of Africa Photobank 263, Images&Stories 21, itdarbs 233, Jacques Jangoux 127, James Kubrick 133, Les Gibbon 72b, Leslie Garland Picture Library 79t, Barry Lewis 146, Mark Bassett 265, Mark Boulton 275, Michelle and Tom Grimm 244b, Gianni Muratore 150, Chris Pancewicz 91, Andrew Parker 260r, Pearl Bucknell 239, Purple Marbles 100, Robert Harding World Imagery 285tl, Sara Blancett 137, Skyscan Photolibrary 76, Doug Steley B 130, Homer Sykes Archive 149, Thomas Cockrem 7b, 147, Thomas Lee 213 (Flats), Tim Gartside London 227, Tom Gilks 171t, Tom Hanley 266, Worldwide Picture Library 47t, 47b, Ariadne Van Zandbergen 270; **Andrew Stacey:** 83; **Andy Palmer:** 73, 85; **Bridgeman Art Library Ltd:** Guildhall Art Gallery, City of London 31; Corbis: Bennett Dean / Eye Ubiquitous 260l, Carlos Cazalis 253, Du Huaju / Xinhua Press 66t, 66b, Gary K Smith 230, Imagebroker / Martin Siepmann 274, Imagine China 180, Jana Renee Cruder 249, Jason Hawkes 74, Jim Sugar 19b, Jose Fuste Raga 126, Kendra Luck / San Francisco Chronicle 132, Kenty Kobertsteen / National Geographic 254, Konrad Wothe / Minden Pictures 264, Matt Mawson 256, Michael T. Sedam 16, Olivier Pitras / Sygma 124, Paul A. Souders 179, Rickey Rogers / Reuters 48, Rudy Sulgan 59t, Skyscan 77, Strauss / Curtis 128, William Taufic 168, The Irish Image Collection 261b, Elizabeth Whiting & Associates

273r; **DK Images:** 33r, Kim Sayer 72t; **Ecoscene:** Simon Grove 7t, 105; **GeoScience Features Picture Library:** 131l, 131r; **Geoslides/ Geo Aerial Photography:** 90l; Getty Images: Art Wolfe / Stone 29 (Volcano), Banaras Khan / AFP 19t, Charles Thatcher / Stone 145, Christopher Furlong 207, Dan Kitwood 248, Daniel Berehulak 94, DEA / C.BEVILACQUA / De Agostini Picture Library 13b, Gandee Vasan / Photographer's Choice 4b, 231, Haywood Magee / Hulton Archive 148, Jason Edwards / National Geographic 61, Jyrki Komulainen / Gorilla Creative Images 273l, National Geographic / Mike Theiss 285br, Oli Scarf 243l, Paula Bronstein 247, Photofusion / Universal Images Group 222t, Ramzi Haidar / AFP 69, Randy Olson / National Geographic 125, Sam Panthaky / AFP 44, Simon Maina / AFP 268, Solar and Heliospheric Observatory 29 (Sunspot), Stone / Will & Deni McIntyre 285tr, The Image Bank / Bjorn Holland 285bl; **Imagestate Media:** Michael Duerinckx 246; **iStockphoto:** Andrew Howe 51 (Willow Warbler), Lubomir Jendrol 51 (Turtle Dove), Lurii Konoval 51 (Yellow Wagtail); **Lucia Oritiz:** 197; **NASA:** Goddard Space Flight Center. Scientific Visualization Studio 110t, 110b; **Natural History Museum Picture Library:** Michael Long 33l; **Pearson Education Ltd:** Lord and Leverett 229t; **Phil Wood:** 297, 309, 297, 309; **PhotoDisc:** 4t; **Photofusion Picture Library:** Dorothy Burrows 5b, 293, Dorothy Burrows 5b, 293; **Practical Action/ZUL:** Steve Fisher 201; **Press Association Images:** David Jones / PA Wire 208, Michael Stephens / PA Archive 209; **Reuters:** Phil Noble 113; **Rex Features:** Dobson Agency 78, Eye Ubiquitous 13t, 60, Geoff Robinson 218, Jane Sweeney / Robert Harding 62, Keystone USA-ZUMA 171b, Mood Board 172, Sipa Press 50, View Pictures 213 (Museum); **Science Photo Library Ltd:** David A. Hardy, Futures: 50 Years in Science 29 (Asteroid), Planet Observer 59b; **Shutterstock.com:** Francesco Carucci 237; **Simon Oakes:** 65, 112, 65, 112; **Skyscan Photolibrary:** 101; **SuperStock:** Radius 11; **Thinkstock:** 51 (Garden Warbler); **Veer/Corbis:** darrenbaker 160, enjoylife25 205, godricl 223, Johan Swanepoel 29 (Earth), Konstantin / Yolshin 224t.

All other images © Pearson Education

The websites used in this book were correct and up to date at the time of publication. It is essential for tutors to preview each website before using it in class so as to ensure that the URL is still accurate, relevant and appropriate. We suggest that tutors bookmark useful websites and consider enabling students to access them through the school/college intranet.

**A note from the publisher**
In order to ensure that this resource offers high-quality support for the associated Edexcel qualification, it has been through a review process by the awarding organisation to confirm that it fully covers the teaching and learning content of the specification or part of a specification at which it is aimed, and demonstrates an appropriate balance between the development of subject skills, knowledge and understanding, in addition to preparation for assessment.

While the publishers have made every attempt to ensure that advice on the qualification and its assessment is accurate, the official specification and associated assessment guidance materials are the only authoritative source of information and should always be referred to for definitive guidance.

Edexcel examiners have not contributed to any sections in this resource relevant to examination papers for which they have responsibility.

No material from an endorsed Student Book will be used verbatim in any assessment set by Edexcel.

Endorsement of a Student Book does not mean that the Student Book is required to achieve this Edexcel qualification, nor does it mean that it is the only suitable material available to support the qualification, and any resource lists produced by the awarding organisation shall include this and other appropriate resources.

# Contents: delivering the Edexcel GCSE Geography B Evolving Planet specification

# Welcome to Edexcel GCSE Geography B Evolving Planet

Why should I choose GCSE Geography?

Because you will:

- learn about and understand the world that you live in
- develop skills that will help you in other subjects and your future career
- get to complete practical work away from the classroom
- learn how to work as a team
- learn by investigating, not just listening and reading.

## What will I learn?

You only have to switch on the news or pick up a newspaper to see that we live in a fast-pace, ever-changing world. GCSE Geography gives you the chance to learn about those changes: from those on your own doorstep to those of global proportions. There are four units:

## Unit 1: Dynamic planet

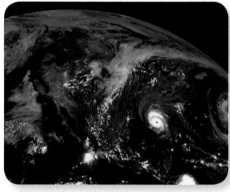

Have you ever wondered...

- How and why climate has changed in the past, and the impact this will have on the future?
- Why water is important to the health of the planet?
- Why conflict occurs on the coast and how these conflicts can be managed?
- What the challenges of extreme climates are?

In this unit you will get a chance to investigate geological processes, ecosystems, the atmosphere and climate, and the hydrological cycle. These topics are interlinked and, although you may study them separately, the unit is designed to show you how physical geography combines to create a 'life support system' for the planet.

## Unit 2: People and the planet

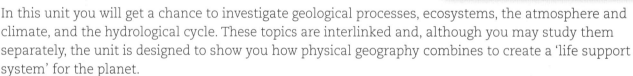

Have you ever wondered...

- How and why the population is changing in different parts of the world?
- What the ingredients of good living spaces are?
- What the environmental issues facing cities are?
- How countries might develop sustainably in the future?

This unit focuses on human geography. In a similar way to Unit 1, it links together to build an overall understanding of how humans interact with the planet. You will study how populations grow and change, where people live and work, and how they exploit and use resources.

There are also options in Units 1 and 2 so you will choose to study some topics in more depth such as rivers or coasts, cities or the countryside, development or economic geography, and oceans or extreme climates.

## Unit 3: Making geographical decisions

This is a decision-making exercise, where you get to stand in the shoes of a real geographer. You will study a specific topic, such as Antarctica, in detail. This is designed to teach you how to make decisions about a specific topic, based on the evidence studied. The skills you will learn in this topic will be valuable in all aspects of this GCSE in Geography, and in the rest of your life.

## Unit 4: Researching geography

This is the unit where you can really get stuck in! It will involve undertaking research, carrying out fieldwork and then writing it up. The research and fieldwork can be undertaken out of class, but the writing up will all be in class time. This means you have to spend less time at home doing your geography coursework!

## How will I be assessed?

- Higher and Foundation examination papers are available.
- Units 1 and 2 exams are resource based. You will have a booklet containing maps, photographs and diagrams to help you answer the questions.
- Units 1 and 2 exam questions will range from short questions up to larger extended-writing questions.
- Unit 3 is a decision-making exercise based on a resource booklet that you will receive in the exam which will outline a geographical problem. Questions will assess your understanding of the resources in relation to sustainable and environmental issues.
- Unit 4 is the controlled assessment unit. You will complete fieldwork and data collection for this unit, and analyse and write up your results in class.

# About this book

**Objectives** provide a **clear overview** of what you will learn in the section. Objectives increase in difficulty from ● to ◉

Clear and accessible diagrams **highlight key concepts and enable skills practice**.

**Results Plus** features help you to understand how to improve, with guidance on answering exam-style questions.

Top Tip These provide handy hints on how to apply what you have learned and how to remember key information.

**Top Tip**

**Watch Out!**

Watch Out! These warn you about common mistakes and misconceptions that students often make. Make sure that you don't repeat them!

## Activity 2

(a) Use the Internet to find another location that suffers from coastal flooding.

(b) Research how this area tries to predict or prevent coastal flooding.

**Activities** provide extra **support** to ensure understanding and opportunities to **stretch** your knowledge.

# Chapter 7 Oceans on the edge

## Objectives

● Describe the impacts of human activities on marine ecosystems such as mangrove forest.

◉ Explain how marine food webs work, and why they become damaged.

◉ Understand that climate change brings new and often unpredictable stresses to oceans.

### Results Plus
**Build Better Answers**

**EXAM-STYLE QUESTION**

**Study the distribution of mangrove swamp shown in Figure 2. Identify the main populated areas where mangrove swamp grows. (3 marks)**

■ **Basic answers** (0–1 marks)
Name only one or two areas, or are very imprecise, writing things like 'there is a lot on the right hand side of the map'.

● **Good answers** (2 marks)
Correctly identify Asia and Central America as named locations.

▲ **Excellent answers** (3 marks)
Also provide more specific details of Asian locations (naming the west coast of India and Indonesia) or mentioning north-west and south-east Africa.

## How and why are some ecosystems threatened with destruction?

The term ecosystem describes a grouping of plants and animals that is linked with its local physical environment – through use of soil nutrients, for example. The oceans, covering two-thirds of our planet, are home to distinctive **marine ecosystem** communities composed of fish, aquatic plants and sea birds – as well as tiny but very important organisms such as krill and plankton.

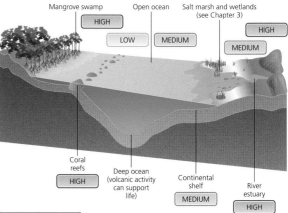

*Figure 1: Biodiversity in the oceans' ecosystems*

**Coral reef** ecosystems have an incredible level of plant and animal species). Although they cover less than 1% of the Earth's surface, they are home to 25% of all marine fish species (see Figure 12, page 114). Biodiversity is also high in waters close to the edges of the world land masses – above the **continental shelf**. There, species enjoy shallow warm water, enriched with silt nutrients from river **estuaries**. Even in the deep ocean, where light cannot penetrate, unique ecosystems are found. Underwater volcanic activity can create densely populated sites of plant and animal life. Here, often at great depth, life has evolved that can survive truly extreme conditions of heat and pressure.

**Key terms** are highlighted in the text, summarised at the end of each chapter, and are detailed in full in the glossary at the end of the book to enable you to **develop your geographical language**.

## The global pattern of mangrove swamps

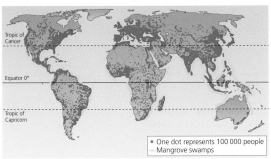

*Figure 2: The global pattern of mangrove swamp and populated areas*

- One dot represents 100 000 people
- Mangrove swamps

Mangroves are areas of swampy forest, originally found in estuaries and along marine shorelines in around 120 tropical and subtropical countries. Many are densely populated, especially in coastal Asia (Figure 2). Mangrove plants have evolved to tolerate daily tidal flooding and high salinity. Long twisting roots anchor the trees against a constantly ebbing and flowing tide (Figure 3). The roots trap mud, making a **habitat** for lobster and prawn.

### Global threats to mangrove swamps

These species used to be fished in **sustainable** ways that preserved the **mangrove swamp**. However, prawns can be more intensively harvested from ponds dug in mud on cleared land. Global demand for prawns has caused widespread removal of mangrove vegetation. Mangrove swamp naturally covered 200,000 km² of the Earth's surface but only half that area now remains, often because of this prawn **aquaculture**. Vietnam's Mekong Delta is a typical site where mangrove trees have been removed, leaving a flat muddy plain studded with blue plastic-lined ponds. Other major areas of loss include Thailand (half lost since 1960), the Philippines and Ecuador.

Mangrove removal creates problems of coastal erosion and loss of nursery grounds for fish. Crocodiles, snakes, tigers, deer, otters, dolphins and birds all lose an important habitat. Carbon dioxide stored over centuries in the rich mud beneath the swamp is released when the trees are removed.

Mangrove swamp is also nature's defence against tsunamis. The enormous ocean wave that struck the coast of south-east Asia in 2004 killed 230,000 people. Where they were still in existence, mangrove trees shielded lives and property. In Thailand, recent conversion of mangrove habitat into prawn farms and tourist resorts close to Phuket contributed significantly to the catastrophic losses experienced there.

Prawn aquaculture can also cause serious **pollution** problems. Antibiotics and pesticides used in the prawn pools frequently leak into the delicate ecosystem of neighbouring areas.

*Figure 3: Mangrove forest*

Engaging photos **bring geography to life**.

**Quick notes (Mangrove swamp):**
- Because of commercial pressures, **unsustainable** exploitation now takes place, such as removal to create space for aquaculture.
- Of 200,000 km² originally, only 100,000 km² now remain.

**Quick notes** pull out the key information in examples and case studies for **quick revision** reference.

Skills Builder exercises will **develop your geographical skills** and understanding in a specific topic.

## Skills Builder 1

Study Figure 3.

(a) Name one helps the environme

(b) Describe mangrove activities.

### Case Study: Singapore's 'Have three or more' policy

Since the mid-1960s, the Singapore government has controlled the size of its population. First, it wanted to reduce the rate of population growth, because it was worried that the small island would soon become overpopulated. This policy was so successful that in the mid-1980s the government was forced to completely reverse the policy. The old family planning slogan of 'Stop at two' was replaced by 'Have three or more – if you can afford it'. Instead of penalising couples for having more than two children, they now introduced a whole new set of incentives to encourage them to do just that. These include:

- Tax rebates for the third child and subsequent children
- Cheap nurseries
- Preferential access to the best schools
- Spacious apartments.

Pregnant women are offered special counselling to discourage 'abortions of convenience' or sterilisation after the birth of one or two children.

**Case study quick notes:**
- Governments are able to control population numbers in a variety of ways.
- Control is usually achieved by a 'stick and carrot' approach.

Real-life case studies show the **theory in practice!** Each case study includes a set of quick notes that pull out the key points.

examzone

Throughout this student book you will find a dedicated suite of revision resources.

We've broken down the six stages of revision to ensure that you are prepared every step of the way.

**Zone in:** How to get into the perfect 'zone' for your revision.

**Planning zone:** Tips and advice on how to effectively plan your revision.

**Know zone:** All the facts you need to know and exam-style practice at the end of every chapter.

**Chapter overview:** Outlines the key issue that the chapter examines. Keep this issue in mind as you work through the Know Zone pages.

**Key terms:** A matching exercise to ensure that you can **understand and apply important geographical terminology**.

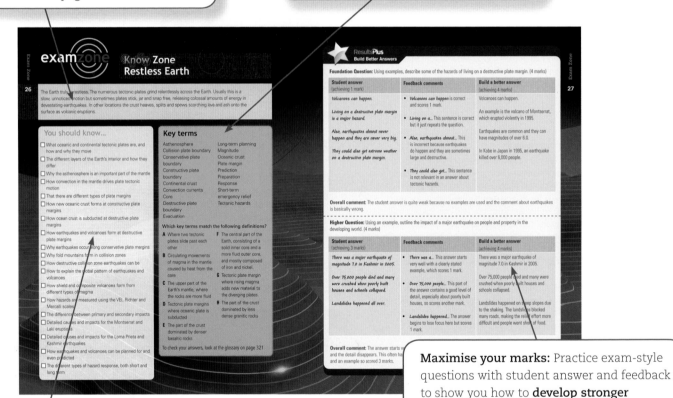

**You should know:** A check-yourself list of the concepts and facts that you should know before you sit the exam. Use this list to **identify your strengths and weaknesses** so you can plan your revision wisely.

**Maximise your marks:** Practice exam-style questions with student answer and feedback to show you how to **develop stronger answers** (see next page).

**Don't panic zone:** Last-minute revision tips for just before the exam.

**Exam zone:** Some exam-style questions for you to try, an explanation of the assessment objectives, plus a chance to see what a real exam paper might look like.

**Zone out:** What do you do after your exam? This section contains information on how to get your results and answers to frequently asked questions on what to do next.

# ResultsPlus

These features help you to understand how to improve, with guidance on answering exam-style questions. Some are based on the actual marks that students have achieved in past exams.

## ResultsPlus
### Exam Question Report

**REAL EXAM QUESTION**

Choose a stretch of coastline or coastal area you have studied where cliff recession is occurring or has occurred. Describe the effects of this cliff recession. (3 marks, June 2007)

**How students answered**

Most students answered this question poorly. They often only gave one effect when effects were required by the question. The question also required specific effects in the location studied rather than general effects of recession.

68% (0–1 marks)

Many students gave two effects, but still did not link these directly to the location that they had studied.

20% (2 marks)

Some students answered this question well. They gave at least two effects and referred to the studied location, using names of roads or buildings that had been affected by cliff recession.

12% (3 marks)

**Exam question report:** These show previous exam questions with details about how well students answered them.

- Red shows the number of students who scored low marks (less than 35% of the total marks)
- Orange shows the number of students who did okay (scoring between 35% and 70% of the total marks)
- Green shows the number of students who did well (scoring over 70% of the total marks).

## ResultsPlus
### Build Better Answers

**EXAM-STYLE QUESTION**

Suggest reasons why some countries are more successful than others in recycling waste. (4 marks)

■ **Basic answers** (0-1 marks)
Only give a description of differences in recycling, with no explanation.

● **Good answers** (2 marks)
Offer some explanatory statements relating to the role of government and/or level of economic development.

▲ **Excellent answers** (3-4 marks)
Provide full explanations of how some governments, such as Germany, encourage recycling by education, legislation and financial incentive.

**Build better answers** These give you an opportunity to answer some exam-style questions. They contain tips for what a basic ■, good ● and excellent ▲ answer will contain.

Exam questions are clearly flagged throughout this book to show you which are real exam questions

**REAL EXAM QUESTION**

and which are exam-style questions.

**EXAM-STYLE QUESTION**

## ResultsPlus
### Build Better Answers

**Foundation Question:** Study Figure 6 on page 17. Describe the site of Budleigh Salterton. Use map evidence in your answer. (3 marks)

| Student answer (awarded 1 mark) | Feedback comments | Build a better answer (awarded 3 marks) |
|---|---|---|
| It is on the sea. It is on a hill. It is south-west of Sidmouth. | • It is on the sea and It is on a hill are both vague and inaccurate but are enough for 1 mark together. Data from the map could be added.<br><br>• It is south-west of Sidmouth is incorrect because it refers to the *situation* instead of the *site*. | It has a settlement of about 2 km, some of it next to a beach and some on the cliffs.<br><br>The town spreads inland for a little over a kilometre, rising from sea level to about 50 metres spot height 48 metres at 063827.<br><br>The gentle slope on which it is built faces southwards. Some of the town appears to be built in a small valley as at Little Knowle, 058822. |

**Overall comment:** The student answer is basic. Most students can describe what they see on maps, but to score well you must use the data that the map offers when asked to – heights, distances and grid references are all useful in this answer.

# Unit 1 Dynamic planet

## Your course

This unit investigates geological processes, ecosystems, the atmosphere and climate, and the hydrological cycle. These topics are interlinked and show you how physical geography combines to create a 'life support system' for the planet. There are three sections:

**Section A** topics are compulsory and introduce you to the main areas of our planet: the geosphere, atmosphere, biosphere and hydrosphere. You will study **all** topics:

- Topic 1 (Chapter 1): Restless Earth
- Topic 2 (Chapter 2): Climate and change
- Topic 3 (Chapter 3): Battle for the biosphere
- Topic 4 (Chapter 4): Water world

**Section B** will cover how aspects of our planet work on a small scale and you will study **one** topic:

- Topic 5 (Chapter 5): Coastal change and conflict
- Topic 6 (Chapter 6): River processes and pressures

**Section C** will cover how aspects of our planet work on a large scale and you will study **one** topic:

- Topic 7 (Chapter 7): Oceans on the edge
- Topic 8 (Chapter 8): Extreme environments

## Your assessment

- You will sit a 1-hour 15-minute written exam worth a total of 78 marks. Up to 6 marks are available for spelling, punctuation and grammar.

- There will be a variety of question types: short answer, graphical and extended answer, which you will practice throughout the chapters that you study. You will answer **all** the questions in Section A, **one** question from Section B and **one** question from Section C.

- **Section A** contains questions on the compulsory topics.

- **Section B** contains questions on the two small-scale topics.

- **Section C** contains questions on the two large-scale topics.

**Remember** to answer the questions for the topics that you have studied in class!

---

Study the photograph of a desert in Australia.

(a) Explain why droughts may become more frequent in some countries.

(b) Describe two impacts of drought on the economy of a country.

# Chapter 1 Restless Earth

## Objectives

- Be able to identify the main features of the Earth's layered structure.

◉ Describe the distribution of earthquakes and volcanoes.

◎ Understand how the distribution of different types of volcanic activity and earthquakes relate to different types of margin.

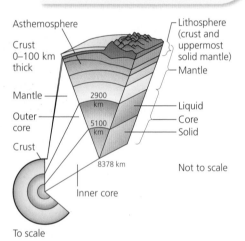

Figure 1: A cross-section view of the Earth

## Skills Builder 1

Study Figure 1. Calculate the thickness of the mantle.

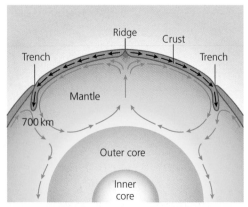

Figure 2: The convectional currents in the mantle

## How and why do the Earth's tectonic plates move?

### The structure of the Earth

The Earth is made up of a series of distinctive layers that are sometimes compared with onion skins (or even a Scotch egg). The solid crust is made up of relatively low density, allowing it to rest on the more flexible and fluid mantle. A real 'journey to the centre of the Earth' would not get very far because temperatures rise quite quickly, reaching 200–400°C at the boundary with the upper mantle.

As you can see in Figure 1, the Earth's crust is very thin compared to the diameter of the planet. The crust is made up of sections, called plates, ranging from the very small to the size of an entire continent. **Continental crust** is much thicker than **oceanic crust** but is made of material that is less dense than oceanic crust. The Earth's surface and its interior are both constantly moving, but you will never notice this movement – unless you are unlucky enough to be caught up in an earthquake.

Beneath the Earth's crust lies the mantle, the upper part of which is solid, just like the crust. The crust and this solid top layer of the mantle are together known as the 'lithosphere'. Below the lithosphere the sticky, viscous dense rock of the mantle slowly moves in great convectional currents (see Figure 2). This process is a bit like a pot of thick soup that is heated to boiling. The heated soup rises to the surface, spreads and begins to cool, and then sinks back to the bottom of the pot, where it reheats and rises again. In the upper part of the mantle – called the **asthenosphere** – the flows affect the lithosphere above, causing earthquakes and creating volcanoes.

Because of the heat and pressure that builds up beneath the surface, the crust is constantly being stressed, which breaks it up. These large-scale processes within the Earth's crust are known as plate tectonics. The processes were suspected for many years but they were only finally accepted as geological fact about fifty years ago. When two plates are being pulled away from each other, deep cracks are opened through the crust. This allows magma to rise to the surface and then, when it cools, it forms new crust in the shape of a ridge (Figure 2).

Plate size can vary greatly, from a few hundred to thousands of kilometres across. The Pacific and Antarctic Plates are two of the largest.

How do these massive slabs of solid rock remain on the Earth's surface? Why don't they sink into the mantle? The answer lies in the composition of the rocks. Continental crust is composed of granitic rocks which are made up of relatively low density minerals (Figure 3). By contrast, oceanic crust is composed of basaltic rocks, which are more dense (Figure 4). When a tectonic plate composed mostly of oceanic material meets a plate composed mostly of continental material, it is the denser oceanic plate that is forced downwards. The variations in plate thickness are nature's way of partly compensating for the imbalance in density of the two types of crust. Because continental rocks are less dense, the crust under the continents is much thicker (as much as 100 km) whereas the crust under the oceans is generally only about 7 km thick.

The **convection currents** in the mantle are themselves driven by the heat of the **core**. That heat is partly created by the pressure of overlying material but also by the radioactivity of the core material itself. As long as there is a temperature difference with depth, there will be a cycle of rising and sinking material. (A lava lamp is a perfect illustration of convection.)

Temperatures in the core are probably much the same as on the surface of the Sun. The other basic facts about the core are:

● It is mostly made up of iron and nickel

● Just over half the diameter of the Earth

● One-sixth of the volume of the Earth

● One-third of the Earth's mass

● The outer core is liquid

● The inner core is solid

● The currents in the outer core generate the Earth's magnetic field.

## The four types of boundary between tectonic plates

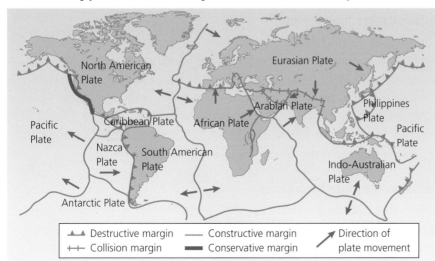

Figure 5: The Earth's tectonic plates and their boundaries

*Figure 3: Granite is made up of large crystals. The magma which formed these granitic rocks cooled slowly, which explains the very large crystals.*

*Figure 4: Basalt is fine-grained with no visible crystals. It ranges from black to dark grey in colour. Most basalts occur in lava flows and sheets.*

### Activity 1

Describe the differences between the rocks that dominate continents and those that dominate oceanic crust.

### Activity 2

1. State two differences between oceanic and continental crust.

2. Explain how tectonic plates move.

### Skills Builder 2

Study Figure 5.

1. Describe the distribution of plate boundaries.

2. Identify the type of plate boundary found between: (a) the Nazca Plate and the South American Plate; (b) the Arabian Plate and the Eurasian Plate; (c) the Eurasian Plate and the North American Plate.

## Activity 3

Study Figure 6.

(a) What is subduction?

(b) Explain why it is the oceanic plate that subducts.

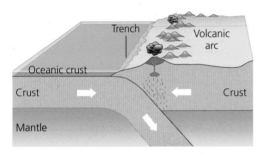

Figure 6: A destructive plate boundary

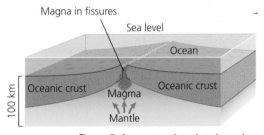

Figure 7: A constructive plate boundary

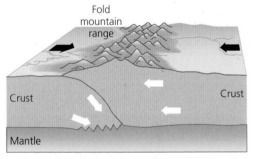

Figure 8: A collision plate boundary

## Activity 4

Study Figure 8.

(a) Explain why continental crust is thickest at collision boundaries.

(b) Explain why no volcanoes are found on this type of boundary.

There are four types of boundaries between tectonic plates – **destructive**, **constructive**, **collision** and **conservative**. Each of them is associated with characteristic tectonic events and landforms.

### Destructive plate boundaries

Destructive plate boundaries are found where two plates are moving together and oceanic plate material is destroyed (Figure 6). The boundary between the Nazca Plate and the South American Plate would be an example.

Destructive plate boundaries are associated with frequent earthquakes and volcanoes. The collision of the two plates buckles the leading edge of the continental plate forming fold mountains – and causing earthquakes. The denser basaltic oceanic plate is dragged downwards below the lighter granitic continental plate – a process known as 'subduction'. This creates an ocean trench and the melting of this material creates molten material (magma) as it is dragged deeper and deeper into the upper mantle. This magma rises through weaknesses in the overlying continental crustal material, some of which is inevitably melted by the rising magma, forming volcanoes on the surface. This is what has happened at the boundary between the Nazca and South American Plates.

### Constructive plate boundaries

Constructive plate boundaries (Figure 7) are found where new basaltic material rises to the surface, forcing plates apart (for example the mid-Atlantic boundary between the Eurasian and North American Plates).

Rising convection currents in the mantle cool and spread outwards as they near the surface. This pulls the crust apart and creates fissures and faults through which molten magma can reach the surface. The vast majority of this creation of new crust takes place in the oceans, forming a ridge and chains of submarine volcanoes. But sometimes, as in Iceland, these volcanoes reach the surface to form islands. It is believed that constructive plate boundaries evolve from rising magma splitting apart continental crust and creating 'new' oceans. This process is thought to be happening in the modern Red Sea and the African Rift valley.

### Collision plate boundaries

Collision plate boundaries are found where two continental plates move towards each other. Neither is destroyed but buckling takes place (Figure 8). An example would be between the Indo-Australian Plate and the Eurasian Plate. Earthquakes are very common (e.g. Pakistan 2005) but because no material is being subducted and melted no volcanoes are formed. The buckling has led to the formation of the world's biggest mountain range, the Himalayan chain, and the Tibetan plateau to the north of it.

### Conservative plate boundaries

Conservative plate boundaries occur where plates are 'sliding' past one another, rather than moving together or away from each other. There is no plate being created or destroyed – because there is no magma and no subduction. But the sliding is not smooth – there is friction between the two plates, and extreme stresses build up in the crustal rocks.

When this is eventually released the result is an earthquake. The best example of this is along the west coast of North America, where a series of faults mark the boundary between the North American Plate and the Pacific Plate (Figure 9). The size of the earthquakes relates to the frequency of movement, so if there is a long period with no earthquake activity the pressure builds up and the eventual movement will be much greater, generating much more energy. On this boundary the plates are actually moving in the same direction but at different rates. Some of the faults are visible on the surface but many are not – and the results can be devastating when there is a sudden movement, as there was along the previously unknown Northridge fault in 1994.

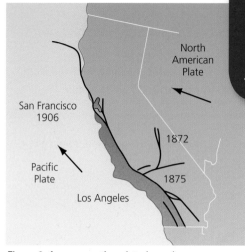

Figure 9: A conservative plate boundary

## Earthquakes and volcanoes

Figures 10 and 11 show the global distribution of earthquakes and volcanoes. By comparing these with Figure 5 it is obvious that there is a very close relationship between **plate margins** and tectonic activity. However there are some important exceptions.

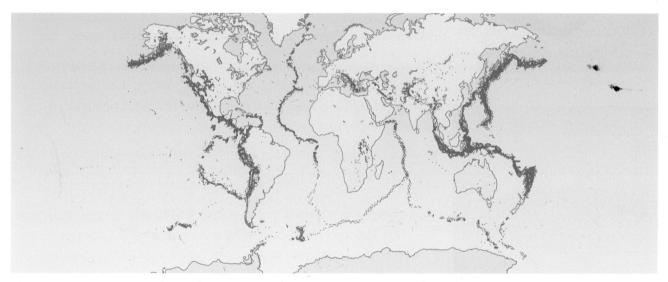

Figure 10: Global distribution of earthquakes (as shown by the red dots)

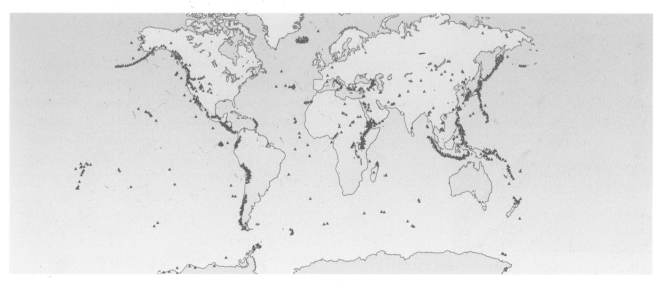

Figure 11: Global distribution of volcanoes (as shown by the red triangles)

## Activity 5

Study Figure 9.

Describe the movements of the North American and Pacific plates.

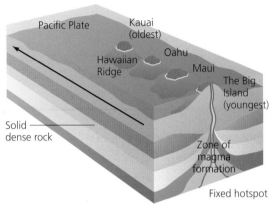

Figure 12: The formation of the Hawaiian chain of islands, according to the hotspot theory

Figure 13: The shield volcano, Mauna Loa, Hawaii – the largest volcano in the world

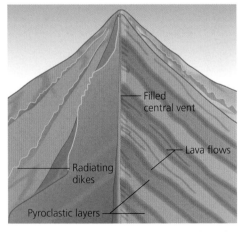

Figure 14: A cut-away view of a composite volcano

The Hawaiian Islands, which are entirely of volcanic origin, have formed in the middle of the Pacific Ocean – more than 3,200 km from the nearest plate boundary. This exception to the usual pattern has been explained by the 'hotspot theory' – the theory that there are fixed spots in the mantle where magma rises to the surface. As the crust is moved over these fixed spots, volcanoes are created, eventually forming a chain of volcanoes, as shown in Figure 12. There are many other hotspots on the Earth, some of which have been extremely destructive in the past and pose threats for the planet in the future (probably a very long time in the future). The best known example of this is the so-called 'supervolcano' under Yellowstone National Park in the USA.

There are many different types of volcano. Experts recognise 539 volcanoes that have erupted in recorded history. These are classified as 'active'. There are a further 529 volcanoes that have not erupted in historic times, but which exhibit clear evidence of the capacity to do so again. These are classified as 'dormant'. The rest are classified as 'extinct'.

In general, the shape and structure of volcanoes – and the explosive threat that they pose – is related to the type of magma that created them:

● Basaltic magma is usually found on constructive margins. It is high temperature, very low in silica, with low gas content. This type of magma produces fluid lava flow with relatively little explosive activity when it reaches the surface.

● Andesitic magma is formed at destructive margins where continental rocks are melted by rising magma. (It gets its name from the Andes mountains.) It is lower in temperature, has more silica and a lot of dissolved gases. As a result, this magma is much less fluid than basaltic magma, and is more likely to explode when it reaches the surface. (It is much harder for the gas bubbles to escape from a viscous magma than from a less viscous one.)

● Granitic magma is relatively low temperature, with a high silica content. These magmas are usually so viscous and 'sticky' that they get stuck before they reach the surface. Granitic magmas are much more likely to cool deep below the ground and become igneous rocks.

The shape of volcanoes is a result of the type of magma that creates them and the frequency of their eruptions. Volcanoes formed by magma that is largely basaltic tend to form very large, gently sloping shapes, known as shield volcanoes (Figure 13). On the other hand, volcanoes made up of andesitic lavas tend to form composite volcanoes which are steep-sided and made up of layers of lava and ash (Figure 14). The ash is formed out of the material destroyed in explosive eruptions which often blow the top of the mountain to pieces. These pieces of ash and rock debris are known as pyroclastics.

## Mauna Loa, Hawaii, a shield volcano

Mauna Loa is the largest volcano on the planet. Its long sides descend to the sea floor to add an additional 5,000 metres to its above sea-level height of 4,170 metres. This makes it well over 9,000 metres in height in total – more than Mount Everest. This enormous volcano forms half of the Big Island in the Hawaiian Islands (and by itself its volume is almost as much as all the other Hawaiian Islands combined). Mauna Loa is shaped like a shield, with very gentle slopes (Figure 13), because its basaltic lava is extremely fluid (it has low viscosity). This lava means that eruptions are rarely violent – the most common form being lava 'fountains' feeding lava flows. Typically, at the start of an eruption, a long rift up to several kilometres long opens towards the summit of the volcano with lava fountains occurring along its length in a 'curtain of fire'. After a few days, activity normally becomes concentrated at one or two fountains only.

Mauna Loa is one of the Earth's most active volcanoes, having erupted 33 times since its first recorded eruption in 1843. Its most recent eruption was in 1984. Mauna Loa is certain to erupt again, and the volcano is constantly monitored for warning signs.

## Mount Pinatubo, Philippines, a composite volcano

In the Philippines, volcanoes provide one of several threats to the population. The volcanoes here are almost all composite because the magma is made up of sticky andesitic lava. This frequently blocks the vent, leading to a build up of pressure. Eventually, the top of the mountain may be blown off in a large explosion.

In 1991 one of the most dangerous Philippine volcanoes erupted, for the first time in 500 years. This eruption produced at least 10 km$^3$ of erupted material – ten times the size of the Mount St Helens eruption of 1980 – putting it at level 6 on the 'Volcanic Explosivity Index' (Figure 15). The eruption had been predicted, and thousands of people had been evacuated from the surrounding areas. Many lives were therefore saved, but the region was damaged by flows of lava and ash.

Large eruptions have an effect on the climate because huge quantities of dust and gases are ejected into the atmosphere. Pinatubo's eruption led to a drop in global temperatures of about 0.5 °C.

### Activity 6

Study Figure 13.

(a) Identify the main characteristics of shield volcanoes.

(b) Explain how the type of magma affects the shape that you have described.

**Results Plus**
**Build Better Answers**

**EXAM-STYLE QUESTION**

**Describe the main features of composite volcanoes. (3 marks)**

■ **Basic answers** (1 mark)
Give a generic reference to a conical shape.

● **Good answers** (2 marks)
Identify one specific feature – alternate layers of ash and lava, steep slopes, etc.

▲ **Excellent answers** (3 marks)
Clearly describe two or more features – alternate layers, steep concave slopes, secondary cones, etc.

| | VEI | Volume of erupted rock and ash | Examples |
|---|---|---|---|
| Non-explosive | 0 | 0.00001 km³ | |
| Small | 1 | 0.001 km³ | |
| | 2 | 0.01 km³ | |
| Moderate | 3 | 0.1 km³ | |
| Large | 4 | 1 km³ | Mount St. Helens, May 18 1980 (1 km³) |
| Very large | 5 | 10 km³ | |
| | 6 | 100 km³ | Pinatubo, 1991 (10 km³) |
| | 7 | | Tambora, 1815 (100 km³) |
| | 8 | | Long Valley Caldera, 760 000 years ago (600 km³) |
| | | | Yellowstone Caldera, 600 000 years ago (1000 km³) |

Figure 15: The Volcanic Explosivity Index (VEI). The erupted volumes of the examples are estimates.

## Objectives

- Be able to identify the various impacts of tectonic hazards on people and property.

- Distinguish between the primary and secondary impacts of a tectonic hazard.

- Understand the role of various agencies in relieving the social and economic impact of various hazards.

## Activity 7

1. Define the term 'vulnerability'.

2. Explain why some areas are more vulnerable to hazards than others.

# What are the effects and management issues resulting from tectonic hazards?

## The effects and impacts of volcanic hazards

Earthquakes and volcanoes are good examples of **tectonic hazards**. Hazards pose a threat to us, but not all hazardous events are disasters. That depends on a number of other factors:

- The type of hazard

- The place's vulnerability to hazards

- The ability or 'capacity' to cope and recover from a hazardous event.

Not all natural hazards are equally devastating. The size and the 'type' of event is crucial. An eruption of Mauna Loa, for example, is seldom threatening to life because the volcano is not explosive. This is not the case for Mount Pinatubo.

Not all of the Earth's inhabitants are at equal risk from natural hazards. Despite the exceptions referred to on page 16, unless people live close to a plate boundary, it is very rare for them or their property to be damaged by earthquakes – they are not 'vulnerable' to earthquakes.

The idea of vulnerability can also be applied to volcanoes. If you do not live close to an active volcano, then you are not likely to be threatened by lava flows. However, you may very well be affected by clouds of volcanic ash, which can significantly alter the climate of places many miles, even continents, away from their point of origin.

'Capacity' refers to the ability of a community to absorb, and ultimately recover from, the effects of a natural hazard. The Japanese increase their capacity to cope with the effects of earthquakes by regularly practising how to respond. In theory, this means that they will have a better chance of coping with a large earthquake if one happens. Compare this capacity to cope with the situation in a sprawling slum in the developing world, where dwellings have been hastily constructed from poor-quality materials, and where there is neither the time nor money to commit to a large-scale community training programme, let alone improve the quality of the buildings.

It is important to distinguish between the primary impacts and the secondary impacts of disasters. Primary impacts are those that take place at the time of the event itself, and are directly caused by it. Secondary impacts are those that follow the event, and are indirectly caused by it. The social and economic impact of volcanic eruptions can be very considerable. In extreme cases it might mean the entire **evacuation** of the area.

What are the effects and management issues resulting from tectonic hazards?

19

## The Kashmir earthquake, 2005

This earthquake occurred in one of the most remote regions in the world – one that is difficult to reach even in normal conditions. This poor, largely agricultural, mountainous region is disputed between Pakistan and India, but is currently administered by Pakistan. Mountainous areas are fragile environments and earthquakes are likely to cause very considerable damage because of landslides and falling rock. The earthquake happened on a Saturday morning – a normal school day – and many of the dead and injured were children. What made it worse was that because it was during Ramadan (the Muslim period of fasting during daytime) many people were sleeping after getting up early to have a pre-dawn meal.

A considerable relief effort by the Pakistani government and international relief agencies was able to prevent the secondary disaster that threatened to overwhelm the region. With winter drawing in and conditions becoming more and more difficult, it became a race against time. But, using helicopters, the agencies managed to fly blankets, tents, basic provisions and medical supplies into the area.

## The Loma Prieta earthquake, California, 1989

The Loma Prieta earthquake (named after a local mountain) is sometimes known as the 'San Francisco 'quake of '89'. It was caused by a slip of several metres on the San Andreas fault and the other faults that mark the boundary between the Pacific Plate and the North American Plate. In common with the rest of the Californian coastal area, earthquakes are expected in this area – but **prediction** of time and **magnitude** is not possible. The earthquake took place during the evening rush hour, and offices were mostly empty or emptying. Purely by chance it was an exceptionally quiet rush hour because the two local baseball teams were competing for the World Series in Candlestick Park in San Francisco. Many people were either at the match or had made an early trip home to watch this local derby on television. The earthquake caused more property damage in the Bay Area than its strength and location suggested likely. This was because the clay soils of that area shook so much that they liquefied, causing properties to sink, gas mains to burst as they broke and fires to break out. (This was a smaller-scale reminder of the great San Francisco earthquake of 1906, when it was the fire after the earthquake that caused most of the destruction of the city.)

Figure 17: The collapsed Cypress Street Overpass – the scene of 41 deaths in the Loma Prieta earthquake

### Quick notes
**(The Kashmir earthquake):**
- Date: Saturday 8 October 2005
- Magnitude: 7.6 on the Richter Scale
- Epicentre: Muzaffarabad, the capital of Pakistani-administered Kashmir
- Death toll: 75,000
- Injured: 75,000
- Homeless: 2.8 million
- Property cost: $440 million.

Figure 16: In Patikka, 17 km from the epicentre of the Kashmir earthquake, villagers have to cross this ruined bridge to get to the Red Cross emergency unit, where they can pick up tents and blankets.

### Quick notes
**(The Loma Prieta earthquake):**
- Date: Tuesday 17 October 1989, at 17.04
- Magnitude: 6.9 on the Richter Scale
- Epicentre: in a mountainous part of Santa Cruz County, 96 km south-east of San Francisco
- Death toll: 63
- Injured: 3,757
- Homeless: 12,000
- Property cost: $10 billion.

## Montserrat, 1995: fleeing the volcano

Montserrat is a tiny Caribbean island of about 100 km² that is a very small leftover of the British Empire. The Chances Peak volcano in the south of the island was dormant – it had not erupted since the seventeenth century. This was because its very viscous lava had blocked the vent but, as with any dormant volcano, from time to time there can be violent eruptions. And between 1995 and 1997 it erupted huge quantities of lava, ash and extremely dangerous pyroclastic flows – high-speed avalanches of hot gases, ash and rock fragments which moved at speeds of 100–150 km/h.

### Impacts

Before the 1995 eruption, Montserrat had a population of about 10,500. Now it has about 5,000. Throughout the twentieth century, Montserratians migrated from their island because of a lack of employment opportunities. Recently, however, since the eruption of Chances Peak approximately two-thirds of the island's population have left the island for very different reasons. In the short term, they were evacuated by the government for their own protection. Some left the island altogether, moving to Antigua and further afield, whilst the remaining evacuees and others living in the south have since had to relocate permanently to the north of the island. The primary impact on the economy was devastating. This was particularly true for two of the mainstays of the economy – agriculture and tourism. Happily, both have slowly recovered.

## The management of tectonic hazards

The management of volcanic and earthquake hazards is expensive. But the costs of a disaster can be much worse. The pattern of volcanic and earthquake risk is fairly well known and, apart from the occasional totally unexpected events, communities that are located in tectonically active areas can develop management strategies to cope.

### Preparedness – being ready

**Preparation** means that governments, communities and individuals are ready to respond rapidly when disaster strikes and cope with the situation effectively. These measures include the formulation of emergency plans, the development of warning systems and the training of personnel. The measures may include evacuation plans for areas that may be at risk from a disaster and training for search and rescue teams. Preparedness therefore encompasses those measures taken before a disaster event which are aimed at minimising loss of life, disruption of critical services, and damage when the disaster occurs.

### Mitigation – reducing the impact

Mitigation measures are taken to reduce both the effect of the hazard and the vulnerability to it, in order to reduce the scale of a disaster. They can be focused on the hazard itself or on the elements exposed to the threat. Hazard-specific measures include relocating people away from the hazard-prone areas and strengthening structures – and using hazard-resistant design to reduce damage when a hazard occurs.

---

**Quick notes (Montserrat):**

- Eruption of a 'dormant' volcano in 1995
- Very viscous lava that gets 'stuck', causing violent eruptions
- Tiny island, so there was nowhere to 'hide'
- Two-thirds of population emigrated
- Devastating social and economic impact.

## Activity 8

Read the case studies on Montserrat and Laki.

(a) Describe the differences in the impact of these eruptions on the population.

(b) Explain why some volcanic eruptions cause much more loss of life than others.

What are the effects and management issues resulting from tectonic hazards?

21

## Iceland's Laki eruption, 1783–84, and beyond

The Laki eruption was one of the most devastating eruptions in human history. Iceland lies on the mid-Atlantic ridge and its volcanoes pose a constant threat, although very few of them produce violent eruptions because the magma is usually basaltic and relatively free-flowing. In 1783–84, a major eruption from the Laki fissure poured out an estimated 14 km$^3$ of basaltic lava and clouds of poisonous compounds. The volcano is located in a remote part of Iceland and no one was killed by the event itself. However, the secondary effects were devastating because the poisonous cloud killed over half of Iceland's livestock, leading to a famine which killed approximately a quarter of the population. At that time, there was no system of international relief in Iceland. The dust cloud (which was much larger than that caused by the eruption of Mount Pinatubo) is thought to have reduced temperatures in Europe for several years, causing poor summers, reduced harvests and, as a result, social unrest. Some historians believe that it helped trigger the French Revolution in 1789.

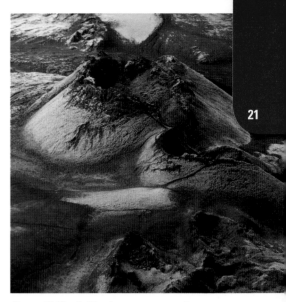

Figure 18: The Laki volcano cones as they appear now

Today, despite its banking collapse in 2008, Iceland is now one of the richest countries on the planet and the impact of volcanic eruptions is very different. The recent eruptions of Mount Hekla illustrate this. It is very active, erupting most recently in 1970, 1980, 1981, 1991 and 2000. The last eruption started on 26 February 2000 and lasted for 12 days, producing lava that covered approximately 18 km$^2$. Mount Hekla is very carefully monitored by the Icelandic authorities, using seismometers to record any movement of magma. They also measure the gas that the volcano is constantly leaking, watching carefully for any sudden changes. The area is very sparsely populated but in the last eruption, 200 families were evacuated and the island ring-road was closed. A future eruption of Hekla might interfere with flights to and from Iceland which would impact on its tourist industry.

Some Icelandic volcanoes are under its ice-sheets and these pose an interesting secondary threat. During their under-ice eruptions they melt vast quantities of water that can be released suddenly in massive floods known as 'jokulhaups'.

Most preparation and mitigation measures are expensive and beyond the budgets of poor communities, unless they can get help from international aid agencies. This explains the difference in impacts of disasters in different parts of the world. A community's ability to cope with a disaster depends on how well prepared they were and how much they had been able to spend on mitigation. But lack of preparedness is not just a feature of poor countries. Democratically elected leaders frequently make promises to potential voters that they will reduce taxes and cut public spending – almost always a popular message. In San Francisco after the Loma Prieta earthquake there was a call for greater spending on the emergency services but, with memories fading, it isn't obvious that the city is any more ready for the next earthquake.

### Quick notes (Laki eruption):
- Huge eruption in a remote area
- Population unable to escape
- No aid possible at that time (1783–84)
- No direct deaths but vegetation died, so animals died – and then people died
- Impact on the climate of Europe.

### Activity 9

In relation to the management of tectonic hazards, define and illustrate the terms 'preparation' and 'mitigation'.

## How ready is San Francisco?

San Francisco is bound to be struck by the 'Big One' – an earthquake many times more powerful than the Loma Prieta. Many experts say that the city is poorly prepared.

Although the city has firefighters and police who train regularly everything is not 'ready'. Experts say many critical structures in the city may also not have been adequately upgraded to withstand earthquakes, including:

- The city's biggest hospital, San Francisco General, where many of the injured would be taken, as well as many of the schools.

- The Bay Bridge Connecting Oakland and San Francisco damaged in 1989 and very vulnerable to a larger earthquake.

- The famous Golden Gate Bridge – if both bridges are damaged San Francisco becomes very isolated with many of its firefighters and hospital staff living on the wrong side of the bridges.

The city has evacuation plans and websites that aim to inform the citizens about the correct procedures. The Hurricane Katrina disaster forced many cities to take a long look at their plans in the light of what happened in New Orleans, which was widely thought to be poorly organised, with the poor and disadvantaged being neglected.

As a result, the Red Cross, fire department and other agencies have trained many citizens in emergency **response**. They urge residents to prepare an emergency kit that will last them three days with food, water and other supplies, while bilingual students offer training to older foreign residents who may not have the language skills to understand the procedures.

San Francisco is famous for its old Victorian buildings, which add to the city's charm. But many are at risk in a major earthquake. High-rise buildings put up in the 1950s and 1960s are also at risk. Earthquake scientists say that many buildings are not being made safe because owners have little incentive to do so, and in a city with a desperate shortage of housing the last thing the authorities want is housing to become even less affordable.

Scientists say there is nearly a 'two in three' chance that another major earthquake will strike San Francisco by 2032, and that the only way to get through disaster is to be ready for it. Although popular attention is fixed on the San Andreas fault, San Francisco is also threatened by the Hayward fault.

As Figure 19 shows the whole region is criss-crossed with faults. Residents can log-on to the seismic net online to check local earthquake activity. The San Andreas fault runs close to the coast and across the entrance to the bay, while the Hayward fault runs in the same south-east/north-west orientation but through Berkeley and Santa Rosa.

US Geological Survey Senior Seismologist Tom Brocher said: 'The Hayward Fault is a tectonic time bomb. [It is] the single most dangerous fault in the entire Bay Area, because it is ready to pop and because nearly 2 million people live directly on top of it.'

**Quick notes (San Francisco):**
- A disaster waiting to happen
- Many faults threaten the city
- San Andreas and Hayward faults are the most dangerous
- Costly to make old buildings safe
- Tension because city government wants to reduce taxes and also be responsible.

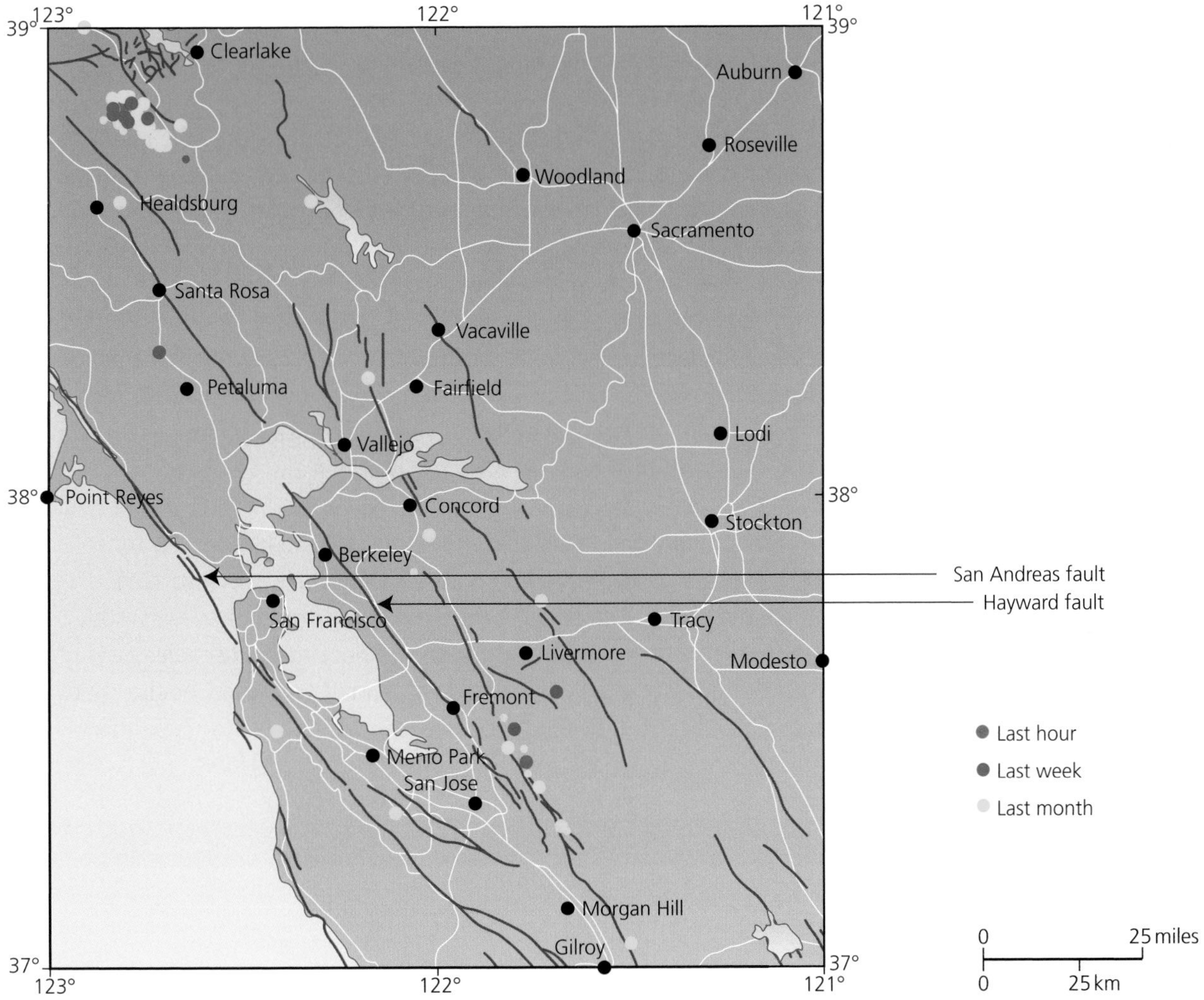

23

Figure 19: The San Francisco seismic net online

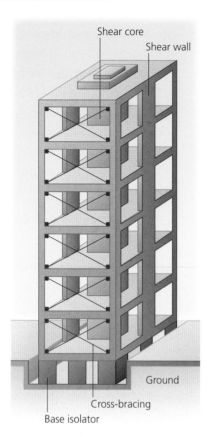

Figure 20: Earthquake-resistant design for tall buildings

## Building design

One of the main ways of mitigating the impact of an earthquake is to improve building design. Engineers have developed ways to build earthquake-resistant structures – not just houses, but office blocks and bridges too. The methods range from extremely simple to complex. For small- to medium-sized buildings, the simpler reinforcement techniques include bolting buildings to their foundations and providing support walls, called 'shear walls'. Shear walls are made from concrete that has steel rods embedded in it to help strengthen the structure and help resist rocking forces. The centre of a building can be constructed to form what is called a 'shear core'. Walls may also be reinforced and supported by adding diagonal steel beams in a technique called cross-bracing (Figure 20).

Medium-sized buildings are constructed using devices that act like shock absorbers between the building and its foundation. These devices – known as 'base isolators' – are usually bearings made of alternating layers of steel and an elastic material, such as synthetic rubber. Base isolators absorb some of the sideways motion that would otherwise damage a building.

Skyscrapers need special construction to make them earthquake-resistant. The foundations need to be very deep. They need a reinforced framework with stronger joints than those used in an ordinary skyscraper. Such a framework makes the skyscraper strong enough and yet flexible enough to absorb the energy of an earthquake – flexibility is the key.

All heavy appliances, furniture and other structures should be fastened down to prevent them from falling. Gas and water lines must be specially reinforced with flexible joints to prevent breaking. Fire, fuelled by broken gas pipes, is a real risk after earthquakes.

In the developing world, all of these methods are used, especially for important buildings in the Central Business Districts (CBDs) and in the richer areas of cities. Elsewhere in the developing world, however, money is rarely available for expensive engineering solutions, and simpler ways of strengthening existing buildings and new buildings are used, as the tables show.

## Strengthening of existing buildings

| Problem | Strengthening methods used |
| --- | --- |
| Heavy roof | Removal of mud overlay on top. |
| Poor timber frame connections | Adding diagonal bracing to the frame, using timber (cheaper) or steel (may not be locally available). |
| Thick walls without 'through-stones' to bind walls together | Installation of through-stones. This needs training of local artisans (new skills) and must be performed very carefully. |
| Separation joint at wall corners | Strengthening of wall corners, using wire mesh and cement overlay (though welded wire mesh not always available in rural areas). |
| Walls moving outwards | Installation of a ring beam (band of concrete) at the roof level. |
| Shaking of stones/bricks from exterior walls | Pointing of exterior walls with cement mortar. |

## Strengthening of new buildings

| Part vulnerable to earthquakes | Strengthening provisions used |
|---|---|
| Walls | Use cement/sand mortar and shaped stones (including through-stones) in construction. Construct concrete ring beam at the roof level. |
| Roof | Limit the thickness of mud overlay to 200 mm. |
| Timber frame | Install 'knee-braces' to reinforce the vertical/horizontal connections. |

Tectonic disasters stimulate both **short-term relief** and **long-term planning** to improve preparedness and mitigation for any future events. Immediate aid is needed to keep people alive, especially if, as in Kashmir in 2005, there are few local resources to fall back on. The aid needed urgently usually includes:

- Tents
- Blankets
- Garbage bags
- Antibiotics
- Baby food
- Milk
- Canned food
- Generators
- Tranquilisers
- Jerry cans
- Prefab toilets
- Disinfectants
- Mobile field kitchens.

## Izmit eathquake, 1999

Volunteers are also required, especially from the professional rescue services and health services. Disaster relief is by no means always smooth. The Turkish authorities, for example, were widely criticised after the 1999 Izmit earthquake – not just for an absence of preparedness and mitigation in a country that is far from the poorest, but also for bureaucratic blunders after the shock. Relief efforts were hindered by a serious underestimation of the size of the earthquake (because of poor quality and inaccurate sensors) which led to the Red Crescent sending far too little aid. The government also seemed slow in responding to help from the aid agencies. Generous aid from the United States and Europe did pick up the slack from NGOs. But as the disaster ran its course, the government mishandled this aid as well. Foreign-aid workers were required to pay customs duties on equipment and wait days before entering the disaster area. Failure to provide maps, interpreters and information on Islamic burial practice also impeded foreign workers' efforts.

### ResultsPlus
### Build Better Answers

25

**EXAM-STYLE QUESTION**

**Describe two ways in which buildings in developing countries can be made more resistant to earthquakes. (2 marks)**

■ **Basic answers** (0 marks)
Confuse developing with developed countries or discuss volcanic eruptions and not earthquakes.

● **Good answers** (1 marks)
Offer one method, such as bracing the frame.

▲ **Excellent answers** (2 marks)
Add a second method, such as installing a ring beam to prevent the walls moving outwards.

## Decision-making skills

Imagine you were involved in assembling aid packs for after an earthquake.

Which of the thirteen bullet point items would be your essential choice for emergency aid? You may choose up to six. Provide clear instructions for local aid workers as to how your six could best be used.

## Decision-making skills

Why are there sometimes conflicts between the following people after a major tectonic disaster has occurred:

- Local people
- Specialist rescue services
- Local leaders
- Local charities
- International charities
- Central government
- Armed forces

# exam zone

## Know Zone
## Restless Earth

The Earth truly is restless. The numerous tectonic plates grind relentlessly across the Earth. Usually this is a slow, unnoticed motion but sometimes plates stick, jar and snap free, releasing colossal amounts of energy in devastating earthquakes. In other locations the crust heaves, splits and spews scorching lava and ash onto the surface as volcanic eruptions.

## You should know...

- [ ] What oceanic and continental tectonic plates are, and how and why they move
- [ ] The different layers of the Earth's interior and how they differ
- [ ] Why the asthenosphere is an important part of the mantle
- [ ] How convection in the mantle drives plate tectonic motion
- [ ] That there are different types of plate margins
- [ ] How new oceanic crust forms at constructive plate margins
- [ ] How ocean crust is subducted at destructive plate margins
- [ ] How earthquakes and volcanoes form at destructive plate margins
- [ ] Why earthquakes occur along conservative plate margins
- [ ] Why fold mountains form in collision zones
- [ ] How destructive collision zone earthquakes can be
- [ ] How to explain the global pattern of earthquakes and volcanoes
- [ ] How shield and composite volcanoes form from different types of magma
- [ ] How hazards are measured using the VEI, Richter and Mercalli scales
- [ ] The difference between primary and secondary impacts
- [ ] Detailed causes and impacts for the Montserrat and Laki eruptions
- [ ] Detailed causes and impacts for the Loma Prieta and Kashmir earthquakes
- [ ] How earthquakes and volcanoes can be planned for and even predicted
- [ ] The different types of hazard response, both short and long term

## Key terms

Asthenosphere
Collision plate boundary
Conservative plate boundary
Constructive plate boundary
Continental crust
Convection currents
Core
Destructive plate boundary
Evacuation

Long-term planning
Magnitude
Oceanic crust
Plate margin
Prediction
Preparation
Response
Short-term emergency relief
Tectonic hazards

### Which key terms match the following definitions?

**A** Where two tectonic plates slide past each other

**B** Circulating movements of magma in the mantle caused by heat from the core

**C** The upper part of the Earth's mantle, where the rocks are more fluid

**D** Tectonic plate margins where oceanic plate is subducted

**E** The part of the crust dominated by denser basaltic rocks

**F** The central part of the Earth, consisting of a solid inner core and a more fluid outer core, and mostly composed of iron and nickel.

**G** Tectonic plate margin where rising magma adds new material to the diverging plates.

**H** The part of the crust dominated by less dense granitic rocks

To check your answers, look at the glossary on page 321.

**Foundation Question:** Using examples, describe some of the hazards of living on a destructive plate margin. (4 marks)

| Student answer (achieving 1 mark) | Feedback comments | Build a better answer (achieving 4 marks) |
|---|---|---|
| Volcanoes can happen.<br><br>Living on a destructive plate margin is a major hazard.<br><br>Also, earthquakes almost never happen and they are never very big.<br><br>They could also get extreme weather on a destructive plate margin. | • *Volcanoes can happen* is correct and scores 1 mark.<br><br>• *Living on a...* This sentence is correct but it just repeats the question.<br><br>• *Also, earthquakes almost...* This is incorrect because earthquakes do happen and they are sometimes large and destructive.<br><br>• *They could also get...* This sentence is not relevant in an answer about tectonic hazards. | Volcanoes can happen.<br><br>An example is the volcano of Montserrat, which erupted violently in 1995.<br><br>Earthquakes are common and they can have magnitudes of over 6.0.<br><br>In Kobe in Japan in 1995, an earthquake killed over 6,000 people. |

**Overall comment:** The student answer is quite weak because no examples are used and the comment about earthquakes is basically wrong.

- - - - - - - - - - - - - - - - - - - - - - - - - - - - - - - - - - - - - - - - - - -

**Higher Question:** Using an example, outline the impact of a major earthquake on people and property in the developing world. (4 marks)

| Student answer (achieving 3 marks) | Feedback comments | Build a better answer (achieving 4 marks) |
|---|---|---|
| There was a major earthquake of magnitude 7.0 in Kashmir in 2005.<br><br>Over 75,000 people died and many were crushed when poorly built houses and schools collapsed.<br><br>Landslides happened all over. | • *There was a...* This answer starts very well with a clearly stated example, which scores 1 mark.<br><br>• *Over 75,000 people...* This part of the answer contains a good level of detail, especially about poorly built houses, so scores another mark.<br><br>• *Landslides happened...* The answer begins to lose focus here but scores 1 mark. | There was a major earthquake of magnitude 7.0 in Kashmir in 2005.<br><br>Over 75,000 people died and many were crushed when poorly built houses and schools collapsed.<br><br>Landslides happened on steep slopes due to the shaking. The landslides blocked many roads, making the relief effort more difficult and people went short of food. |

**Overall comment:** The answer starts very well, with some good, accurate detail. However, it becomes much more general and the detail disappears. This often happens when students have not revised in depth. The answer had some good impacts and an example so scored 3 marks.

# Chapter 2 Climate and change

## How and why has climate changed in the past?

### Natural climate change over time

The average temperature of the Earth's atmosphere has changed a great deal in the past. For example, 100 million years ago – at the time of the dinosaurs – conditions were much hotter than they are today. There have also been many cold phases, called **ice ages**. Our last major cold period, the Pleistocene, started 2.6 million years ago and ended just 10,000 years ago. Since then, conditions have been warmer. This most recent 10,000 years is called the Holocene. The Pleistocene and the Holocene are part of the **Quaternary Period** of Earth history.

We know – from a range of evidence – that temperatures were colder during much of the Pleistocene. The most important evidence is found in ice cores extracted from polar ice caps. Snow has been falling and building up into thick ice there for many thousands of years. Ice cores allow us to look back in time – they are like cross-sections drilled down through the snow and ice. One core was cut down through 3 kilometres of ice, to where the ice was 500,000 years old. The ice was taken to a laboratory and melted, releasing bubbles of ancient air. Changes in air content were then analysed, showing scientists how temperatures have warmed and cooled over time (Figure 1).

Another important data source for science is the fossil record. This shows whether animals preferring warm or cool conditions were alive and thriving at different times in the Earth's past.

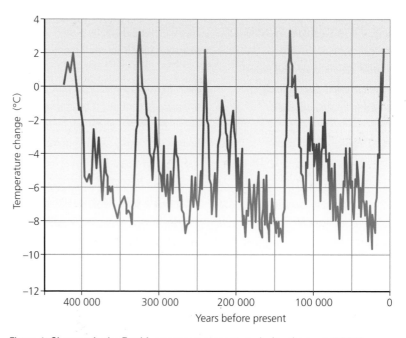

*Figure 1: Changes in the Earth's average temperature during the past 450,000 years*

### *Natural causes of climate change*

The Earth's atmosphere is affected by changes in the lithosphere as well as by cosmic influences such as the Sun's activity or variations in the Earth's orbit. These are **natural causes** of climate change. As shown below, some changes take place over a short timescale and these can explain the particular years or decades of warming or cooling. Other changes are far more significant and can last for thousands, or even millions of years.

| Long-term changes (may last for many centuries) ⬇ Short-term changes (lasting just a few years) | | |
|---|---|---|
| **Orbital changes** Changes in how the Earth moves around the Sun are believed to cause ice ages. According to Milankovitch (a Serbian physicist), every 100,000 years or so the Earth's orbit changes from a circular to elliptical (egg-shaped) pattern. This changes how much sunlight we receive. He also identified that the Earth's axis moves and wobbles about, changing over 41,000 and 21,000 year cycles. This also affects how much sunlight is received. Put all of this together and the history of ice ages can be explained! |  |
| **Solar output** The Sun's output is not constant. Cycles have been detected that reduce or increase the amount of solar energy. The most well-known phenomenon is sunspot activity, when uneven temperatures develop on the Sun's surface. These can be seen as tiny black spots on photographs of the sun taken by experts (never try this yourself). Sunspots seem to come and go following an irregular cycle that lasts about 11 years. Interestingly, temperatures are greatest when there are plenty of spots – because it means other areas of the Sun are working even harder! | 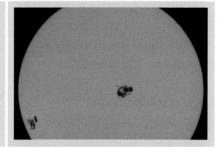 |
| **Volcanic activity** Major volcanic eruptions lead to a brief period of global cooling, due to ash and dust particles being ejected high into the atmosphere, blanketing the earth. The 1883 explosion of Krakatoa is believed to have reduced world temperatures by 1.2 °C for at least one year afterwards. The most recent explosion to have a similar effect was Pinatubo (1991). Sunlight reaching Earth was reduced by 10%. World temperatures fell by nearly half a degree in the following year. |  |
| **Asteroid collisions** A large asteroid or meteorite colliding with the Earth would lead to a short period of cooling because of the enormous cloud of dust that would be thrown up by the impact, blocking out sunlight. The extinction of the dinosaurs, 65 million years ago, may have been caused by an asteroid impact. |  |

## The impact of natural climate change on people and the environment

So far we have only considered the **climate changes** that took place many thousands of years ago. We now need to analyse more recent changes. In just the last few centuries, there have been noticeable changes in world temperatures (see Figure 2 on page 30).

The world warmed slightly about one thousand years ago. This was around the time of the Norman invasion of England. However, it began to cool again during the thirteenth and fourteenth centuries. By 1600, when Elizabeth the First was on the English throne, a notably cooler phase had begun. Icy winters became far more common in Europe. This period is called the **Little Ice Age**.

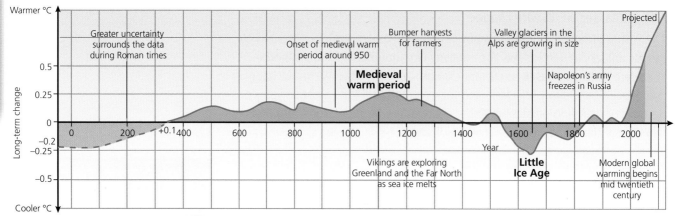

Figure 2: Climate changes since Roman times

## ResultsPlus
### Build Better Answers

**EXAM-STYLE QUESTION**

**Explain one natural cause of climate change in the past. (2 marks)**

■ **Basic answers** (0 marks)
Are about the impact humans have had on climate, instead of natural causes.

● **Good answers** (1 mark)
Correctly describe a natural cause of climate change, such as volcanic activity, but do not explain why this causes climate to change.

▲ **Excellent answers** (2 marks)
Not only describe a natural cause, such as volcanic activity, but also explain that ash clouds ejected into the atmosphere can cut out the Sun's rays, causing temperatures to fall.

## Skills Builder 2

Study Figure 2.

(a) When did the Medieval Warm Period start and end?

(b) Describe one possible natural cause of the onset of the Little Ice Age.

(c) Describe one disadvantage and one advantage of the Little Ice Age for people alive at that time.

Of course, modern technology did not exist to accurately record these events when they were occurring. Worldwide temperature readings (using reliable thermometers) have only been recorded in the past couple of centuries, and photography only really got going in the late nineteenth century. Luckily, however, we do have all kinds of other evidence to help us build a case for climate change before modern scientific readings started to be made. Diaries, folklore, paintings, books and very old newspapers can all provide a pretty reliable picture of the Little Ice Age and the years that followed. Some scientists study old wood, examining tree rings (because a cold year brings less new growth and thinner rings). We also get clues from old coral reefs (because when seas were cooler, coral grew more slowly).

Evidence for the warmer world that preceded the Little Ice Age has also been pieced together. For instance, archaeological evidence shows that the Vikings sailed to many northerly places after they left their homes in Norway and Denmark around AD 900. They even settled and farmed in icy Greenland. Many historians believe that there must have been less Arctic sea ice at this time to allow the Vikings to move around so easily.

Modern **global warming** is widely believed to be the result of pollution caused by humans since 1800. So these much older climate changes must have been natural in origin. Sunspot activity, as we have already learned, can result in warmer or cooler temperatures on the Earth. One explanation for the Little Ice Age is that sunspot activity was much lower than today. Another is that it coincided with a more active period of volcanic activity.

### The Little Ice Age

Charles Dickens wrote his novels as the Little Ice Age was drawing to a close. Many of his books, including *A Christmas Carol*, describe very cold and snowy winters in London. But these days, London and south-east England seldom experience a 'white Christmas'. Were winters actually colder back in the seventeenth, eighteenth and early nineteenth centuries? Plenty of data sources suggest so.

Scientists have examined roof timbers in houses built at this time. They have found that tree-ring growth was often slow during winter in the 1700s. Backing up this scientific fact-finding is evidence gathered by art and book historians. Some landscape paintings from the Middle Ages show a winter landscape strikingly different from today. There are, for example, many illustrations of the River Thames covered with ice. Very cold temperatures are required to freeze moving water, especially in a salty tidal river like the Thames. But it is clear that between 1607 and 1814 Londoners enjoyed 'ice fairs' and skating competitions on the frozen river – events never seen today (Figure 3).

There is evidence for the Little Ice Age from all over Europe. Farming records tell us that the thermal growing season was often a little shorter than today. In a cool year, with spring arriving late, planting was often delayed. When autumn came early, so too did the harvest. Some of the most useful and reliable records are grape harvest dates from winemaking regions. These have been used to reconstruct summer temperatures in Paris from 1370 to 1879, and a clear decline in temperature can be seen through much of the period after 1500.

Cold winters often played a part in wars at this time – which are well documented. During Napoleon's retreat from Russia, for example, thousands of his troops froze to death in the winter of 1812. Changes in ecosystems and animal populations are also recorded. Seal populations appear to have been badly affected by a fall in their fish food supplies during this period. Most importantly of all, surviving records tell us that our European ancestors adapted well to a cooler world (otherwise we would not be here today). At the simplest level, leisure patterns changed as people took to skating – on the Thames, for example. In Iceland there was a more serious impact. When farmers suffered cereal crop failures because of the cold, Icelandic society adapted by turning to a fish-based diet.

## Watch Out!

The changes associated with the Little Ice Age and Medieval Warm Period were not major events in the Earth's history. Look carefully at Figure 2 and you will see that variations in average annual temperature never exceed half a degree either way. This was enough to make winters colder, or to make sea ice melt a little – but was in no way comparable with the arrival or departure of proper ice ages.

## Activity 1

1. How reliable are art and literature, compared with scientific data?

2. Do you trust the evidence shown in a painting (Figure 3) as much as a photograph or a thermometer reading?

Quick notes (The Little Ice Age):
- The world may be warming today but in the past it has sometimes been colder.
- Data showing such changes come from both scientific and artistic sources.
- Humans successfully adapted to these changes in their environment.

*Figure 3: Paintings of the Little Ice Age include ice fairs on the River Thames*

*Figure 4: Megafauna that became extinct near the end of the last ice age – sabre-toothed tiger (top)and a woolly mammoth*

## Geological climate events and their impacts

Significant changes taking place over millions of years are called **geological climate events**. This is because they form major chapters in the Earth's history (and the study of the Earth's history is called 'geology').

Whenever the Earth starts a new chapter of its history, enormous climatic changes are usually taking place. In the past, these had drastic knock-on effects for flora and fauna (plants and animals). The fossil record is what alerts scientists to big changes that took place in the distant past, such as the extinction of the dinosaurs 65 million years ago. Sometimes known as the Cretaceous–Tertiary (K–T) extinction event, the cause is likely to have been a massive asteroid impact or increased volcanic activity (see page 29).

Humans have also been affected by big climate events and changes in the past. Although modern humans evolved just 200,000 years ago, we have been around in one shape or another for about 5 million years. One of our oldest ancestors (who scientists named 'Lucy') lived in Ethiopia in Africa at the end of a geological period called the Pliocene, when the world was a little cooler and drier. The world's thermometer has moved up and down many times since then. Sometimes our ancestors have had to migrate away from warming or cooling regions. It may be the reason why Lucy's descendants left Africa.

### The extinction of megafauna

The most recent major change for plant and animal life was the extinction of certain large animals called **megafauna** at the end of the last Pleistocene ice age. At its coldest, climate change had left the UK covered in ice as far south as what is now the Midlands. It was not just the UK that was affected of course, as changes were global. What are now the central states of the USA were ice-bound too. However, a warming of the Earth's climate was well underway by 10,000 years ago, probably linked to the orbit changes discussed earlier. The ice that covered the UK and much of Northern Europe retreated to its present-day Arctic limit.

During the Pleistocene ice ages, very large mammals lived in Europe and North America (Figure 4). These included woolly mammoths, sabre-toothed tigers, large wolves and giant beavers (the size of bears). However, within just a few centuries of the ice melting, these animals – as many as 135 species – were all extinct. Why?

Scientists think that they were unable to adapt to new conditions. Weather and plant life were changing, affecting whole **food chains**. Some places were left drier once the glaciers – a source of meltwater – retreated north. If food was scarce, some megafauna would have died out naturally.

There is an opposing argument that says that humans were to blame, rather than changes in the climate. This theory claims that towards the end of the ice ages, humans migrated around the world in large numbers and that wherever they went they hunted native megafauna species such as the mammoth – until they eventually became extinct. Remains of mammoths have been found at many settlement sites dating from this time.

# What challenges might our future climate present us with?

## The climate of the UK today

The UK has a cool, temperate **maritime** (coastal) climate (see Figures 5, 6 and 7). The mean average temperature for southern England is 17 °C in July, compared with around 5 °C in January, showing marked seasonality. The **thermal growing season** ranges from 7 to 9 months at lower altitudes. A warm ocean current known as the **Gulf Stream** keeps much of the west coast warmer than might be expected during winter (compared with other places found as far from the equator).

Precipitation (rain and snow) in the UK ranges from 700mm to 2500mm annually, depending on location. Rainfall is generally high along the west coast, which frequently receives **frontal rainfall** brought by **Atlantic depressions** (unstable weather systems that develop over the ocean and convey large volumes of warm, moist air in a westerly direction over Europe). Precipitation also increases with altitude due to lower temperatures, which can be a trigger for condensation to occur, leading to **orographic** (relief) **rainfall**.

The movement and constant jostling of **air masses** is a further feature of the UK's climate. An air mass is a large body of air that originally develops over a source region such as Scandinavia (cold) or north Africa (hot). The air mass may later move away from its source region, driven by complex global-scale atmospheric processes. Summer drought is associated with the arrival of **tropical continental** air masses from north Africa. Very cold dry winters are linked with the movement of **polar continental** air masses from Scandinavia and Eurasia. Both of these events are characterised by high pressure, clear skies and low rainfall.

### Objectives

- Recognise that human activities produce greenhouse gases.

- Explain how this results in an enhanced greenhouse effect and a changing climate.

- Understand that people everywhere will face climate change challenges in the future.

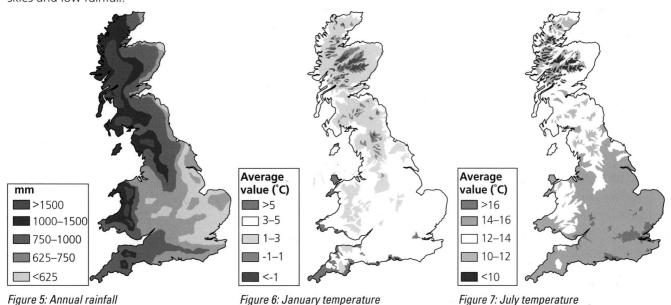

**mm**
- \>1500
- 1000–1500
- 750–1000
- 625–750
- <625

*Figure 5: Annual rainfall*

**Average value (°C)**
- \>5
- 3–5
- 1–3
- -1–1
- <-1

*Figure 6: January temperature*

**Average value (°C)**
- \>16
- 14–16
- 12–14
- 10–12
- <10

*Figure 7: July temperature*

The UK's climate may already be changing in complex ways that are hard to predict. This is due to the large number of influences on the UK's climate described above, any and all of which could be modified in a range of different ways by global warming. For instance:

- A warmer climate would provide Atlantic depressions with even more energy, increasing the strength and frequency of rainfall, especially in western parts of the UK.

- The movement of air masses could become less predictable, bringing longer or unexpected periods of drought.

- However, a weakening of the Gulf Stream, possibly as an effect of melting Arctic ice, would mean a colder climate for the UK.

## The human causes of modern climate change

We have learnt about the extinction of the megafauna. Today it seems humans may be guilty of starting a new mass extinction. This time it is linked to a new era of climate change – one that we are very probably responsible for. The Earth's temperature is rising due to the human addition of extra **greenhouse gases** to the atmosphere. By some estimates, a quarter of the world's species may not survive the worst predicted changes.

Greenhouse gases naturally help to warm our atmosphere and make the Earth habitable. However, if *extra* greenhouse gases are added then the Earth begins to get unnaturally warmer. What are these greenhouse gases? What kinds of human activities are boosting their supply?

### Greenhouse gases and the activities that produce them

Carbon dioxide ($CO_2$) is a greenhouse gas that is naturally produced when humans and other animals respire (breathe out). Plants do the opposite, taking in carbon dioxide, while producing oxygen. For this reason, trees and other plants are called a carbon store. However, **deforestation** releases all that stored carbon kept locked up in the trees (see Chapter 3). Other human activities have also boosted $CO_2$ levels well beyond their natural level. The burning of fossil fuels, such as oil and gas, contributes greatly. Other significant activities that release $CO_2$ are cement making and steel manufacturing.

In addition to carbon dioxide, there are other greenhouse gases that human activities produce.

| Methane | Methane emissions result from the raising of livestock. Recently, scientists estimated that wind given off by cattle makes up 11% of Australia's carbon footprint. Methane is also emitted during the production and transport of coal, natural gas and oil. The decomposition of organic wastes (such as uneaten food) in landfill sites also contributes methane. |
| --- | --- |
| Nitrous oxide | Nitrous oxide is emitted during agricultural and industrial activities, as well as during the combustion of solid waste and fossil fuels. |

**ResultsPlus**
**Build Better Answers**

**EXAM-STYLE QUESTION**

**Describe how human activities produce two different types of named greenhouse gas. (4 marks)**

■ **Basic answers** (0–1 marks)
Name carbon dioxide but do not link its production with an activity such as deforestation or car driving.

● **Good answers** (2 marks)
Also name a second gas, such as methane. However, still do not say how human activities produce the gases.

▲ **Excellent answers** (3–4 marks)
Not only describe a second greenhouse gas, usually methane, but also accurately link the growth of this gas with more cattle being reared by people (the best answers even state that this was a response to rising demand for meat in Asia).

## The growth of greenhouse gases over time

Since pre-industrial times, atmospheric concentrations of many greenhouse gases have grown significantly. This is due to economic development processes spreading around the world. As countries are industrialised, their citizens become:

● consumers of energy and goods

● producers of air pollution (through burning fossil fuels).

Large economically advanced countries, like the USA, produce enormous amounts of new $CO_2$ each year. The world current level of $CO_2$ is the highest reached for 650,000 years (some think it is the highest for 20 million years). Even worse, it is increasing now at a rate 200 times faster than at any time in the last million years. Just since the 1800s, carbon dioxide concentration has increased beyond its natural amount by about one-third.

The growth rates of methane and nitrous oxide also fluctuate from year to year, but their long-term trend is again upwards. Methane concentration has more than doubled since the 1800s. It is well over 100% higher than at any time in the past 900,000 years. Nitrous oxide concentration has grown less, but has still risen by 16%. The table below shows one estimate of the rise for all three greenhouse gases.

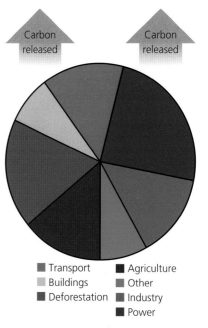

Figure 6: Activities that produce greenhouse gases

## An estimate of the rise of the three main greenhouse gases

| Year | 1800 | 1820 | 1840 | 1860 | 1880 | 1900 | 1920 | 1940 | 1960 | 1980 | 2012 |
|---|---|---|---|---|---|---|---|---|---|---|---|
| Carbon dioxide (parts per million) | 280 | 282 | 283 | 287 | 291 | 295 | 299 | 310 | 323 | 347 | 396 |
| Methane (parts per billion) | 700 | 713 | 727 | 756 | 796 | 835 | 871 | 980 | 1430 | 1656 | 1800 |
| Nitrous oxide (parts per million) | 270 | 271 | 275 | 276 | 283 | 286 | 290 | 294 | 304 | 311 | 318 |

## Which places produce the most greenhouse gases?

We have already looked at the activities that produce greenhouse gases, but which nations are the biggest polluters? You can probably guess the names of some of the main culprits straight away – the United States and Europe. Both regions experienced an industrial revolution back in the 1800s and have therefore been polluting for a long time, along with Japan.

The emerging superpowers of India and China have recently caught up with these older polluters. In the 1990s, both countries began to 'take off' at breakneck speed. Their economies have been growing at around 10% per year since then – and so have their greenhouse gas emissions. China is now the world's largest single polluter.

### Skills Builder 3

Look at the table of greenhouse gases.

(a) Draw a graph to illustrate changes in one of the gases shown.

(b) Identify the decade when major changes first started to occur.

(c) Suggest reasons why major changes began around this time.

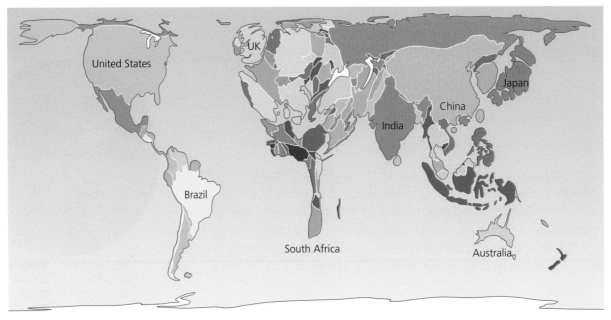

*Figure 7: A map showing the size of countries in proportion to how much $CO_2$ they emit*

## Skills Builder 4

Study Figure 7.

(a) Who are the biggest polluters?

(b) Where are the smallest polluters found?

(c) What are the strengths and weaknesses of mapping data in the way shown here?

## Top Tip

Students who know that the greenhouse effect is a *natural* phenomenon will score well. Without greenhouse gases, the atmosphere would not trap the Sun's heat, leaving the Earth too cold for human life. Recently, however, human activity has been boosting the process, trapping *too much* heat.

At first, China and India were not put under much pressure to deal with their pollution. The view was that both places needed to industrialise, in order for health and welfare to improve. With a combined population of 2.5 billion people, many of whom still live in poverty, it is hard to imagine progress *not* producing a lot of greenhouse gas emissions. However, the global situation has become so urgent that both India and China now acknowledge that they need to do more to tackle their emissions.

### What do greenhouse gases do?

Don't forget that the greenhouse effect is, in its natural state, a good thing for life on Earth. Without it, conditions would be up to 30 °C cooler – too cold for humans to exist. The way the greenhouse effect works is not especially complicated. Sunlight arrives at the Earth's surface and heats up the ground. Some of this warmth then escapes back into space, just like heat leaving a room when the window is left open. However, greenhouse gases act like the glass in a greenhouse and trap the escaping heat energy (Figure 8). They keep the Earth warm.

Another way of explaining it is to think of the natural amount of greenhouse gases as being like a blanket that helps you sleep at a constant, comfortable temperature. You wouldn't want to take the blanket off (and, similarly, we wouldn't want to lose the greenhouse gases in the Earth's atmosphere). But neither would you want to put an extra blanket on the bed – because it would make you too hot. The human-produced greenhouse gases are like this extra blanket, resulting in what is called the **enhanced greenhouse effect**. And if more people around the world do not change their lifestyles, the greenhouse effect will be enhanced further, with more 'blankets' being added to the atmosphere – a sure recipe for sleepless nights!

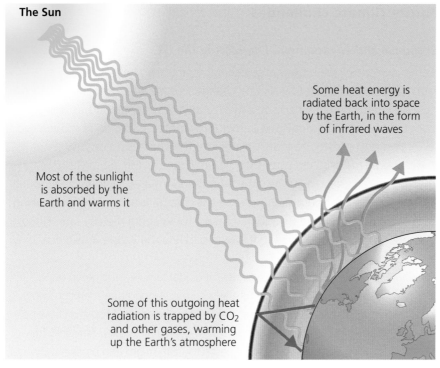

The Sun

Some heat energy is radiated back into space by the Earth, in the form of infrared waves

Most of the sunlight is absorbed by the Earth and warms it

Some of this outgoing heat radiation is trapped by $CO_2$ and other gases, warming up the Earth's atmosphere

*Figure 8: The greenhouse effect*

## Uncertain projections for the future

In 2011, world carbon dioxide concentration passed 390 ppm (parts per million). In 1800, it was just 280 ppm. The figure is currently growing by 2 ppm or more every year, despite the best efforts made by people and governments to reduce energy use, increase efficiency and introduce 'green solutions' like solar or wind power. Worldwide fossil fuel combustion and rates of deforestation have not slowed down enough yet to make a big dent in total carbon emissions. Methane production is also rising again. With wealth increasing in Asia, more people want a meat and dairy diet. This means more cattle being raised – resulting in more methane from bovine flatulence (windy cows).

Many of the world's most knowledgeable climate change scientists belong to the Intergovernmental Panel on Climate Change (IPCC). This group believes that greenhouse gas emissions need to level out below 550 ppm. They see this limit as a 'tipping point', beyond which events could spiral out of control (see below). However, scientists are uncertain whether or not this tipping point will actually be reached. There are just too many variables, including uncertainty over the future success of global climate change agreements and unreliable economic growth forecasts.

**ResultsPlus**
**Build Better Answers**

**EXAM-STYLE QUESTION**

**What is the enhanced greenhouse effect? (3 marks)**

■ **Basic answers** (0–1 marks)
Offer only a simple (and possibly inaccurate) description of the Earth getting warmer over time.

● **Good answers** (2 marks)
Accurately describe an increase in global temperatures due to the human production of greenhouse gases.

▲ **Excellent answers** (3 marks)
Also explain that human activities are increasing greenhouse gases beyond a natural (and desirable) level, thus making the warming effect of greenhouse gases even more powerful than it used to be.

| | |
|---|---|
| **Below 550 ppm** | If greenhouse gases rise no higher than 550 ppm, then the global temperature rise should not exceed 2 °C. However, this could still be enough to cause widespread melting of glacier ice and a world sea level rise of nearly one metre, submerging low-lying areas. There would also be more storms and hurricanes due to warmer sea temperatures. Globally, many species might become extinct, although warmer conditions could encourage tree growth and greater biodiversity at high latitudes (Chapter 3). So there could be some winners as well as losers. |
| **Above 550 ppm** | If greenhouse gas emissions rise above this critical level, then conditions will rapidly worsen. Vicious circles will develop. For instance, as the ice caps melt, their bright white surface is lost. This ice surface usually reflects some sunlight back into space. Without it, more energy will be absorbed by the Earth – and temperature will rise even faster. The worst predictions suggest that the Earth could one day be ice-free if nothing is done to stop a global temperature rise of 6 °C or more. Billions of humans will lose their homes (because of sea level rises) or their fresh water supplies if this happens. |

**Results Plus**
**Exam Question Report**

### REAL EXAM QUESTION

**Explain one possible good effect of global warming and one possible bad effect of global warming. (4 marks, June 2006)**

#### How students answered

Most students answered this question very poorly. The question requires two developed answers that explain why the effect is seen as *good* or *bad*. The majority of students could only write statements like 'warmer temperatures' or 'ice caps melt'.

80% (0–1 marks)

A few students gave two effects, and said *why* one of these effects was good or bad.

10%    (2 marks)

A few students answered this question really well. They gave two effects and in both cases showed what the positive or negative impact on people or places would be. 'Melting of ice caps causes coastal flooding, damaging cities' would have been a very clear *negative* effect, for example. While 'warmer temperatures allow more crops to be grown' scores two marks as a *positive* effect.

10% (3–4 marks)

## Future climate challenges

### *Economic and environmental impacts in the UK*

Richer countries have money and technology that can help them adapt to higher sea-levels brought by climate change. For instance, Holland has long built enormous embankments to protect the low-lying areas it has reclaimed from the sea. The Thames Flood Barrier (TFB) protects London from high sea levels that drive North Sea water inland into the Thames estuary. Without it, occasional major storm surges would flood central London. With climate change, these surges might become more frequent and dangerous, although there is uncertainty about how extreme any change may be. The TFB is being used more than ever and is vitally important for keeping the city safe. Flooding of London cannot be allowed as it would paralyse the whole British economy.

The entire UK faces change and challenges if global temperatures keep rising (Figure 9 below). In a low-emission future (with the temperature rise kept below 2 °C) there may be winners as well as losers. One winner would be the UK wine industry, which is currently experiencing record growth in profits, especially for sales of fizzy champagne-style wines. Coastal tourism might also prosper. However, there will be many losses. In a high-emissions future, economic costs could be high:

- A complete loss of winter sports, as snow disappears from highland areas

- More cases of tropical diseases like malaria

- More severe storms would cause costly damage and disruptive transport problems

- The economic cost of helping climate change refugees who migrate from poor countries to the UK

- Major changes for fishing industries, especially if ocean currents are disrupted by melting ice in Greenland, making British waters turn colder (though this would not happen before 2100).

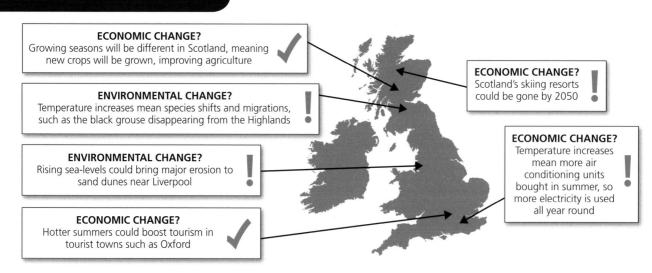

*Figure 9: Rising temperatures and climate change create economic and environmental uncertainty for the UK*

### Economic and environmental impacts in poorer nations

Many nations are at serious risk of sea-level rise, and the poorest of them lack the resources needed to adapt to climate change. In a high-emissions future, some places could be abandoned entirely:

| The Maldives | Most of the tiny islands that make up the Maldives are less than two metres above sea level. The 400,000 people that live there may soon become 'climate change refugees'. In 2008, the Maldives President asked neighbours India and Sri Lanka if he could buy some of their land! |
|---|---|
| Tuvalu | Half of the 10,000 residents of the Pacific island state of Tuvalu live within three metres of today's sea level. Many islanders think they may soon need to migrate elsewhere. |

It is not just rising sea-level that brings new risks. If climate change permanently melts glaciers in the Himalayas, millions of people in Asia will lose their water supply (Chapter 4). Drought and famine would follow. Saharan Africa is also highly threatened by future water shortages. Projections show that naturally low levels of rainfall could be further reduced.

### Bangladesh

Bangladesh is the world's seventh most populous country, with 160 million people. It is also one of the poorest nations on Earth, with an average income of just over one pound a day. Bangladesh is also a vulnerable nation, with much of its land at or near sea-level on the delta of the River Ganges – and this land is naturally sinking and subsiding. Flooding is already frequent and damaging.

A small rise in sea-level would leave large areas of Bangladesh permanently under water (Figure 10). Climate change could also help drive tropical storms further inland, causing more farmers to lose crops (because when storms temporarily raise the water level, sea salt deposits are left behind, killing plants).

One of the many problems that flooding brings is long-term interrupted schooling when schools are flooded. If the country is to be lifted out of poverty, then safeguarding education is vital. Bangladesh needs well-educated citizens if it is to move further forwards.

Britain recently gave £75 million to help Bangladesh tackle the worst impacts of climate change. The money will be spent on special adaptation measures to protect schools, rebuilding them on raised stilts and platforms. Flood waters may then pass safely beneath the buildings. Farmers will also be helped with the introduction of crops that are more tolerant to salt.

### Skills Builder 5

Study Figures 9 and 10.

(a) Describe how different parts of the UK could be affected by climate change.

(b) Describe how much of Bangladesh would be lost to (i) a two-metre rise, or (ii) a five-metre rise.

(c) Suggest how a sea-level rise could affect the economy of a poor nation like Bangladesh.

### Decision-making skills

Look at the bulleted list on page 38. Make a table to list the environmental and economic costs. Rank them in what you think is the order of severity of impact.

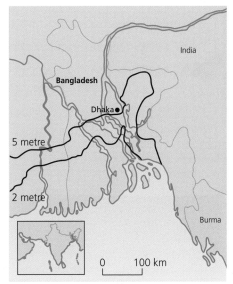

Figure 10: Land that could be lost to potential sea-level rises in Bangladesh

### Quick notes (Bangladesh):
- Poorer countries are more vulnerable to natural hazards than rich countries.
- Natural hazards, like flooding, can interfere with a country's long-term economic development.
- Climate change means that the risk of flooding and storm damage is increased for Bangladesh.

# examzone

## Know Zone
## Climate and change

The Earth's climate may seem stable, but in fact it has changed in the past and is likely to change in the future. Climate can, and does, change on a timescale of a few years and many thousands of years. In the past climate changed quite naturally, but today there is increasing evidence that humans are changing the climate.

## You should know...

- ☐ How climate has changed in the past
- ☐ The timescales of climate change
- ☐ How scientists know about past climate changes
- ☐ The natural causes of climate change
- ☐ How climate change has affected people and the environment since Roman times
- ☐ How large, long-term climate change contributed to extinction at the end of the last ice age
- ☐ What the UK's climate is like today
- ☐ How the greenhouse effect works
- ☐ The types of human activities increasing greenhouse gases
- ☐ How greenhouse gas levels have changed over time
- ☐ Who the main producers of greenhouse gases are
- ☐ What scientists think might happen to climate and sea-levels in the future, and why they are uncertain
- ☐ How the UK's climate could change in the future
- ☐ The challenges our changing climate might bring
- ☐ How climate change might affect people in the developing world.

## Key terms

Air masses
Climate change
Deforestation
Depressions
Ecosystems
Enhanced greenhouse effect
Extinction
Geological climate events
Global warming

Greenhouse gases
Ice age
Little Ice Age
Megafauna
Natural causes
Orbital changes
Orographic rainfall
Quaternary period
Solar output
Thermal growing season
Volcanic activity

### Which key terms match the following definitions?

**A** A trend whereby global temperatures rise over time, linked in modern times with the human production of greenhouse gases

**B** Those gases in the atmosphere that absorb outgoing radiation, hence increasing the temperature of the atmosphere

**C** A community of plants and animals that interact with each other and their physical environment

**D** A period in the Earth's past when the polar ice caps were much larger than today

**E** The most recent major geological period of Earth's history, consisting of the Pleistocene and the Holocene

**F** Long-term changes in global atmospheric conditions

**G** The chopping down and removal of trees to clear an area of forest

**H** Very large mammals, such as those that lived during the last ice age

**Foundation Question:** Describe two human activities which are increasing the amount of greenhouse gases in the atmosphere. (4 marks)

| Student answer (achieving 2 marks) | Feedback comments | Build a better answer (achieving 4 marks) |
|---|---|---|
| Fossil fuel.<br><br>Methane is increasing.<br><br>This is because there are more cows grown for their meat and they give it off. | • *Fossil fuel* is on the right lines, but this is not an activity so it does not score any marks.<br><br>• *Methane is increasing* is correct and scores 1 mark.<br><br>• *This is because...* scores 1 mark. This part of the answer describes an activity and links well to the previous point. | Fossil fuels are burnt in power stations and this gives off carbon dioxide.<br><br>Methane is increasing.<br><br>This is because more cows are bred for their meat and they produce methane. |

**Overall comment:** The student identified fossil fuels but did not develop this idea to describe how they are linked to greenhouse gases.

**Higher Question:** Describe two challenges the UK might face in the future due to global warming. (4 marks)

| Student answer (achieving 2 marks) | Feedback comments | Build a better answer (achieving 4 marks) |
|---|---|---|
| In the future there could be more rainfall, leading to floods.<br><br>There were severe floods in 2007 in Gloucestershire.<br><br>The floods were very costly and many people had to be rescued.<br><br>Flooding is a major challenge. | • *In the future...* scores 1 mark. In this part of the answer a challenge, flooding, is identified.<br><br>• *There were severe...* This is a specific example which supports the challenge of flooding. This part of the answer scores 1 mark.<br><br>• *The floods were...* Although this is an extension point, it is still about floods so does not gain any marks.<br><br>• *Flooding is a...* This repeats the point about flooding and does not identify a second challenge. As such, it does not score any marks. | In the future there could be more rainfall, leading to floods.<br><br>There were severe floods in 2007 in Gloucestershire, which were very costly and many people had to be rescued.<br><br>Higher temperatures could mean much hotter summers.<br><br>These could bring more forest fires and cause crops to die. |

**Overall comment:** Although there is some good information and examples included, the answer covered only one challenge – flooding. Look carefully at questions and spot how many different points are being asked for.

# Chapter 3 Battle for the biosphere

- Learn what ecosystems, biomes and the biosphere are, and why they matter.

- Explain how precipitation and temperature affect the global distribution of biomes.

- Understand that the biosphere and its people are interdependent on one another.

## What is the value of the biosphere?

### The world biome map

An **ecosystem** is a community of plants and animals that interact with each other and their physical environment. For instance, an area of woodland where trees draw nutrients from the soil and return nutrients too, as leaf-fall. The Earth is home to very large ecosystems called **biomes** (such as tropical rainforest or savannah). Together, these biomes make up our **biosphere**.

Latitude is a very important influence on air temperature because of the shape of the Earth. Sunlight arriving in the Tropics is highly concentrated whereas at the poles it is spread thinly. The reduced concentration of solar energy nearer the poles means less energy for photosynthesis. As a result, the rate of vegetation growth at higher latitudes is less than in the Tropics.

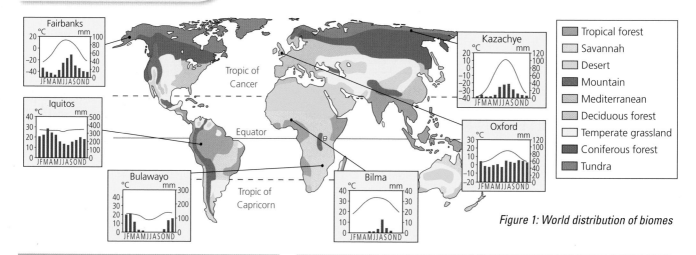

*Figure 1: World distribution of biomes*

### Tropical rainforest

Tropical rainforest lies mostly in a band either side of the Equator. Here, the Sun's rays are concentrated, heating moist air and causing it to rise. Heavy rainfall is the result – perfect conditions for evergreen rainforest.

### Deserts

Deserts are found close to the Tropic of Cancer in the northern hemisphere and close to the Tropic of Capricorn in the southern hemisphere. The air that rises over the Equator travels polewards after losing its moisture but sinks back down at the Tropics (due to complex air patterns caused by the Earth's rotation). The Sun's rays are still highly concentrated at this latitude and because the air is dry it brings desert conditions to places like the Sahara.

### Deciduous

Deciduous forest grows in higher latitudes. It is found in the UK and other places along the coast of continents where rainfall is high. The Sun's rays are less strong at this latitude, and cooler winter temperatures encourage trees to shed leaves at the end of autumn (this is what 'deciduous' means).

### Coniferous forest

Coniferous forest dominates by 60° north. Temperatures are so cold that trees have evolved with needle leaves that reduce moisture and heat loss. Snow slides easily off their sloping branches.

### Tundra

Tundra (cold desert) is found at the Arctic Circle. The Sun's rays have little strength here, and temperatures are below freezing for most of the year. Only tough short grasses can survive.

Precipitation is the other significant climatic influence on biome distribution. World patterns are very complex, but rainfall generally tends to be high in coastal and highland regions.

## Local factors affecting biomes

The world map of biomes (Figure 1) does not show the variations occurring at a local level. For example, the UK is shown as entirely covered with deciduous forest – which clearly is not true. Physical factors – especially drainage – affect local conditions. In parts of Scotland, where soil conditions are especially wet there are peat bogs rather than forests. But it is human factors – deliberate clearance of the original forests – that have caused the most change, wiping out most of the natural biome.

**Altitude** and distance inland are important influences too. Temperatures fall by about half a degree for every 100 metres increase in altitude. As a result, tough grasses quickly replace trees on higher mountain slopes (Figure 2). In the USA and in Asia, inland areas isolated from the sea suffer from low rainfall because winds blowing off the oceans quickly lose moisture – especially if the air passes over high mountains. The drier lands found further inland are said to be in a 'rain shadow'. On a smaller scale, the same effect is found in Great Britain.

Finally, geology and **soils** have an influence. For instance, limestone bedrock weathers to leave behind thin soil, which can store little water. Percolating water can also pass through limestone rock relatively easily. As a result, in the UK trees are rarely found in limestone areas, due to limited water availability.

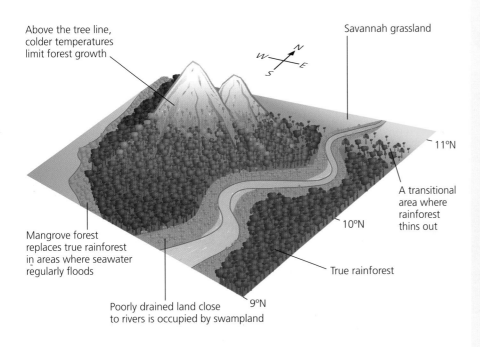

Figure 2: Local factors affecting where tropical rainforest can grow

**EXAM-STYLE QUESTION**

**The Earth's biosphere provides vital goods and services for people. Describe how the tree in Figure 4 could be used. (2 marks)**

■ **Basic answers** (0 marks)
Describe a tree, but not how it is used.

● **Good answers** (1 mark)
Correctly state that the enormous tree shown in Figure 4 gives water.

▲ **Excellent answers** (2 marks)
Recognise that the baobab tree is used by people as a source of water for vital services like drinking or washing.

*Figure 4: A baobab tree with its water-storing trunk*

## The biosphere life-support system

The biosphere acts as a life-support system for the planet, helping to regulate the composition of the atmosphere, maintaining soil health and regulating the **hydrological** (water) **cycle** (Figure 3).

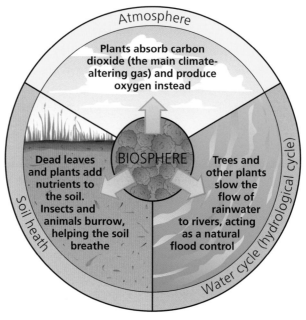

*Figure 3: The biosphere's interaction with other parts of the physical world*

As we learned in Chapter 2, the biosphere is an important carbon store. Plants take in and store carbon dioxide while producing oxygen. Tropical rainforest is sometimes described as 'the world's lungs'. This is why **deforestation** is a major cause of global warming.

Healthy soils require constant inputs of new nutrients from falling leaves and rotting plant remains. In the absence of human interference, nutrients such as nitrogen and potassium are continually recycled and exchanged between plants and the soil beneath them. Insects and animals that live on the forest floor also play a vital role by digging and burrowing in the soil. This is very important for soil health because it allows fresh air to circulate. When forest cover is removed, soil health can quickly deteriorate.

The biosphere also interacts with the hydrological cycle. Trees and plants 'catch' rainwater and slow down the speed at which it travels to the nearest river. They also take up water through root networks, later releasing vapour back into the atmosphere via a process called 'transpiration'. You will learn more about this in Chapter 4.

### The biosphere and people

The biosphere naturally interacts in good ways with the Earth's climate, soil and water. It also supports people. Plants and animals play a vital role in maintaining human life and aiding economic development.

The biosphere provides us with a wide range of **goods** (Figure 5), including:

| | |
|---|---|
| **Food** | When population numbers are low, natural ecosystems can be sustainably harvested, for example through berry and fruit picking. With larger populations, there is more pressure to replace natural vegetation with commercial crops. A good example of this is cereal production in the central parts of the USA. 'America's bread basket' is a region where naturally occurring grasses have been replaced by wheat and corn – increasingly used for the production of **biofuels** as well as food. |
| **Medicines** | Many naturally occurring substances act as medicines and remedies. For example, quinine is a plant extract found in the tropical rainforest. Native tribes paralyse animals with darts dipped in quinine. We call this natural store of medicines the **gene pool**. Genes are the building blocks of life found in the cells of plants and animals. |
| **Raw materials** | The biosphere provides raw materials for industrial activities. The most important of these is wood for the construction of houses and boats. Roofing material for thatched roofs also came from plants. In the dry savannah grasslands of countries like Kenya, trees have evolved to store water so they can survive long periods of drought. The baobab tree (Figure 4) has a trunk that stores water like a barrel. It is a vital source of water for people living in rural areas where rainfall is rare. |

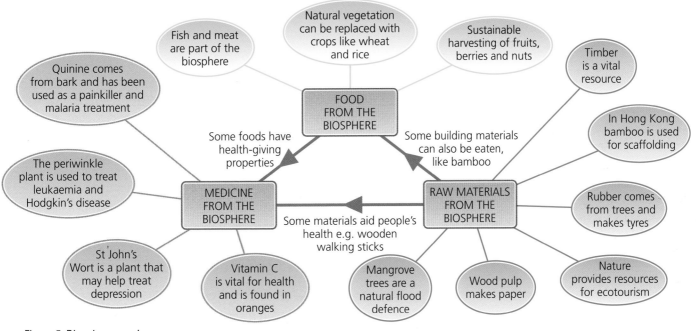

Figure 5: Biosphere goods

### Biosphere protection

The biosphere helps protect people's health, thanks to its gene pool. Vegetation also saves people's lives in other ways. For example, it has long been recognised that India's tropical rainforest provides an important service by stopping the river Ganges from flooding. In Hindu mythology, the Lord Shiva protects people from the power of the river goddess Ganga. He ties her to the land with his hair. In this legend, Shiva's hair represents rainforest trees which intercept rainfall and stop the river Ganga (the modern Ganges) from flooding the surrounding plains.

In the UK today, natural salt marsh vegetation is sometimes seen as a better defence against the sea than artificial sea walls. Faced with rising sea-levels, the UK government has abandoned some low-lying agricultural land to the sea. The idea is to create new salt marshes. At Wallasea Island in Essex, protective ditches and dykes have been deliberately breached, allowing sea-water to spill on to agricultural land behind where it deposits silt and builds new salt marsh. This 'managed retreat' from the sea is a relatively cheap and effective way of protecting more important areas further inland. Wave energy is quickly reduced by friction when high tides move across the new salt marsh. This is a good way of using the biosphere to deliver an important service. It is also a sensible local adaptation to climate change. (See soft engineering methods, page 83, Chapter 5.)

### The importance of forests to human life

- In Britain, wood from the deciduous forests helped make and fuel the factories that drove the Industrial Revolution. Centuries earlier, the sustainable practice of coppicing produced flexible strips of wood for longbows, while oak timber ships defeated the Spanish Armada. Today, remaining areas of forest are a resource for UK tourism.

- Coniferous forests in Scandinavia and North America provide much of the world's softwood. Pine is an especially popular material for lightweight furniture. In paper mills, the softwood is ground up and immersed in caustic soda to separate its fibres and create the pulp used to make paper.

- Tropical rainforest is a vital resource for poor countries in South America, Asia and Africa. Wood is used in many ways to generate incomes and build homes. Forest can also be cleared to provide land for other uses. However, many of these actions are unsustainable, as they do not leave behind forested areas for future generations. The environmental impacts of tropical rainforest removal are costly, both to local people and for the Earth as a whole. (Important medicines and raw materials can be derived from sustainably managed forests.)

Ten facts about the Amazonian rainforest and the River Amazon can be seen on the next page.

### Activity 1

In 2005, environmental campaigner Dorothy Stang was murdered after criticising powerful loggers in Amazonia. This shocking incident shows that very real battles are fought between the people who want to save the biosphere and others who use it in unsustainable ways.

Find out more about the people who try to save the rainforest at: http://ngm.nationalgeographic.com/2007/01/amazon-rain-forest/wallace-text

**1** Many tropical plants are used as industrial raw materials. Plant oils, gums, resins, tannin, rubber and dyes all come from the rainforest.

**2** Hardwood timber is a valuable resource that can be exported to help feed the growing population of rainforest countries like Brazil.

**3** The rainforest tree canopy protects the soil. If it is removed, there is a risk of soil being washed away by the convection rain that falls daily in Amazonia.

**10** Trees intercept and slow down rainwater as it moves through a river's drainage basin. This helps lower flood risk. Cutting down trees can result in greater flooding.

**4** If soil is removed by rain following deforestation, it gets washed into the River Amazon. Water temperatures rise as a result of this sediment being added, making survival hard for some fish.

## 10 facts about the Amazonian rainforest

**9** The population of South America is expected to increase by about 25% in the next 40 years. South American people want to clear their forest to gain living space.

**5** The Amazon forest has been described as 'the world's lungs'. Its trees produce much of the world's oxygen, which humans breathe, while also soaking up carbon dioxide.

**8** The tropical rainforest is a gene pool that the health of the entire planet is dependent upon. Millions of plant, insect and animal species live in Amazonia. They contain genetic material that can provide vital medical resources for the fight against diseases like leukaemia.

**7** Soils are protected from the drying effect of direct sunlight by the shade that trees provide. Soils suffer from the formation of a hard crust called laterite if the forest canopy is removed.

**6** The rainforest is a vital store of carbon. If it were all to disappear, global carbon dioxide levels would rise considerably – causing runaway climate change and faster temperature and sea level rises.

### Windsor Forest

The UK was once covered with deciduous forest. However, over-use meant that very little was left, especially after 1750, when more farmland was required to feed the rapidly growing population. Timber was also needed for housing, shipbuilding and to prop up mine shafts. By 1919, only 5% of England's natural forest remained. Since then, efforts have been made to protect what is left and to restore the forest in some places.

Windsor Forest is one small sustainably managed area of ancient woodland that has survived centuries of change in Britain. Once a medieval Royal hunting forest, the 3,100 hectares of woodland holds more than 900 oak and beech trees over 500 years old. More than 2,000 species of invertebrate and 1,000 species of fungi have been found here, displaying the rich **biodiversity** of the native deciduous forest.

The government agency Natural England now manages Windsor as a 'post-industrial forest', meaning that it is used for recreation and **conservation** rather than timber production.

*Figure 6: Windsor Forest*

Quick notes (Windsor Forest):
- The biosphere provides important goods, such as timber.
- The biosphere can also be used to provide services, such as tourism and leisure.

## Objectives

- Recognise that tropical rainforest is being degraded by human actions.

- Explain the range of indirect effects that pollution and climate change have on the biosphere.

- Understand that management is needed at a variety of scales if human use of the biosphere is to become more sustainable.

*Figure 7: The destruction of Amazonia*

# How have humans affected the biosphere and how might it be conserved?

Few places on Earth remain free from human interference. Even wilderness regions, far from settlement, are vulnerable to climate change and the sea-level rise that it threatens. Air and water pollution are problems that have 'gone global' over the past hundred years. You can be standing on a remote wilderness beach in Alaska and still find rubbish washed up along its shore. Over the next few pages, we will look at the effects of pollution and climate change on the Earth's biosphere – after first taking an in-depth look at the destruction of Amazonia.

## The destruction of tropical rainforest

Tropical rainforest is found between 10° north and 10° south of the Equator. Along with constant high temperatures of around 28 °C, there are daily falls of convection rain (the result of frequent intense thunderstorms). This ever-wet climate provides optimum conditions for plant growth, and the biodiversity of trees, vines, shrubs and plants reaches almost unbelievable levels. One study of the Napo region of Peru found 283 species of trees in an area of forest no larger than a football pitch.

But tropical rainforest in the Amazon and elsewhere is under threat. The rainforest is suffering from **degradation** – its character and quality are being constantly lowered – caused by human actions. The rate of rainforest clearance recently reached record levels. Now that deforestation is known to be a major cause of climate change, what drives countries like Brazil to cut down 100,000 square kilometres of precious forest each year?

Several large-scale land use changes have resulted in rainforest removal in Amazonia and elsewhere:

| | |
|---|---|
| **Mining** | The Grande Carajas development programme brought iron mines and aluminium plants to places where virgin forest once stood in Brazil. In order to transport mined materials, new roads must always be constructed. The building of the Trans-Amazon Highway led to enormous losses of primary forest. |
| **Timber** | The commercial clearance of tropical hardwood like mahogany and teak has been occurring for centuries (Figure 7), fuelled by demand amongst the world's rich for expensive furniture and flooring. This trade continues today, despite attempts to restrict imports by countries like the UK. |
| **Agricultural land** | Landless farmers migrate into Amazonia along the new roads. They then cut down forest for firewood or clear land to grow crops. As a result, the pattern of deforestation often follows the road network (see Figure 8). Huge areas of forest have been cleared for commercial agriculture. Major crops such as soya beans are grown on old rainforest soils in Brazil. However, other countries have taken political decisions to conserve rainforest, whilst also trying to meet economic goals. In Australia, the Skyrail Rainforest Cableway ferries tourists high above the canopy of the Queensland rainforest. 'Debt for nature' swaps have allowed some developing nations (e.g. Peru, Bolivia) to reduce their debts (for development) by agreeing to introduce new conservation measures. |

| | Rivers |
|---|---|
| | Roads |
| | Areas of deforestation |
| ooo | Urban zones |

0 ⊢———⊣ 150 km

*Figure 8: The deforestation and degradation of Amazonia (adapted from National Geographic Map)*

## The soya craze

In recent years, removal of Brazil's forest can be linked with the rise of soya bean production. Half of all forest clearance in 2005 took place in the state of Mato Grosso – whose governor Blairo Maggi also runs the world's largest soya bean production company. Maggi is known locally as 'O Rei da Soja' (King of Soya). During 2003, his first year as governor, the rate of deforestation in Brazil rose by 40%. In 2006, environmental campaigners Greenpeace awarded Maggi their 'Golden Chainsaw Award' for being the Brazilian who most contributed to the destruction of the Amazon rainforest!

Growth in soya cultivation is a response to increased global demand. Cattle ranchers around the world, especially in Europe, are switching to soya as a safe and healthy food source for their animals.

More people now want a diet that is meat-rich. Demand for meat is soaring in emerging **superpower countries** like China, whose citizens have become richer in recent years. This means that more crops like soya need to be grown to feed the ever-growing number of cattle. In other countries, such as South Sudan, reported 'land grabs' have taken place. An American company gained control of a large area of South Sudan's Central Equatoria State, including unlimited rights to harvest hardwood.

## People pressure

Future population growth in Brazil will put even more pressure on Amazonia. Already numbers have increased from 72 million in 1960 to nearly 200 million today. With a population growth rate of 1.5% per annum, more land is needed each year to provide room for farming and housing. In recent years, as many as 20 million extra people have also migrated to live in settlements along the Amazon and its tributaries. On the fringes of settlements such as Manaus, virgin forest is cleared every day to make way for more shanty dwellings.

**Results Plus**
**Build Better Answers**

### EXAM-STYLE QUESTION

**Study Figure 8. Describe the impacts that human activities are having on Amazonia. (4 marks)**

■ **Basic answers** (0–1 marks)
Neglect to mention human activity at all.

● **Good answers** (2 marks)
Correctly identify that the yellow areas are places where deforestation is taking place and emphasise that the area lost is now very large (using the scale to suggest the number of square kilometres that have been removed).

▲ **Excellent answers** (3–4 marks)
Also mention other impacts, such as urbanisation, or suggest that there is a pattern to the deforestation (you can see it follows the road and river routes through the region).

Over-fishing of the River Amazon's once well-stocked waters is another problem that Amazonia faces because of people pressure. The worst effects have been seen in the Pantanal, a giant wetland bordering the forest. Populations of the Pacu fish are in serious decline. This is worrying because the Pacu fish, in turn, helps disperse the seeds of some rainforest trees. Knowledge of this kind of species interdependency is important if the biosphere is to be managed and protected properly.

## Pollution and climate change bring stress and change

The deliberate removal of forest to create space for agriculture is an example of direct human actions damaging the biosphere. Degradation can also be caused by indirect means, such as pollution. The greatest indirect threat comes from air pollution in the form of carbon dioxide. Human-produced emissions of carbon dioxide (and other equivalent gases such as methane) are widely agreed to be the cause of global warming.

What effects will a warmer world have on the distribution of biomes and the lives of plants and animals? Scientists are observing the changes now taking place in environments all over the world. Studies of changing species habitats show that nature is already on the move. Polar bears are extending their hunting range inland as Arctic ice melts. In places that already suffer from water shortages, the consequences of climate change may be severe for the biosphere. Areas that currently enjoy a Mediterranean climate could experience a shift towards even drier conditions. This would result in frequent wildfires which could devastate ecosystems (Figure 9).

## Changes for the tropical rainforest

The flora (plants) and fauna (animals) of the tropical rainforest are reportedly already under stress from a warming climate:

- In recent years, there have been droughts reported in parts of the Amazon rainforest that normally experience little in the way of seasonal weather changes, other than lower rainfall in August and September. But in 2005 and 2010, longer-lasting drought affected a large area covering the northwest, central and southwest Amazon, including parts of Colombia, Peru and northern Bolivia. The cause was an unusual northerly shift in rain-bearing winds. Both droughts correspond well with some climate-change models that project a drying-out of the Amazon.

- Rainforest amphibians, reptiles and frogs are highly sensitive to environmental change. Frogs are viewed by some scientists as being the 'canary in the cage' of climate change. Scientists studying the rapid decline of the world's frog populations suspect that temperature changes brought on by climate change might be making rainforest frogs more vulnerable to disease.

*Figure 9: Wildfires destroy a forest during a drought*

## Activity 2

Study Figure 1 on page 42.

1. Consider what the effects might be of the world warming up by 1 or 2 degrees.

2. Consider what the effects of an even larger rise of 3 or 4 degrees might be.

3. What would happen to the distribution pattern? Remember that precipitation trends could change too. A warmer world could also be wetter in some places, due to increased evaporation over oceans and more storms.

How have humans affected the biosphere and how might it be conserved?

51

## Changes for the UK

The global map of biomes (Figure 1 on page 42) shows the British Isles as originally an area of deciduous forest. However, future changes in temperature and precipitation patterns for the south of England could make it more likely that grass will become the natural vegetation, rather than trees. Unusual and severe heatwaves have already affected the UK, most recently in 2006. There are also fewer frosts and winter cold spells now than in the 1960s. If these trends continue, then trees could become stressed wherever water supplies are limited (for instance in very well-drained areas with highly permeable soils).

If temperature rises between 2 °C and 4 °C, there will be major changes in the geographical distribution of British forests and other plant communities. Bird and animal species that live here would be affected by change too. Half of the UK's migratory birds have already experienced a severe decline in numbers in recent decades (Figure 10). Scientists believe that this may reflect difficulties these birds face in adapting to changing conditions, both in the UK and in Africa (where species such as the swallow spend winter).

However, while global warming will certainly bring changes to the biosphere in Britain, it may not lower our total biodiversity, because there could be winners as well as losers:

- Changed growing season – The growing season for plants in central England has already lengthened by about one month since 1900. Exotic species such as yucca and olive now thrive in the warmer and sometimes drier conditions experienced by southern England.

- New marine wildlife – There are more sightings of whales, sharks and sea turtles in British waters. These changes in marine biodiversity are thought to be a result of warming surface water temperatures.

- Moving tree-line – Some scientists think that the tree-line is moving upwards into higher areas in the Scottish Highlands. This should bring greater biodiversity to some upland areas.

## Biosphere conservation at the global scale

With nearly 200 countries in the world needing to sign agreements, global-scale attempts at conservation are quite a challenge – especially when people in poorer countries still rely heavily on farming, fishing, hunting and forestry to make a living. Despite these difficulties, important progress has been made.

Notable conservation framework successes include:

- CITES (Convention on International Trade in Endangered Species) – Throughout the world, harsh penalties now exist for poachers caught shooting endangered species, such as the rhino and tiger.

- Rainforest conservation – The global importance of rainforest has led to the introduction of international 'debt-for-nature' swap agreements. The US reduced the size of Guatemala's foreign debt repayments in exchange for promises that less rainforest will be cut down.

*Garden warbler* — down 21%

*Yellow wagtail* — down 70%

*Turtle dove* — down 82%

*Willow warbler* — down 60%

Figure 10: Is climate change to blame for Britain's vanishing birds?

### Skills Builder 2

Study Figure 10.

(a) Which bird has been worst affected?

(b) Describe the changes observed among the UK's migrant bird population.

(c) Explain ways in which global warming could be causing these changes.

## Decision-making skills

Which of the conservation frameworks is the most successful? Think about the extent of their coverage (scale and focus), cost and management.

### ResultsPlus
**Build Better Answers**

**EXAM-STYLE QUESTION**

**Choose a local example of biosphere management. Explain the methods used to make it more sustainable. (4 marks)**

■ **Basic answers** (0–1 marks)
Do little more than name a place (such as Scotland) where improvements have been made, but do not give any details.

● **Good answers** (2 marks)
Provide specific details (e.g. of wild boar coming back to the Caledonian Forest) and identify that this was important.

▲ **Excellent answers** (3–4 marks)
Also show proper understanding of sustainability and identify that all the plant and animal species that make up the biosphere should be preserved and passed on for future people to use and enjoy.

### Quick notes
(The Caledonian Forest):
- Sustainable living involves protecting the natural environment so that future people may enjoy it too.
- Where damage has been done in the past, it is still possible to make repairs.
- Controversial decisions are sometimes taken by the people responsible for protecting the natural environment.

- National Parks – Covering 13% of the land surface of the Earth, National Parks are a conservation success story. Globally, they are overseen by agencies including the United Nations.

- World Heritage Sites – Over 950 important sites have been awarded official recognition by the UN since 1972, many on account of their biosphere credentials.

- Wetland management – The Ramsar Convention has achieved real success, protecting wetlands everywhere from unsustainable over-exploitation.

### Global wetland management

Wetlands are areas where water is the primary factor controlling the environment and the associated plant and animal life. They occur where the **water table** lies at or near ground level, or where the land is covered by shallow water. Five major wetland types are generally recognised:

- Coastal wetlands, including coastal lagoons and coral reefs

- Estuaries, including river deltas and mangrove swamps

- Lakes and their edges

- Wetlands along rivers and streams

- Marsh, swamp and bogs.

In addition, there are human-made wetlands such as fish and shrimp ponds, farm ponds, irrigated agricultural land, salt pans, reservoirs, gravel pits, sewage farms and canals.

The **Ramsar** Convention on Wetlands is a global treaty established during 1971 at a meeting in the Iranian city of Ramsar. It was the first modern global treaty on the conservation and sustainable use of natural resources. 162 countries from all parts of the world have signed the treaty and 2,060 wetlands now have special protection. Ramsar sites cover 2 million square kilometres – a larger area than France, Germany, Spain and Switzerland combined.

## Sustainable biosphere management at the local level

### The Caledonian Forest

The remaining fragments of Scotland's Caledonian Forest are an important biological resource. They are all that remains of the original forest that first colonised Scotland at the end of the last Ice Age (10,000 years ago). The forest is an important 'environmental inheritance' for people living in the Scottish Highlands. The European Union has provided funding for the restoration of this internationally important habitat for the benefit of future generations. One of the objectives is to bring back lost animal habitats and populations within a part of the forest known as Glen Affric.

How have humans affected the biosphere and how might it be conserved?

53

The aim is to restore biodiversity to its natural level by adopting a countryside management strategy known as environmental stewardship. One of the first decisions taken was to reintroduce wild boar to Glen Affric, a species that has been hunted to extinction over the years. A close relative of pigs, wild boars feed on all sorts of forest floor organisms (such as fungi and insects). They disturb forest floor vegetation in ways that can aid the growth of trees. Forest managers describe the boars as a very useful 'ground disturbance force'.

## The challenge of achieving a sustainable outcome

An 'outcome' is the end-result of something. Sustainable development, in its fullest sense, is an approach which pledges that future generations of people should have the same economic and social opportunities as those alive today, while protecting the environment too. This is shown diagrammatically as a three-legged stool, with each of the three strands – environmental, economic and social **sustainability** – shown as being essential to the welfare of future generations (Figure 12).

However, the dual pressures of population growth and rising resource consumption (due to poverty reduction, especially in large countries like Brazil, China and Indonesia) make sustainable development hard to achieve. For instance, wetlands (Figure 11), as we have learned, are internationally important, highly ecologically productive environments, with high concentrations of birds, mammals, reptiles, amphibians and fish. Yet wetlands are also important water sources for people and so are threatened by population change. Climate change will worsen the situation further by bringing higher temperatures and water shortages to some vulnerable regions. The Zhalong wetlands in China's north-eastern Heilongjiang province, for example, are severely threatened by falling water levels and increased human activity. It seems unlikely that Zhalong has a sustainable future.

Elsewhere in the world, sustainable development programmes provide headaches for policymakers when the needs of different **stakeholders** come into conflict with one another. In the name of environmental sustainability, wolves and bears have been reintroduced into some wilderness areas in Europe and North America. However, these are still regarded as dangerous animals by the public, which makes the idea highly controversial. Walkers might not like meeting potentially dangerous animals! Tourist revenues for an affected area may fall as a result.

Finally, Brazil, as we have learned, is an especially interesting and important case. Half of its 200 million population are now 'middle-income' and the rest would probably like to join them. Brazil's government, like national governments throughout the world, hopes to keep building on this economic progress, encouraged by influential organisations such as the International Monetary Fund. In the short-term, one of the easiest ways for Brazil to increase its wealth is by more and more clearance of the rainforest – which would clearly not be sustainable. Do you think it is actually possible for Brazil's government to lift its entire population out of poverty whilst also safeguarding the survival of the rainforest? If so, how can it be done?

*Figure 11: Wetlands have rich biodiversity*

*Figure 12: The 'sustainability stool' is a visualisation of sustainable development*

# examzone

## Know Zone
## Battle for the biosphere

The biosphere contains a huge diversity of plant and animal species. This biodiversity is an important resource, as it provides humans with important goods and services. These vital resources are increasingly threatened and degraded by human activity, and urgent action is needed to conserve them for future generations.

## You should know...

- [ ] How to describe the distribution of biomes across the Earth's surface
- [ ] How climate (temperature and precipitation) influences the distribution and types of biomes
- [ ] How local factors, such as altitude and soil, also influence biome distribution
- [ ] How the biosphere provides important services to humans
- [ ] How the biosphere provides goods (resources) that humans use
- [ ] How goods and services support human life on Earth
- [ ] How humans are directly degrading biomes by their actions, such as deforestation
- [ ] An example of tropical rainforest degradation
- [ ] How global climate change is affecting biomes
- [ ] Why humans need to conserve biomes and biodiversity
- [ ] How to define sustainable and unsustainable
- [ ] Why humans need to use biomes more sustainably in the future
- [ ] How global actions and agreements could help make this possible
- [ ] Examples of global actions and agreements
- [ ] How local and national management could conserve biomes
- [ ] Examples of local and national management
- [ ] Why sustainable management is a challenge

## Key terms

| | | |
|---|---|---|
| Biodiversity | Degradation | Services |
| Biofuels | Ecosystem | Superpower |
| Biome | Gene pool | countries |
| Biosphere | Goods | Sustainability |
| CITES | Hydrological | Unsustainable |
| Conservation | cycle | Water table |
| Deforestation | Ramsar | Wilderness |

**Which key terms match the following definitions?**

**A** The chopping down and removal of trees to clear an area of forest

**B** A naturally occurring process or event which has the potential to cause loss of life or property

**C** Uncultivated, uninhabited and inhospitable regions

**D** The world's most powerful and influential nations – the USA and, increasingly, China and India

**E** The number and variety of living species found in a specific area

**F** The ability to keep something (such as the quality of life) going at the same rate or level

**G** A plant and animal community covering a large area of the Earth's surface

**H** The level in the soil or bedrock below which water is usually present

To check your answers, look at the glossary on page 321.

**Foundation Question:** Describe some of the goods that the biosphere provides humans with. (4 marks)

| Student answer (achieving 2 marks) | Feedback comments | Build a better answer (achieving 4 marks) |
|---|---|---|
| The biosphere gives humans goods and services.

Plants add oxygen into the air.

Humans might use the biosphere for food.

It could be to do with floods. | • **The biosphere gives...** This part of the answer just repeats the question so does not score any marks.
• **Plants add oxygen...** is correct and scores 1 mark.
• **Humans might use...** scores 1 mark but food is not very specific.
• **It could be...** does not score any marks because it is not linked to goods or services. | People can cut trees down for timber, which are goods.

Plants add oxygen into the air.

Humans can hunt animals for food.

Forests can stop flooding from happening. |

**Overall comment:** The student would have scored more marks if they had focused more carefully on goods and services. Also, remember that there is no need to repeat the question in your answer.

---

**Higher Question:** Describe **two** goods that the biosphere provides and explain why they are important. (4 marks)

| Student answer (achieving 3 marks) | Feedback comments | Build a better answer (achieving 4 marks) |
|---|---|---|
| The biosphere regulates water movement around the hydrological cycle.

This can prevent flooding from destroying homes.

It also makes the atmosphere have the right amount of gases.

This is really important. | • **The biosphere regulates...** This is a good start to the answer as the student describes a good and uses some good geographical terminology. This part of the answer scores 1 mark.
• **This can prevent...** This also scores 1 mark because it links to a good already described and explains why flooding is important.
• **It also makes...** This describes another good and scores 1 mark, but it is only partly correct.
• **This is really important.** No explanation is given so this part of the answer does not score any marks. | The biosphere regulates water movement around the hydrological cycle.

This can prevent flooding from destroying homes.

It also regulates the amount of carbon dioxide and oxygen in the air.

This means the air is breathable for humans and animals. |

**Overall comment:** Although there is some good information and examples included, the answer covered only one challenge – flooding. Look carefully at questions and spot how many different points are being asked for.

# Chapter 4  Water world

## Objectives

- Recognise the main stores and transfers of the hydrological cycle.

- ◉ Consider how hydrological systems might respond to climate change.

- ◎ Explain how changes in water supplies can impact on people and ecosystems.

## Why is water important to the health of the planet?

### The hydrological cycle

The Earth's water is continuously on the move. It is constantly recycled, moving from place to place via key processes called **water transfers**. Sometimes it remains held for a period of time in a **water store**, such as glacier ice, before continuing its journey again. The complete picture of what happens is called the global **hydrological cycle** (Figure 1). The key transfers are detailed in the table below.

**Key:**

**13** Water store. The number of cubic kilometres of water that is stored.

↑ Water transfer. Movements of water through the atmosphere, biosphere and lithosphere.

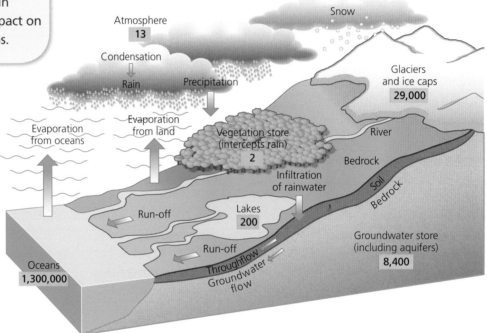

Figure 1: The global hydrological cycle, showing the links between the atmosphere, biosphere and lithosphere

| | |
|---|---|
| **Evaporation** | Water will return to the atmosphere in vapour form, once it has been warmed by sunlight. |
| **Condensation** | When water vapour cools down again, tiny water droplets form, often becoming visible as cloud. |
| **Precipitation** | Tiny water droplets collide and grow, building bigger droplets that then – under gravity – fall to the ground as rain or snow. |
| **Run-off** | Some rainfall flows quickly over the surface of the ground to reach rivers and lakes. **Run-off** is often responsible for causing flash floods. Also called overland flow. |
| **Throughflow** | Sometimes rainwater has time to soak into the soil (this is called **infiltration**). The water is then transferred slowly through the soil until it reaches a river. |
| **Groundwater flow** | Once it has entered the soil, some rainwater soaks into the rock below. It moves very slowly through tiny air spaces in the rock, called **pores**, or it flows along cracks, called **joints**. |

## Water stores

While moving through the stages of the hydrological cycle, water sometimes stops flowing for periods of time. Instead, it becomes temporarily stored in lakes, rivers, ice caps or the oceans. 99% of the Earth's water is kept in the last two of these – the ice caps and the oceans. Transfer movements through the cycle begin again when the Sun's energy triggers evaporation or ice melting.

The Earth's biosphere also captures and stores water. Vegetation leaves are a temporary storage space where water sits for a time, until it evaporates or drips to the ground. In vegetated drainage basins, therefore, the transfer of rainwater is regulated by plants. This means that river levels do not rise and fall too quickly, even when sudden bursts of heavy rainfall arrive. For this reason, forest is an excellent flood defence. It intercepts rainfall that might otherwise be transferred to the river too quickly and cause a flash flood.

The Earth's lithosphere also acts as a water store. Soils and rocks can hold water for a period of time. The actual volume that can be stored is determined by how **permeable** substances are, along with other important geological factors such as slope angle and jointing. Sometimes a layer of permeable chalk lies on top of a different rock type that is **impermeable** – it will not let water pass through. Rainwater can soak down into the tiny pores (air spaces) found in the chalk, but the impermeable layer below acts as a barrier, and a water store – called an **aquifer** – develops.

### Skills Builder 1

Study Figure 1.

(a) Name the two largest water stores.

(b) Describe the fastest transfer route that rain falling over land can take to reach the ocean.

(c) Explain why tree planting slows water transfers through the hydrological cycle.

(d) Explain why some clouds do not produce precipitation.

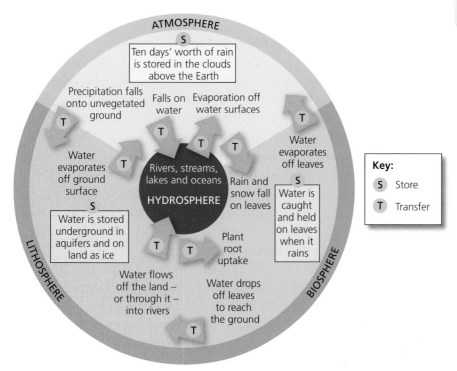

*Figure 2: The links between the hydrosphere, atmosphere, biosphere and lithosphere*

## Climate change and the hydrological cycle

How might climate change and global warming impact on the world's hydrological systems? Some places where water is naturally scarce may become even more drought-stressed in the future, as we have already learned. Places that lack enough water to meet local social and economic needs suffer from **water insufficiency**. Climate change could lead to more places being affected by this condition. However, a warmer world will also be one where more evaporation takes place over the oceans – and what goes up must come down! Climate change scientists therefore believe that some already wet places will become even wetter, although not always at the time of the year when we might expect it (resulting in **water unreliability**).

There are reports of more rain-bearing storms forming over the Atlantic, close to the UK. Heavy inputs of water into Britain's drainage basins were seen in 2004 (the flash flooding of Boscastle), 2007 (widespread urban flooding), 2008 (flooding in Wales and north-east England) and June 2012 (the wettest ever recorded). Many scientists see these events as symptoms of a changing hydrological cycle. They think that in future, the UK will be faced with:

- Longer, hotter summers, especially in London and the south-east (like the extraordinarily dry summers of 1976 and 2006 that stretched UK water supplies to their limit)

- Warmer, wetter winters, with more rain-bearing storms predicted to pass over all parts of the UK, and not just northern regions.

### Climate change and water supplies in America

Climate change may bring reduced rainfall inputs to arid places where water is already naturally scarce for all or part of the year. People and ecosystems in such areas could become seriously threatened. One such imperilled region is the south-west of the United States. Already far too dependent on irrigation from the Colorado River (see pages 64–65), this region is naturally very dry. Yet despite this arid climate, people like the hot weather there, and the state of Arizona saw a 40% rise in population in the 1990s. Just across the border, in Nevada, the growing city of Las Vegas (Figure 3) doubled its consumption of water between 1985 and 2000. However, severe drought conditions have recently begun to affect the entire region. For Las Vegas, 2002, 2004 and 2007 were some of the very driest years on record.

The city's authorities have responded by removing the dried-up remains of some sports pitches and re-covering the ground with plastic turf. Drought-tolerant 'desert landscaping' is now encouraged instead of planting grass lawns in front of new buildings. This could be bad news for Las Vegas wildlife such as snakes and grey squirrels, who have taken up residence in the artificially irrigated areas.

Las Vegas originally grew due to local supplies of groundwater. Artesian spring water provided travellers with a stopping point, back in the 1800s. Today, however, local water supplies have been depleted, sometimes leading to ground subsidence. Almost 2 million residents in Las Vegas Valley, and tourists too, are now dependent on water piped in from the Colorado River.

**ResultsPlus**
**Build Better Answers**

### EXAM-STYLE QUESTION

**Describe two ways in which climate change could impact on water supplies. (4 marks)**

■ **Basic answers** (0–1 marks)
State that climate change would make the sea level rise. People do not take their drinking water from the sea, so this is not a good answer about changing water supplies.

● **Good answers** (2 marks)
Describe two climate change impacts, one of which is well connected with the idea of changing water supplies. For example, the disappearance of glaciers might deprive some places of their regular drinking supplies from annual meltwater.

▲ **Excellent answers** (3–4 marks)
Provide two developed impacts. For example, if more storms start passing over the UK, it could add water to reservoirs (meaning that climate change could actually help water supplies in some cases).

*Figure 3: Fountains in Las Vegas. Can this kind of water use continue in the Nevada desert?*

## Climate change and water supplies in Spain

The Ebro is the longest river in Spain. It rises in the Cantabrian Mountains and flows south through semi-arid lands for nearly a thousand kilometres before reaching the Mediterranean Sea. Over the past 50 years, many dams have been put in place to help build water stores for the local people in this dry climate. These dams reduce the level of silts carried by the river, causing shrinkage of the depositional landform at the mouth of the river – the Ebro Delta. The delta is a valuable agricultural area and a huge nature reserve of international importance.

With less water reaching the river estuary, due to its retention in reservoirs, one further problem, 'salt wedge growth', has also developed. Because of the reduced river flow, salty sea water now pushes further upstream into the estuary, harming freshwater fish and other organisms that cannot tolerate saline conditions.

Climate change now threatens to intensify all of these problems further:

● Changing rainfall patterns could result in even lower river flow, leading to further shrinkage of the delta.

● If seas rise by one metre by 2100, this will lead to most of the Ebro delta becoming submerged. Every year, 5 mm in height is already being lost, due to a combination of sea-level rise and the ground sinking.

● The salt water wedge will penetrate even further inland in the future, bringing greater harm to freshwater **ecosystems**.

*Figure 4: The Ebro Delta*

## Unreliable and insufficient water supplies

The transfers of the hydrological cycle vary over different time scales, especially in places that naturally experience a fluctuating or seasonally changing climate. These fluctuations can result in a state of water unreliability, meaning that dry conditions are found at a time of year when water is needed (even if the average annual rainfall total looks to be enough).

| Seasonal variations | Some climates have distinct wet and dry seasons. |
| --- | --- |
| Longer natural cycles | Natural weather cycles bring clusters of drier or wetter years in some regions of the world, for reasons not always fully understood. |
| Climate change | Year-on-year temperature rises are being recorded across the world. These data suggest global warming is taking place, affecting all areas. |

The Sahel region (see page 121), where rainfall is notoriously unreliable, suffers from all three problems (Figure 5). The climate here naturally has a marked wet and dry season. The wet season is short, and when the rain does fall, it comes in high-intensity bursts. Torrential rains in August and September 2007 affected thousands of people throughout the region and 50,000 people were left homeless in Niger. However, the rainwater ran over the land so quickly that little was captured or stored, so water availability did not actually increase for many local people and remained insufficient.

Since the 1970s, the region has also been in the grip of a longer-term drought. Figures suggest rainfall in some parts of the Sahel has fallen by 30% over this time. During most years of the past three decades, the Sahel countries – Burkina Faso, Chad, Mali, Mauritania, Niger and Senegal – have suffered chronic water and food shortages.

In the past, these countries coped with unreliable water supplies. Long-term records tell us that the Sahel's climate has always been variable and that cyclical drought is a fact of life in this part of world. But local people used to be more resilient to drought when it arrived. Farmers were often nomadic and moved to find new water if local supplies ran short. Changes in agriculture, however, have meant that people cannot do this any more. Food is increasingly grown for export by big businesses – using all the best land and local water supplies. The expansion of large-scale cotton and peanut production has permanently pushed many subsistence farmers into marginal lands that are too dry for most farming. Hunger, thirst and malnutrition follow as a result of insufficient water supplies.

The consequences of the changes is a growing number of people suffering from **water insecurity**. This means that there is insufficient availability of water to meet people's health and economic needs, possibly combined with issues of water safety too (such as polluted water sources).

## Activity 1

Local people can experience water shortages in places like the Sahel, even when rain has fallen recently.

Think about how each of the following can impact on water availability for local people:

- lack of technology
- civil war
- cash crops
- population increases.

Is it possible that water shortages actually owe more to human rather than physical factors?

*Figure 5: The Sahel suffers from seasonal variations, longer natural cycles and climate change*

## Case study: Australia's unreliable water

Australia is a naturally arid continent (Figure 6). However, water supplies have recently become even more unreliable than usual. Since the 1990s, Australia has experienced a decade-long dry spell. Rainfall has fallen to 25% of the long-term average. In 2007, the Murray River reached its lowest level in over 100 years of records. Some Australian geographers say that this is the worst drought for 1,000 years. In 2009, the country experienced terrible bush fires.

There are several reasons why water shortages are even greater than usual. El Niño is a complex global weather pattern that periodically reduces Australian rainfall. It has caused further drying-up of this already arid continent. Over-use of water by agriculture and industry has made the situation even worse. It may also be that we are seeing the first major effects of climate change and global warming in Australia. Impacts of increased aridity include:

- Wheat crop yields have fallen by around one half and people face big food price rises.
- Hundreds of winemakers could be forced out of their AU$2.6 billion business.
- The government says it will pay severely affected farmers AU$150,000 each to walk away from their land. This 'exit with dignity' strategy will help farmers retrain or relocate.
- Local ecosystems are suffering. Wells dug by sheep farmers are drying up, and local native animals like kangaroos lose their use of them too.

A new technology called electromagnetic imaging is helping some farmers survive by finding new hidden water stores beneath their fields. However, if current trends continue, Australia also needs to seriously change its water consumption habits.

Figure 6: The Australian desert, where rain may be starting to fall even less frequently

**Case study quick notes:**
- Unreliable water supplies are a natural feature of life in some places.
- Unreliable water supplies threaten a range of agricultural and industrial activities and seem to be getting worse.
- Farmers must try adapting to the increased aridity or else 'retreat' and change their occupation.

*Figure 7: Pollution in the Kabul River, Afghanistan*

# How can water resources be sustainably managed?

## Unhealthy hydrology – problems with water quality

The human pressure on rivers is enormous – they are used for transport, for industrial processes, for drinking and for sewage disposal. Rivers in industrialising nations like India and China experience especially damaging impacts because, as yet, they have few laws in place to protect their water and ecosystems. **River pollution** in rivers outside Europe and North America is often appalling (Figure 7). Common types of damaging pollution include:

| | |
|---|---|
| **Sewage disposal** | Cholera bacteria spread quickly in river water when contaminated human sewage is added. As long as toilet waste goes untreated, the cycle of cholera infection will continue. |
| **Industrial pollution** | Liverpool's River Mersey used to be hugely polluted. Alkali factories used the polluting Leblanc process to make vital chemicals for the manufacture of textiles. Vast amounts of calcium sulphide by-product were leached into local streams, making conditions poisonous for wildlife. Fortunately, this is no longer the case today. The factories have mostly disappeared, while the UK Environment Agency can issue heavy fines to any remaining polluters. |
| **Plastic bags** | In China and Bangladesh, thin plastic bags (less than 0.025 mm thick) are prohibited because they can block waterways and sewers. This worsens flooding problems, especially in the monsoon season. Plastic bags also contribute to health problems in many countries. They get trapped by vegetation along river banks, providing pools of still warm water where malaria-bearing mosquitoes will breed. |
| **Intensive agriculture** | Nitrate fertiliser feeds the growth of algae when it is washed into rivers and lakes by rainwater run-off. Later, the algae die. The decomposition process exhausts oxygen in the water (eutrophication), resulting in the death of fish and other aquatic organisms. |

## Disrupted water supplies

It is not just water quality that can be harmed by human actions. Water supplies are also affected. The transfers of the hydrological cycle can be speeded up or slowed down, and natural water stores such as lakes may shrink because of over-use by people – the Aral Sea is a good example.

### Changing transfers

Run-off is a very important hydrological transfer, as Figure 1 showed. Too much overland flow (run-off) may result in an over-supply of water to rivers, possibly leading to flooding. One common cause for accelerated run-off is **deforestation**. Without trees to intercept and store rainfall, run-off is more likely to occur. This is because the interception store naturally slows down the rate at which rain reaches the soil surface (via drip and stem-flow from the vegetation canopy). Another cause is urbanisation. In urban areas, the soil is covered by tarmac and concrete. As a result, rainwater travels much more quickly towards the nearest river via run-off. Drainpipes also speed up urban water flows. Dam and reservoir construction also modifies the hydrology of a catchment area in quite radical ways (see pages 64–67).

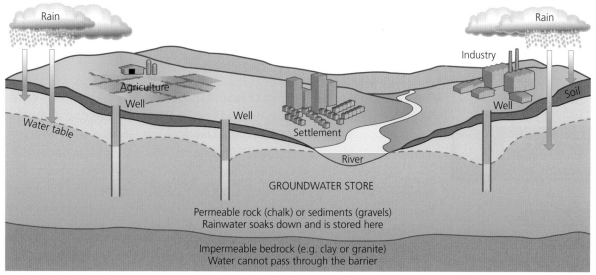

*Figure 8: The features of a groundwater store (or aquifer)*

## Changing stores

Water stores are essential for human societies. However, they need careful management if they are to remain a sustainable resource. Water stores are naturally recharged whenever it rains. However, if too much water is taken from them too quickly then they will dry out before they have a chance to recharge. Many major cities like Mexico City are sited above **groundwater** stores called aquifers. However, **over abstraction** (over-pumping) sometimes takes place. When water is removed from the pore spaces in permeable rock, the rock loses strength. If too much weight has been added overhead by buildings, empty pores close up (imagine pressing down on a Swiss cheese full of holes). Ground beneath the city becomes compacted, leading to a permanent loss of water storage capacity.

## Coca-Cola and Plachimada's groundwater store

In the drought-prone Indian state of Kerala, a large groundwater store used to be found close to the village of Plachimada. In 2000, this water supply attracted the attention of the transnational corporation Coca-Cola. The company was keen to increase sales of its famous drink in India, but the most important ingredient in its production – clean water – was not always easy to find locally. Groundwater supplies were needed.

First, Coca-Cola set up a subsidiary firm called Hindustan Coca-Cola Beverages. Next, they established a bottling plant near Plachimada. Finally, six wells were dug, tapping into the precious groundwater store. Very soon afterwards, water shortages began to be reported. By 2004, news from the region described a desperate situation. Wells used by the Plachimada villagers had dried up.

It is likely that drought in the region would have brought water shortages anyway, even without Coca-Cola's arrival. However, it was probably not a smart move to make industrial use of this groundwater store, given the region's known drought risk. Coca-Cola has since helped the people of Plachimada survive the drought by driving tankers full of fresh water to the village.

## Skills Builder 2

Study Figure 8.

(a) Name one impermeable rock type.

(b) Explain how the presence of impermeable rock helps create a water store.

(c) Suggest how groundwater stores impact on population distribution in arid regions.

## Quick notes (Coca-Cola):

- Economic development is more likely to be located near hydrological stores.
- Water stores in dry areas must be used carefully if they are to remain a sustainable resource for local people.

## EXAM-STYLE QUESTION

**Choose one large-scale water management project. Describe one cost and one benefit of the project. (4 marks)**

■ **Basic answers** (0–1 marks)
Do not recognise that a large-scale project is the focus of the question and score badly.

● **Good answers** (2 marks)
Choose an appropriate example like the Three Gorges or the Colorado River, state that water supplies in the region have been improved and also say why this is the case (river flow has been evened out over the year).

▲ **Excellent answers** (3–4 marks)
Also provide good details of a cost, for example the loss of Colorado squawfish due to cold sediment-free water.

## Large-scale water management: The Colorado River

Some of the world's biggest construction projects have been involved with **water management** on an immense scale. Worthy of an in-depth look is the management of the Colorado River.

The Colorado is one of the most highly managed rivers in the world. Its water feeds the needs of seven US states and Mexico. The natural regime of the river was completely altered during the twentieth century to give US cities and farms reliable water supplies all year round. This is because in its natural state, the Colorado had a very low flow between September and April, whereas during the summer months, snowmelt in the Rockies and Wind River Mountains used to cause a big rise in river levels. In the most extreme years of the early twentieth century, the Colorado's discharge was 13 times higher in mid summer than in winter.

Since then, water management projects have radically changed the regime of the Colorado. Flood peaks have been smoothed out by dam building and reservoir construction. The Hoover Dam was built in 1935. Regarded as a technological marvel at the time, the dam stores the equivalent of two years' river flow in Lake Mead.

The Glen Canyon Dam followed in 1966. It also holds an enormous amount of water. The two lakes build up reliable supplies for thirsty desert cities like Las Vegas. Water is piped along man-made constructions called aqueducts.

However, these direct changes to the annual pattern of river flow have had negative knock-on impacts for river processes, river valley landforms and river ecosystems:

● Silts and sands get trapped behind the dams. Rates of transportation of sediments have fallen nearly to zero since the dams were built. This makes the water colder (silt heats up in sunlight, warming water around it). Four of the Colorado's fish species, including the squawfish, have been lost. Only trout, which prefer clear cool water, have risen in numbers as a result of the changes.

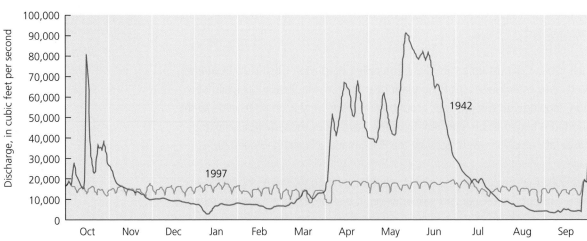

*Figure 9: The Colorado River regime before and after construction of the Glen Canyon Dam (10,000 cubic feet = 283 cubic metres)*

Figure 10: View of the Colorado River and the surrounding area

◉ Sandbanks along the sides of the river in its lower course have been starved of sediment and are reduced in size. Plants and animals that live on the sandbanks – such as the Kanab ambersnail and the Southwestern willow flycatcher – have also declined. The sandbanks were once used for fishing and rafting but they cannot be used any longer.

US government agencies introduced new management measures in 1996 to tackle these problems (Figure 11). Four giant steel pipes were fitted to Glen Canyon Dam in 2004. Every few years they are opened, releasing massive jets of water from the reservoir. This escaping water has high energy levels. It can flush out large amounts of sediment that have built up behind the dam. As the water heads downstream it begins to lose energy and sediments are deposited, helping rebuild the lost sandbanks. In time, scientists believe they can restore river and ecosystem conditions to a more natural state.

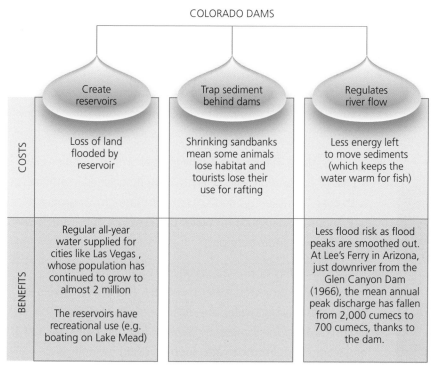

COLORADO DAMS

| | Create reservoirs | Trap sediment behind dams | Regulates river flow |
|---|---|---|---|
| COSTS | Loss of land flooded by reservoir | Shrinking sandbanks mean some animals lose habitat and tourists lose their use for rafting | Less energy left to move sediments (which keeps the water warm for fish) |
| BENEFITS | Regular all-year water supplied for cities like Las Vegas , whose population has continued to grow to almost 2 million<br><br>The reservoirs have recreational use (e.g. boating on Lake Mead) | | Less flood risk as flood peaks are smoothed out. At Lee's Ferry in Arizona, just downriver from the Glen Canyon Dam (1966), the mean annual peak discharge has fallen from 2,000 cumecs to 700 cumecs, thanks to the dam. |

Figure 11: Costs and benefits for the Colorado River

## Decision-making skills

Why are the following four people likely to have conflicting views about the use of the Colorado River?

- Californian farmers who need the irrigation water in huge quantities
- Native Americans on whose land the giant dams are built
- Californian urban dwellers who need huge quantities of water for golf courses and gardens
- Mexicans near the mouth of the Colorado who would also like some water for their farming.

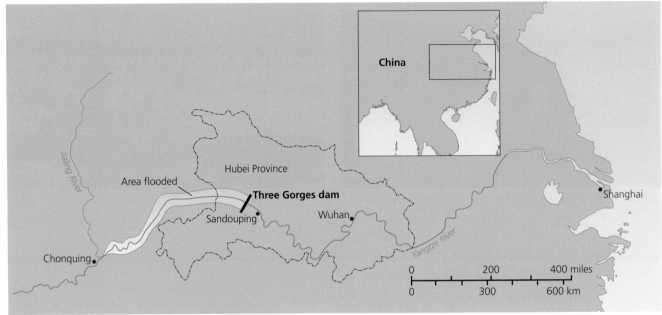

*Figure 12: Location of the Three Gorges dam in China*

*Figure 13: The Three Gorges dam, with the reservoir behind*

*Figure 14: Outflow of water from the Three Gorges dam*

## Large-scale water management: The Three Gorges dam

China is a middle-income developing country that desperately needs to improve its water and energy supplies. An important step towards that goal has been taken with the completion of the Three Gorges dam across the Yangtze River in Hubei Province. The enormous project was completed in various stages between 2003 and 2012 at a cost of between US$22 billion and US$39 billion (estimates vary). The reservoir is over 1 km wide and has flooded a 600-km stretch of the river valley. The dam is the world's largest and is also the site of the largest **hydropower** project on the planet.

The Three Gorges project was undertaken to help China meet the changing needs of its population:

● China is home to 1.3 billion people, many of whom have recently moved to urban areas where they are earning higher wages than in rural areas. As a result, individual water consumption is growing (with more people owning appliances such as washing machines).

● By 2025, there may be 220 Chinese cities with populations in excess of one million, 8 of which will be megacities – settlements with over 10 million residents. Many of these growing cities have high industrial demands for water (as a coolant, lubricant or raw material for manufacturing).

● The sustainability of China's urban industrial settlements is threatened by projected water shortages. Many major rivers are fed by seasonal meltwater run-off from major glaciers in the region, notably the Himalayan Plateau. Future climate change projections suggest meltwater supplies could dwindle as the climate warms (see page 58).

As well as meeting the water needs of one-fifth of the world's entire population, China's leaders also need to address the country's growing energy needs. Already the world's largest emitter of greenhouse gases, China's energy use is projected to keep increasing for many decades to come. The Chinese leadership says it is committed to limiting future increases in the country's carbon footprint size as far as it can. Increased use of hydropower means that less new energy needs to be generated by burning coal.

Giant turbines are driven by water as it flows out of the Three Gorges reservoir. These turbines power electrical generators that have a combined capacity of 22.5 million kilowatts (22,500 megawatts), the equivalent of fifteen nuclear reactors.

### Benefits and costs

China's government has argued that the environmental benefits of hydroelectric power outweigh its costs. Specifically, the Three Gorges dam is producing enough electricity each year to replace 50 million tonnes of coal use and thus reduce China's greenhouse gas emissions by 100 million tonnes of carbon. There are other key benefits too:

- By regulating water flow, the reservoir and dam can help China meet a range of growing domestic and industrial water needs (see above).

- The dam decreases the risk of flooding at certain times of the year (in the past, millions of people have lost their lives to flooding and water-borne diseases along the Yangtze River).

However, the social and environmental costs of the project have been criticised. These include:

- The forced relocation of 1.4 million people during the initial construction phase, beginning in 1993. (Thirteen cities, 140 towns and 1,350 villages were de-populated and China's government has since admitted that it could have done more to help people re-settle elsewhere.)

- Ongoing relocation of around 100,000 more people who have been left at danger from landslides and bank collapse in settlements upstream from the reservoir. (The water rises and falls depending on the season, making the banks unstable – there have been landslides at 350 sites and another 5,000 are being monitored.)

- Algal blooms have been recorded along the length of the reservoir. (These can lead to de-oxygenation of the water and fish stocks being killed.)

- Increased deposition and sedimentation upstream from the dam. (The flow of the Yangtze has been reduced to a fraction of what it was, causing sediment to settle on the riverbed – local government officials say that increased deposition could soon make parts of the river impassable for shipping.)

Do the costs outweigh the benefits? This is very hard to evaluate, given that a global benefit (lower carbon emissions) needs to be weighed against a host of local problems. What do *you* think?

68

## ResultsPlus
### Build Better Answers

## EXAM-STYLE QUESTION

**Using examples, explain how human activity can interfere with the size of water supplies. (4 marks)**

■ **Basic answers** (0–1 marks)
Ignore the word 'size', and instead deal with pollution and water quality.

● **Good answers** (2 marks)
Clearly describe how water supplies are improved by dam building (e.g. the Colorado) or made worse by over-using underground supplies.

▲ **Excellent answers** (3–4 marks)
Give good details of more than one example, or show sound understanding of physical processes (e.g. mentioning the Plachimada aquifer, or explaining that aquifer use becomes unsustainable if all the water is extracted before new rain refills it).

## Decision-making skills

Draw a diagram to show how small-scale water projects can be environmentally, socially and economically sustainable.

## Small-scale water management

In water-stressed areas, small-scale solutions can be introduced that keep water supplies more constant throughout the year. Poor countries especially can benefit from the use of **intermediate technology** – appropriate, small-scale, practical solutions that local people can apply and maintain themselves. They include the use of:

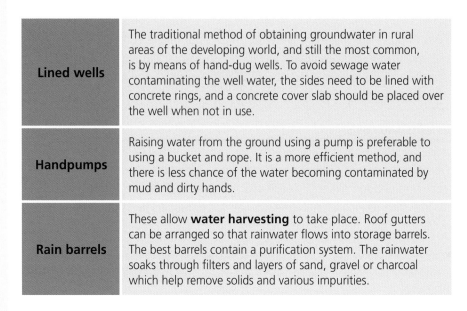

| | |
|---|---|
| **Lined wells** | The traditional method of obtaining groundwater in rural areas of the developing world, and still the most common, is by means of hand-dug wells. To avoid sewage water contaminating the well water, the sides need to be lined with concrete rings, and a concrete cover slab should be placed over the well when not in use. |
| **Handpumps** | Raising water from the ground using a pump is preferable to using a bucket and rope. It is a more efficient method, and there is less chance of the water becoming contaminated by mud and dirty hands. |
| **Rain barrels** | These allow **water harvesting** to take place. Roof gutters can be arranged so that rainwater flows into storage barrels. The best barrels contain a purification system. The rainwater soaks through filters and layers of sand, gravel or charcoal which help remove solids and various impurities. |

In addition to increased supplies, improved water quality is another common aim of intermediate technology projects (Figure 15). Without improved water quality, cholera or diarrhoea often remain persistent problems for communities. Cholera bacteria find their way into the human small intestine after being swallowed. There they thrive, drawing water into the bowel by osmosis. As a result, the sufferer experiences violent diarrhoea. New generations of thriving bacteria rapidly exit the unhappy victim. Soon they are back in the river or they have leached into unlined wells, once again polluting potential drinking water. This cycle of infection will continue without help from intermediate technology.

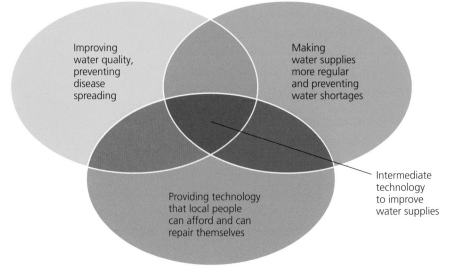

*Figure 15: Intermediate technology for improved water supplies*

### Introducing the Afridev handpump to Tanzania

Tanzania is a desperately poor country in east Africa, where life expectancy is just 46. 70% of Tanzania's rural population and 30% of urban dwellers have no access to safe water. Diarrhoea accounts for at least 20% of infant deaths. It is a common problem in places where water supplies are contaminated with sewage, and can prove fatal, particularly to children.

International agency WaterAid brings intermediate technology to communities needing sewage-free water. Working in partnership with Tanzania's Anglican Church, for example, WaterAid helped the community of Chessa village construct a new well, 24 metres deep, fitted with an 'Afridev' handpump (Figure 16). Fifteen families can now drink safe underground water.

A key principle of WaterAid projects is that communities take ownership of the technology and are responsible for its upkeep. For this reason the technology must be simple, so that ordinary people can fix it when it breaks. The Afridev handpump provided in Chessa is not very sophisticated. Better handpumps exist with a lower rate of breakdown. However, when more advanced machines break down, a specialist engineer needs to be called in – which can leave local people without water while they wait for repairs.

An Afridev handpump may break down fairly often but it can quickly be repaired by a local 'caretaker'. In Chessa, brothers Vincent and Lazaro William volunteered to be the Afridev handpump's caretakers. After one week's training, they were given a toolkit consisting of a spanner and wrench. They take the pump apart once a year to check the whole system, and fix whatever problems arise on a day-to-day basis.

Also helping people in Tanzania is local radio soap show Pilika Pilika (Busy, Busy). WaterAid supports the show and uses it to spread messages on hygiene, sanitation and water management. In one recent episode, a major character had an unpleasant experience when he tripped and fell into a well that had been heavily polluted with toilet water. The audience were reminded that they need to keep their water supplies free of sewage!

**Quick notes (Afridev handpump):**

- Poor water supplies are still a major cause of death and illness for low-income countries like Tanzania.
- Charities like WaterAid do vital work providing aid for improved water supplies in poor countries.
- Intermediate technology provides clean water in a way that allows local people to take ownership of the project.

### Activity 2

WaterAid's website (www.wateraid.org/) contains a wealth of up-to-date case studies (from 17 countries) and technology details appropriate for GCSE students.

You could work in groups and prepare short presentations to share with your classmates, perhaps using PowerPoint.

Figure 16: An Afridev handpump

# Know Zone
# Water world

Water is vital for life on our planet. Water moves around the Earth through an amazing system called the hydrological cycle. This ensures our water supply by regulating the flow of water and keeping water clean. If humans interfere with the hydrological cycle it can threaten our water supply and quality of life.

## You should know...

- ☐ How the hydrological cycle links to the biosphere, atmosphere and lithosphere
- ☐ Why the global hydrological cycle is called a system
- ☐ The key stores and transfers of the hydrological cycle, and their names
- ☐ Why some areas, like the Sahel, have unreliable water supply
- ☐ How changing water supply has affected Australia
- ☐ How future climate change might affect water supplies
- ☐ How too much or too little water can lead to problems in the UK, Asia and America
- ☐ How humans reduce water quality through pollution
- ☐ How human use of stores and transfers can reduce water supply
- ☐ How water supply can be disrupted by human activities, such as deforestation, over-abstraction of groundwater and construction of reservoirs
- ☐ How large-scale water management can interfere with water supply
- ☐ How the Colorado River management has had positive and negative impacts
- ☐ How the Three Gorges dam has had positive and negative impacts
- ☐ What intermediate technology is
- ☐ How intermediate technology might improve water supply in the developing world

## Key terms

| | | |
|---|---|---|
| Aquifer | Joints | Water |
| Deforestation | Over abstraction | harvesting |
| Ecosystem | Permeable | Water insecurity |
| Groundwater | Pollution | Water |
| Hydrological | Pores | insufficiency |
| cycle | Precipitation | Water |
| Hydropower | River pollution | management |
| Impermeable | Run-off | schemes |
| Infiltration | Stores | Water table |
| Intermediate | Throughflow | Water |
| technology | Transfers | unreliability |

### Which key terms match the following definitions?

**A** When water is being used more quickly than it is being replaced

**B** Storing rainwater or used water ('grey water') for use in periods of drought

**C** A community of plants and animals that interact with each other and their physical environment

**D** A technology that the local community is able to use relatively easily and without much cost

**E** Small air spaces found in a rock or other material that can also be filled with water

**F** An underground store of water, formed when water-bearing (permeable) rocks lie on top of impermeable rocks

**G** When moisture falls from the atmosphere – as rain, hail, sleet or snow.

**H** Lines of weakness in a rock that water can pass along

To check your answers, look at the glossary on page 321.

**Foundation Question:** Describe how deforestation could affect or disrupt water supply. (3 marks)

| Student answer (achieving 1 mark) | Feedback comments | Build a better answer (achieving 3 marks) |
|---|---|---|
| Deforestation could increase interception.

There would be more surface run-off.

This will soak in water. | • *Deforestation could...* is incorrect so does not score any marks.

• *There would be...* This sentence is good and includes geographical terminology. This scores 1 mark.

• *This will soak...* This part of the answer is not precise and the meaning of *soak in* is not clear. | Deforestation would decrease interception.

There would be more surface run-off.

The amount of infiltration will fall. |

**Overall comment:** This answer shows how important it is to learn key terms, such as infiltration and interception, and use them accurately. You should avoid using phrases like 'soak in', which are not correct geographical terms.

- - - - - - - - - - - - - - - - - - - - - - - - - - - - - - - - - - - - - - - - - - - - - - - - - -

**Higher Question:** Explain how human activity could change the amount of infiltration. (3 marks)

| Student answer (achieving 2 marks) | Feedback comments | Build a better answer (achieving 3 marks) |
|---|---|---|
| If humans build impermeable surfaces like roads and roofs infiltration falls.

Around some rivers trees are planted to intercept water.

This decreases infiltration. | • *If humans build...* scores 1 mark. It is a good start to the answer and uses an example.

• *Around some rivers...* is also correct and uses a key term, so scores 1 mark.

• *This decreases infiltration* is incorrect. | If humans build impermeable surfaces like roads and roofs infiltration falls.

Around some rivers, trees are planted to intercept water.

This increases infiltration. |

**Overall comment:** This is a good answer, including key terms such as infiltration and interception. The student explains how humans affect transfers, but made a mistake with the last point – possibly because they were rushing.

# Chapter 5  Coastal change and conflict

## Objectives

- Know that there are different types of rock and that they affect the coastline.

- Be able to describe some of the landforms found on coastlines with hard or soft rock.

- Understand how geology and a range of processes impact on the coastal landforms.

*Figure 1: Granite cliffs near Land's End*

*Figure 2: The Yorkshire coast, south of Scarborough*

## How are different coastlines produced by physical processes?

### Coasts and geological structure

Many coasts around Britain are under attack from the waves. Britain's **geological structure** is varied and the coasts are made up of different rocks. In some places, such as Land's End in Cornwall, the rocks are made of granite. Granite is a 'hard' rock – it is resistant to **erosion** – with relatively few points of weakness such as joints and faults (cracks). This hardness means that it is not easy for the waves to erode the rock (wear it away). So the coast here has high, steep cliffs, with few beaches (Figure 1).

The coast looks different in other places, such as the east coast of Yorkshire. South of Scarborough, the coast is made up of softer boulder clays which are more easily eroded by the waves. The result is that the rocks are eroded quickly and the coastline retreats nearly two metres every year. Figure 2 shows what this coast looks like. The table below summarises the differences between the two coasts.

### Contrasts in the cliffs in hard rock areas (e.g. Land's End) and in soft rock areas (e.g. Yorkshire coast, south of Scarborough)

|  | Hard rock coast | Soft rock coast |
|---|---|---|
| Shape of cliffs | High, steep and rugged | May be high but are less rugged and not so steep. |
| On the cliff face | Cliff face is often bare, with no vegetation and little loose rock. | There may be piles of mud and clay which have slipped down the face of the cliff. |
| At the foot of the cliff | A few boulders and rocks which have fallen from the cliff. | Very few rocks, some sand and mud. |
| In the sea at the foot of the cliff | Sea is clear. | Sea is often brown, the colour of material from the cliff. |

### How do waves erode the coast?

Wave erosion involves several different processes, including:

| Hydraulic action | This results from the force of the water hitting the cliffs, often forcing pockets of air into cracks and crevices in a cliff face. |
|---|---|
| Abrasion | This is caused by the waves picking up stones and hurling them at the cliffs and so wearing the cliff away. |
| Attrition | Any material carried by the waves will become rounder and smaller over time as it collides with other particles and all the sharp edges get knocked off. |

## What landforms are created at the coast by wave action?

Although headlands are made of resistant rock, they gradually become eroded. Waves approaching the coastline bend around headland and erode it from both sides. Any points of weakness, such as cracks or joints in the rock, become eroded more quickly than the surrounding rock. This forms small caves on either side. With more erosion taking place over time, the two caves may join to form an arch. More erosion, and **weathering** of the arch roof, leads to the roof collapsing. This leaves an isolated **stack**. The stack itself gradually becomes undercut by wave erosion, eventually collapsing to leave a **stump**, which may only be visible at low tide. All of these landforms are shown in Figure 4.

## Concordant and discordant coasts

Coasts in some places, such as the Swanage area (Figure 5), are made up of both hard and soft rocks. The harder rocks, which here are chalk and limestone, are more difficult for the waves to erode, so they stand out as **headlands**. The softer rocks, which here are clays, are more easily eroded and so form the **bays**. In the Swanage area the alternating hard and soft rocks are at right angles to the coast. This coast is called a **discordant coast**.

In other places, such as the coast near Lulworth Cove, the alternating bands of hard and soft rock run parallel to the coast. Figure 6 on page 74 shows the resulting coastline when the rocks are eroded at different rates. This is called a **concordant coast**. The entrance to the cove is narrow where the waves have cut through the resistant limestone. Then the cove widens where the softer clays have been more easily eroded. At the back of the cove is a band of more resistant chalk, so erosion is slower here.

Swanage has both discordant and concordant coastline (Figure 5). North of Durlston Head, with is Portland limestone, is the softer rock type of greensand in Swanage Bay. North of the Bay, the third rock type is the chalk outcrop of the headland which includes Old Harry Rocks. To the west of Durlston Head the coastline is concordant, with the same rock type along its length. Notice there are fewer bays and headlands than in the discordant coast.

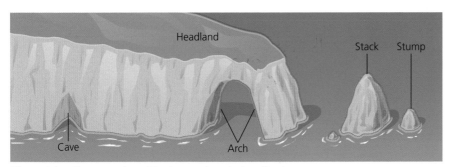

Figure 4: Landforms produced by the erosion of a headland

### Skills Builder 1

Figure 3: Landforms of erosion at Dyrholaey, southern Iceland

Study Figure 3.

(a) Name the landforms labelled A and B.

(b) Describe how these landforms might change over time.

(c) Explain how natural processes might cause these changes.

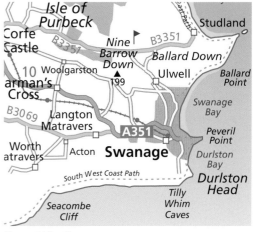

Figure 5: The Swanage coast

## Skills Builder 2

Study Figure 6.

(a) Label the band of limestone rock.

(b) Label the band of clay.

(c) Label the band of chalk.

(d) Explain how the cove was formed.

## Skills Builder 3

(a) Find a photograph of a bay and headland coastline from the Internet.

(b) Label the bays and headlands, using drawing tools.

(c) Annotate to suggest how the bays and headlands have been formed, using text boxes.

Figure 6: Aerial view of Lulworth Cove

## Different types of waves

There are different types of wave that move towards the coast. When waves break, water rushes up the beach, due to the energy from the wave. This is called the **swash**. When the water has lost its energy further up the beach, it runs back down again, under gravity. This is the **backwash**. Big waves break with lots of force and energy and this means they have the power to carry out erosion of beaches or rocks on the coast. These are called **destructive waves** and you can see their main features in Figure 7. As well as being tall and breaking with lots of power, they usually arrive quickly and have a high frequency – a lot of them come in a short period of time. Their backwash is greater than their swash and so sediment is taken away from the beach, back into the sea.

**Constructive waves** are very different. In calm conditions – without much wind – the waves are usually small, weak and with a low frequency (Figure 8). They don't break with much force and they tend to add sand and other sediment to the coastline by **deposition**. The swash is greater than the backwash and so sediment is pushed up the beach, helping to build up the beach.

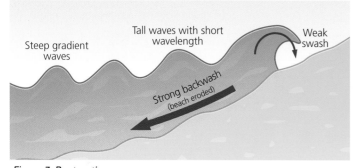

Figure 7: Destructive waves

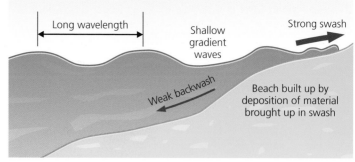

Figure 8: Constructive waves

## Erosional landforms

The landforms that are produced by the processes working on cliffs can be seen in many coastal areas. The wave-cut platform is a fairly flat rocky area at the base of a cliff that is being eroded. Waves cause undercutting at the base of the slope, as Figure 9 shows. This forms a notch that gradually gets bigger. The rock above eventually loses its support and then it collapses. The debris is gradually washed away by the waves.

The process is repeated, and so the cliff slowly retreats backwards and becomes steeper. At Seven Sisters in East Sussex, the cliffs are up to 80 metres high, while the wave-cut platforms extend up to 540 metres out to sea.

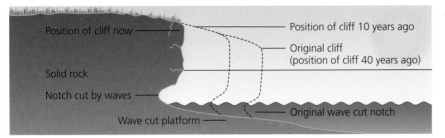

*Figure 9: Retreating cliffs*

## Depositional features

Where waves approach the coastline at an angle, when they break, their swash pushes beach material up the beach at the same angle. The backwash then drags the material down the beach at a 90° angle, due to the force of gravity. This produces a zig-zag movement of sediment along the beach known as **longshore drift**, which you can see in Figure 10. The smallest material, such as fine sand, is easily moved and so ends up furthest along the beach. The largest materials – pebbles perhaps – are heavy and are not moved as far.

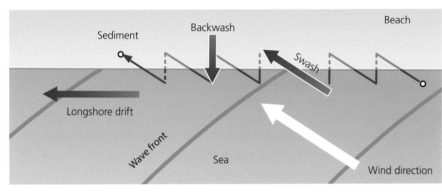

*Figure 10: The process of longshore drift*

## The formation of beaches, spits and bars

Constructive waves add sediment to the coastline and lead to the formation of a beach. This is particularly the case on low-energy coastlines that are sheltered from strong winds and waves, such as in bays between two headlands. A good example is Swanage Bay in Dorset.

Longshore drift can carry beach sediment beyond a bend in the coastline leading to an extension of the beach into the open water, known as a **spit**. The end of the spit often becomes curved because it is exposed to strong winds and waves. Deposition can then happen in the sheltered water behind the spit, forming a salt marsh. You can see both of these features of a spit in Figure 11.

If longshore drift continues to extend the length of the spit, it may join up with the coastline on the other side of an opening, such as a bay. This results in the creation of a 'bar' with a 'lagoon' behind. The bar and lagoon at Slapton, Devon (Figure 12) were formed when sea level was rising. As sea level rose, the water transported the shingle ridge ahead of it. This linked the headland with Torcross and created a freshwater lagoon.

75

**EXAM-STYLE QUESTION**

**Describe and explain the process of longshore drift.**
**(6 marks)**

■ **Basic answers** (Level 1)
Offer ideas only about waves, rather than the movement of sand and sediment along a beach.

● **Good answers** (Level 2)
Accurately describe the movement of sand and sediment along a beach in a zig-zag motion.

▲ **Excellent answers** (Level 3)
Not only describe the zig-zag movement accurately, but explain that it happens because of the wind and wave direction.

 **Top Tip**

Examination questions sometimes ask about how landforms develop over time. With a spit, you could mention that if it continues to grow across a bay or other indentation in the coastline, then it could develop into a bar.

 **Top Tip**

Good quality examination answers usually show a clear understanding of processes. Don't just name the process – explain how it works.

## Skills Builder 4

(a) Draw a sketch of the spit in Figure 11.

(b) Label the spit, the curved end and the salt marsh.

(c) Annotate your sketch to explain how the spit has formed and developed.

Figure 11: Aerial view of the Dawlish Warren spit

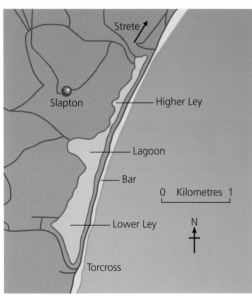

Figure 12: The bar and lagoon at Slapton in Devon

## Weathering

Weathering can happen in many ways. Some common ones in coastal areas are:

● Mechanical weathering – Salt crystal growth, for example, happens because sea water contains salt. When spray from waves lands on rocks, the water can be evaporated leaving the salt behind. The salt crystals grow and create stresses in the rock, causing it to break down into small fragments.

● Chemical weathering – Solution, for example, happens because all rain is slightly acidic, and this can react with weak minerals, causing them to dissolve, and the rock to decay.

● Biological weathering – The roots of vegetation, for example, can grow into cracks in a rock and slit the rock apart.

The term '**sub-aerial processes**' refers particularly to the disintegration of rock through the processes of weathering (such as freeze-thaw and thermal expansion) and the impact of wind and rain.

## Mass movement

There are several different processes of **mass movement**. In coastal areas, the main forms of mass movement are:

● **Rock fall** – This occurs when fragments of rock weathered from a cliff face fall under gravity and collect at the base.

● **Slumping** – This often happens when the bottom of a cliff is eroded by waves. This makes the slope steeper and the cliff can slide downwards in a rotational manner, often triggered by saturation due to rain, which both 'lubricates' the rock and makes it much heavier.

## Climate change and its impact on marine erosion and deposition

Scientists broadly agree that climate is changing. In the UK this means rising sea levels, as warmer temperatures resulting from global warming start to melt glaciers and ice sheets. In addition, climate change is changing storm patterns that affect the UK, with stronger and less predictable storms.

### The impacts on marine erosion

● Wave erosion depends on the energy of the wave, and stronger storms means greater wave energy and so more erosion. So in places where cliffs are being undercut at the moment, the cliff will continue to collapse and retreat – but at a faster rate.

● Abandoned cliffs may come under renewed attack by waves as sea level rises.

● Coastal areas of softer clay will be particularly affected by the changes, and here retreat of the coastline may be much faster.

## The impacts on marine deposition features

● Rising sea levels mean that beaches and bars may be at risk of submergence or erosion. For example, the Studland beach in Dorset (Figure 5 on page 73) which attracts over a million visitors each year, is already being eroded by two or three metres each year. So cafes, beach huts, toilets and a car park have had to be relocated and the sand of the beach replenished by imported sand. This will become worse as sea level rises and other beaches are affected.

● Extreme weather may become more common, and so violent storms (including hurricanes) can destroy or damage depositional features such as a beach or a spit. The rising sea temperature also gives more energy to storms, so spits may be breached, creating a large gap. In 2004, for instance, Hurricane Charley devastated the coast of Florida. The North Captiva spit was breached with a 450-metre gap as a result of high tides and very strong winds.

## Changing sea levels and storm activity

It is widely predicted by scientists that sea levels will rise by about 0.3 metres by the year 2100, mainly due to global climate change. This is going to put many low-lying coastal areas around the world at risk of flooding. These places include Bangladesh, the Netherlands, the Maldives and even parts of eastern England, such as the Thames Estuary. Insurance companies in the UK predict that flooding will be eight to twelve times more frequent here by 2100. These predictions have led governments to take action. In London, although the Thames Barrier is already in place (see Figure 13) to hold back very high tides, an Environment Agency project called 'Thames Estuary 2010' will install a series of new flood walls along the river. There are also plans to leave areas of open space, on to which flood water can go without damaging buildings. Both of these measures should help prevent the potentially serious effects of flooding.

## Storms at sea

Strong winds and storms can increase the height of waves and tides and so cause flooding. The Environment Agency monitors sea conditions over a 24-hour period, 365 days a year. The Storm Tide Forecasting Service provides the Environment Agency with forecasts of **coastal flooding**, surge and wave activity, together with warnings when hazardous situations are seen to be developing. If people are concerned about flooding from the sea, they can contact the Environment Agency's 24-hour Flood Line, or seek advice via their website. They will advise them of what precautions they should take and what action is needed in the event of a flood emergency.

## Watch Out!

Try not to confuse the words 'prediction' – suggesting what flooding is likely to happen in the future – and 'prevention', which relates to stopping the impacts of any flooding.

Figure 13: Aerial view of the Thames Barrier

## Activity 1

1. Use the Internet to find another location that suffers from coastal flooding.

2. Research how this area tries to predict or prevent coastal flooding.

## Objectives

- Know the factors affecting rates of cliff retreat.

- Be able to explain how people and the environment are affected by coastal retreat.

- Understand why methods of engineering used to protect the coast have both advantages and disadvantages.

## Watch Out!

Good examination answers on this topic will typically show awareness that rapid rates of cliff retreat are likely to be the result of two or three different causes acting together on a coastline, rather than just one.

## Skills Builder 5

(a) Use an atlas to draw a sketch map of the British Isles.

(b) Add arrows to show the fetch that would directly affect the south-west of Ireland, the south coast of England and the coastline of eastern England

(c) How would these different fetches affect the coastal processes in these three locations?

# Why does conflict occur on the coast, and how can this be managed?

## Differential rates of cliff retreat

Erosion and retreat of coastal cliffs can happen at very different rates. Some cliffs erode at rates of nearly two metres per year, such as those at Holderness in Yorkshire, whilst others are barely eroded at all. The rate of erosion and retreat is influenced by factors such as the **fetch** (the distance of sea over which winds blow and waves move towards the coastline), the geology and coastal management strategies.

Coasts that face a major ocean, such as the south-west coast of England facing the Atlantic Ocean, have a very long fetch, and the winds are strong and persistent. This produces destructive waves with high energy that can erode cliffs at rapid rates. But the fetch is only one of the factors that affect rates of erosion and retreat. The geology of south-west England is mainly granite, which is a very resistant rock. The actual rates of erosion are, therefore, very slow – typically only a few millimetres per year. One of the reasons for the high rates of erosion and retreat at Holderness is that the geology is very weak clay.

The final factor that affects these rates is that of **coastal management**. If coastal defences such as concrete sea walls protect weak geology then rates of erosion will be much slower.

## The effects of rapid erosion

Cliff retreat can have many impacts on both people and the environment. Many houses, apartments and hotels are built on cliff tops to take advantage of the wonderful sea views. Cliff retreat in many coastal areas has put these properties at risk of collapse into the sea. At Holbeck near Scarborough on the Yorkshire coast, the Holbeck Hall Hotel collapsed in June 1993 because the sea had undercut the cliff on which it stood (Figures 14 and 15).

In the UK, loss of property to cliff retreat is currently not covered by insurance policies. In the USA, about $80 million per year is paid out by a national insurance programme.

*Figure 14: The Holbeck Hall Hotel, near Scarborough*

Why does conflict occur on the coast, and how can this be managed?

79

Figure 15: The collapse of the Holbeck Hall Hotel

## People and coastal retreat

In 1994 Sue Earle, a farmer at Cowden, near Hornsea, East Yorks had to watch while her home of 25 years was demolished because of the rapid erosion of the cliff. Miss Earle fought a losing battle against the waves of the North Sea. When Sue first moved to the farm 39 years earlier, it had been 150 metres from the sea. Unfortunately the soft clay cliffs were eroded over the years by powerful storm waves. So in 1994 – when the farm was just two metres from the cliff edge – she was forced to move out, and the farm was demolished. Sue argued that the problem accelerated when sea defences were built nearby to protect the village of Mappleton.

Another family faced problems from cliff erosion at Barmston, just north of Hornsea. The Norcliffe family owned a bungalow holiday home at Barmston. By 2004 the porch was only 1.5 metres from a 25-metre drop – and the family had to move out. They were furious because this area of coastline is being left to fall into the sea. There are no defences against the waves. The Humber Estuary Coastal Shoreline Management Plan for Barmston concludes that the potential economic damage to caravan parks and isolated farms was not enough to justify the expense of building defences against the sea. The preferred option is to **'do nothing'**. The East Riding Council's coastal management team established a 'rollback' policy for caravan parks. This allows the caravan owners to retreat inland with their caravans, as the sea advances. The problem for the Norcliffe family was that the policy does not apply to houses. Neither Sue Earle nor the Norcliffe family have been able to gain any compensation for the loss of their houses.

Figure 17: The demolition of Sue Earle's farm

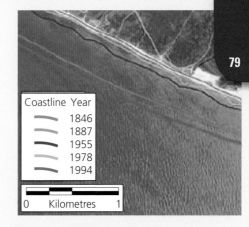

Figure 16: Digital map of Nantucket Island, Massachusetts

### Activity 2

Study Figure 16.

(a) Describe the changes in the coastline between 1846 and 1994.

(b) Name and outline two processes by which waves erode.

(c) Explain two possible reasons why cliff retreat is happening at a fast rate on this coastline.

### Activity 3

(a) Why is the preferred option at Barmston to do nothing?

(b) What are the effects of this policy?

(c) Give reasons why you think Sue Earle should or should not have received compensation for the loss of her farm.

## People, property and the coast in Holderness

The Holderness coast of East Yorkshire is a 60 km long stretch of low cliffs (20–30 metres high). The cliffs are composed of soft, easily eroded boulder clay. The cliff line is retreating at a rapid rate of nearly two metres per year, which is the fastest in Europe. Most of the erosion takes place in storms and tidal surges. The highest recorded loss was six metres of land in a storm in 1967. There is a lot of historical evidence as to the rate of retreat of the cliffs. Over 4 km of land has been lost since Roman times. Dozens of villages have been washed away by the sea, as Figure 18 shows.

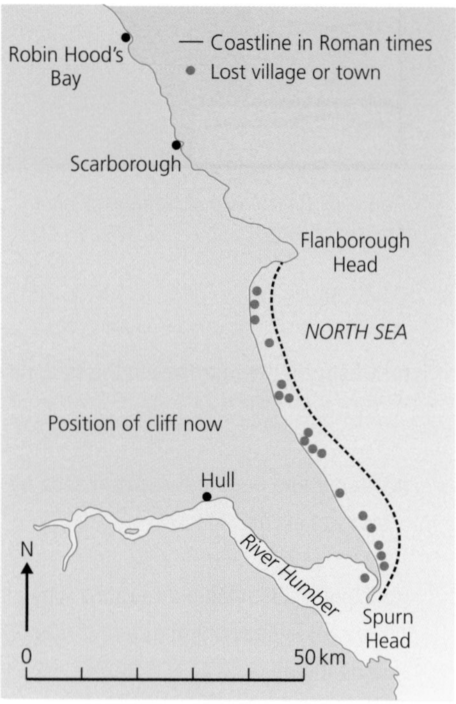

Why is cliff erosion such a problem here?

- The cliffs are made of soft glacial clay.

- The coast is very exposed, and waves have a long fetch over the North Sea.

- As sea-levels rise, more of the cliffs come under attack from the waves.

- The waves are mainly destructive ones.

- Most of the material which falls into the sea as the cliffs collapse is washed out to sea. The rest is moved along the coast by longshore drift. So the beaches in Holderness are narrow and give little protection to the cliffs.

## Why do people have conflicting views about management of the Holderness coast?

People who live on the coast in Holderness have different views about what should be done about the erosion of their cliffs:

Figure 18: The lost villages of Holderness

'Erosion is a natural process. You cannot stop it. We cannot afford to defend the whole coast.' *Chief Engineer of East Yorkshire*

'I have lived here all my life and I do not want to move, but I will have to because my home is getting near the edge of the cliff. I am too old to move.' *Local older person*

'I have lost valuable farmland and received no compensation. It is not fair – it's not my fault the coast is eroding.' *Local farmer (female)*

'My whole business is threatened by the erosion of the cliffs. I cannot afford to buy more land to move the caravans.' *Local caravan park owner*

'Not many people live along the coast, so I do not think we should try to build expensive sea defences for a few people. We should be spending our money in the towns where there are big problems of old houses that need replacing.' *Local politician*

'There is nothing to do here – especially at night – so I don't think it matters if the cliffs are eroded.' *Local young person*

Why does conflict occur on the coast, and how can this be managed?

81

### The options available

There is now a range of strategies in the Shoreline Management Plan which engineers have developed in order to prevent or reduce cliff erosion. These strategies are:

**1**. Do nothing.

**2**. Advance the existing defence line by even more **hard engineering**.

**3**. Hold the existing defence line by maintaining or improving the standard of protection.

**4**. Retreat the existing defence line (this is called **strategic realignment**) and so allow the caravan parks to move inland.

So it is possible to protect the communities of people who live close to the edge of the cliffs. However, these strategies are expensive to put in place, so planners and engineers have to do a cost–benefit analysis to assess whether or not it is worth trying to protect a stretch of coastline. One important part of any assessment of what to do is how sustainable each option may be. Clearly doing nothing is sustainable, in that no additional resources will be needed. Building defences in options 2 and 3 is a lot less sustainable because it would involve building extensive concrete and steel defences against the sea.

Option 4, the strategic realignment, would involve some additional resources such as an extra field for caravan sites and holiday homes, but it would probably be less environmentally damaging than options 2 and 3. The costs in any cost–benefit analysis include things such as:

● The loss of farmland if nothing is done.

● The loss of caravan parks and holiday chalets if nothing is done.

● The costs of building defences against the sea.

● The loss of houses if nothing is done.

● The loss of roads if nothing is done.

The benefits are mainly:

● People do not lose their homes or businesses.

● Erosion of the cliffs stops.

● People do not lose their farms and land.

● The coastline is stabilised.

But, as we have seen, there is disagreement about what to do with the Holderness coast in the future. Some towns and villages have been protected against the sea, but other will not be, according to present plans. More than this, some local people (like Sue Earle) argued that building sea defences in some parts of the coast in fact only made it worse for other places further along the coast, where erosion was even stronger than before.

## Activity 4

(a) What do you think a local ice cream seller would say about the erosion of the coast?

(b) What do you think a person owning a holiday bungalow would say about the erosion of the coast?

(c) What do you think a person owning a nightclub in Scarborough would say about the erosion of the coast?

## Decision-making skills

Read the different opinions on coastal management above.

Arrange these opinions on a scale from *strongly for* to *strongly against* to show the likely views on coastal management.

## Watch Out!

When you are answering questions about the costs and benefits of different methods, you do not have to 'balance' them by giving an equal number of each, but you should give at least one of each.

## Coastal management options – from hard engineering to holistic approaches

Coastal areas can be defended against wave erosion in a variety of different ways. Recently the focus has been more on a **holistic approach** to coastal management. This means considering all the social and economic factors (such as tourism) that affect a coastal area, as well as the physical conditions of the coast, such as the size and scale of the waves, beach and cliffs. In this way all the many factors affecting a coast will be considered in an integrated way. The purpose of all this is to make our management of coastal areas and their resources more sustainable. Many of the methods used in the holistic approach to managing coastal areas are described as being soft engineering, which is generally considered to be more environmentally friendly than more traditional hard engineering.

On pages 84 and 85 there is a case study of traditional coastal engineering structures being employed on the Dorset coast. The following tables show the benefits and costs of all the hard engineering methods and the soft engineering methods.

### Hard engineering methods

**Sea wall** – a long concrete barrier built at the base of a cliff offshore

| Benefits | Costs |
| --- | --- |
| It protects the base of the cliffs against erosion because it is made of resistant concrete. Land and buildings behind it are protected. If it is 'recurved', it can reflect wave energy. | It is expensive to build, and the cost of maintenance is high. It restricts access to the beach and it may be unsightly. |

**Groynes** – wooden, rock or concrete 'fences' built across the beach, perpendicular to the coastline

| Benefits | Costs |
| --- | --- |
| These prevent the movement – by longshore drift – of beach material along the coast. The beach can then build up as a natural defence against erosion – and as an attraction for tourists. | They may look ugly and they do not last very long because the wood rots. Sand is prevented from moving along the coast, and places elsewhere may lose their beach and the natural defence it provides. |

**Rip rap** – large boulders of resistant rock

| Benefits | Costs |
| --- | --- |
| These absorb wave energy and protect weak cliffs behind. They look quite natural. | They can be expensive. They still let some wave energy through. They can restrict access for the very young and the elderly. |

**Revetments** – slatted wooden or concrete structures built at the base of a cliff

| Benefits | Costs |
| --- | --- |
| These absorb and spread wave energy through slats. They do not interfere with longshore drift. | Regular maintenance is needed and they are quite expensive. |

**Off-shore reefs** – rock or concrete barriers built on the sea bed a short distance from the coastline

| Benefits | Costs |
| --- | --- |
| Waves break on the barrier before reaching the coast. These significantly reduce wave energy and allow a wide beach to develop. | They are very expensive to build and can interfere with boats. |

Why does conflict occur on the coast, and how can this be managed?

83

## Soft engineering methods

| Beach replenishment – adding sand taken from somewhere else, often offshore | |
|---|---|
| **Benefits** | **Costs** |
| This looks completely natural. It provides a beach for tourists. The beach absorbs wave energy and protects the land or buildings behind. Quite cheap. | The sea keeps on eroding it away – so it has to be replaced every few years. |

| Managed retreat – people and activities are gradually moved back from the vulnerable areas of coast | |
|---|---|
| **Benefits** | **Costs** |
| Natural processes are allowed to happen. | Compensation has to be paid. There is quite a lot of disruption to people's lives and to businesses. |

| Cliff regrading – making the cliff face longer, so that it is less steep | |
|---|---|
| **Benefits** | **Costs** |
| The angle of the cliff is reduced, making mass movement less likely. This method is relatively cheap. | Other methods need to be used at the base of the cliff to stop it being steepened again by erosion. Properties on the cliff may have to be demolished. |

### *Integrated Coastal Zone Management*

Coastal zones all over the world are changing rapidly. They are subject to the processes of erosion, but they are also subject to other pressures, such as the demand for space to build holiday homes or caravan parks. In some places the demand is to conserve the plants, animals, fish and insects of the area. So now planners everywhere are trying to find more sustainable ways of managing the coast – to reduce the damage to the environment. The main way of doing this is to treat the whole of the coastal area from the shoreline to several kilometres inland as one area for planned development. This approach is called **Integrated Coastal Zone Management (ICZM)**. In the case of the Holderness area in Yorkshire, for example, the East Riding Integrated Coastal Zone Management Plan was produced in 2002. It sets out which areas will be protected from erosion by the sea and in which areas the 'do nothing' approach will be employed. Local people are still very unhappy with this approach and argue that local economic and social needs are being ignored. So debate continues as to the best method to protect coastal areas.

## Activity 5

With a partner, discuss and decide whether you think hard or soft engineering methods are more suitable to protect busy tourist resorts from coastal erosion.

## Activity 6

With a partner, discuss and decide whether you think Integrated Coastal Zone Management is more sustainable than hard or soft engineering.

*Figure 19: Coastal protection at Mappleton, East Yorkshire*

## Decision-making skills

Look at the tables on pages 82 and 83. Summarise the factors which influence the choice of coastal management methods, e.g. whether hard or soft engineering, what type of hard engineering?

## Case study of traditional coastal engineering structures

Swanage Bay and Durlston Bay in Dorset have both suffered from some significant coastal erosion during the last century. Rates of erosion in both places are estimated to have been about 40–50 cm per year.

In Durlston Bay, several methods were used to protect the cliffs from erosion and to safeguard the apartments and houses on the cliff top. Erosion mainly occurred at one particular point, where there was a major weakness in the rock. The methods (Figure 20) included:

- Regrading of the cliff – this extended it forward at the base, making the slope longer and therefore less steep.

- Installing drainage – this removed excess water, so that the slope was not as heavy or as well lubricated after rain.

- Placing rip rap – large granite boulders each weighing about 8 tonnes – at the base of the cliff to resist wave attack.

*Figure 20: Management methods in Durlston Bay*

Why does conflict occur on the coast, and how can this be managed?

85

Swanage Bay needed different methods of defence from Durlston Bay, because the erosion occurred along a considerable length of cliff rather than at one point. On the cliff top, the houses and hotels (such as the Grand Hotel) were losing their gardens and were in danger of collapsing. The methods used here included:

● Sea wall – this was built in the 1920s and provided a promenade (walkway) as well as a barrier to wave attack.

● Cliff regrading – a series of steps were made in the cliff to lower slope angles.

● Groynes – a series of mainly timber groynes were installed in the 1930s, and 18 of them have recently been replaced with new ones. These reduced longshore drift and helped make sure that the beach remained in place to absorb the energy of breaking waves.

● Beach replenishment – 90,000 m³ of sand was dredged from Studland Bay and pumped on to the beach at Swanage. This works with the groynes to ensure a good size of beach.

The cost of the recent new groynes and the beach replenishment was about £2.2 million. You can see some of the methods used at Swanage in Figure 21.

*Figure 21 Swanage Beach and cliffs*

Coasts are packed with physical geography and important processes. They are popular and often crowded places, where physical and human geography often combine to cause conflict.

## You should know...

- [ ] How rock type (geology) and structure influence coastal landforms
- [ ] How landforms such as cliffs and stacks form
- [ ] The key terminology of coastal landforms
- [ ] The difference between erosion, weathering and mass movement
- [ ] How these processes help form coastal landforms
- [ ] The different types of waves, constructive and destructive
- [ ] How wave action moves sediment and changes the profile of beaches
- [ ] How sediment is transported and deposited on coasts
- [ ] How some coasts are threatened by rapid erosion and rising sea levels
- [ ] How erosion can cause conflict
- [ ] The range of management options for coasts
- [ ] Traditional and holistic approaches to coastal management
- [ ] The coasts and benefits of different options for Holderness and Swanage

## Key terms

Backwash
Bay
Coastal flooding
Coastal management
Concordant coast
Constructive wave
Deposition
Destructive wave
Discordant coast
Do nothing
Erosion
Fetch
Geological structure
Hard engineering

Hard rock coast
Headland
Holistic approach
ICZM
Longshore drift
Mass movement
Soft rock coast
Spit
Stack
Strategic realignment
Stump
Sub-aerial processes
Swash
Weathering

### Which key terms match the following definitions?

**A** The dropping of material that was being carried by a moving force

**B** A coastline created when alternating hard and soft rocks occur parallel to the coast, and are eroded at different rates

**C** The distance of sea over which winds blow and waves move towards the coastline

**D** A coastline created when alternating hard and soft rocks occur at right angles to the coast, and are eroded at different rates

**E** Water from a breaking wave which flows under gravity down a beach and returns to the sea

**F** Large, powerful waves with a high frequency that tend to take sediment away from the beach, because their backwash is greater than their swash

**G** The movement of material along a coast by breaking waves

**H** Material deposited by the sea which grows across a bay or the mouth of a river

To check your answers, look at the glossary on page 321.

**Foundation Question:** Describe the advantages and disadvantages of different hard engineering methods used to protect coastlines. (6 marks)

| Student answer (achieving Level 2) | Feedback comments | Build a better answer (achieving Level 3) |
|---|---|---|
| Hard engineering like wooden groynes can be used to build up beaches to protect coasts.<br><br>Some people think they are ugly and stop you moving along the beach easily.<br><br>On some coasts 'do nothing' is used which just lets the coast erode.<br><br>This can be cheaper, but you might have to compensate some people who lost land.<br><br>Sea walls made of concrete are expensive at over £5,000 per metre.<br><br>They can be ugly and might stop people getting onto the beach easily. | • *Hard engineering like...* This is a good example and is correct – groynes do build up beaches.<br><br>• *Some people think...* This includes a disadvantage of groynes so balances the first statement.<br><br>• *On some coasts...* This is not an example of hard engineering so does not score any marks.<br><br>• *This can be...* Although this part of the answer is correct, it is not answering the question.<br><br>• *Sea walls made...* This is a good example with appropriate facts on costs.<br><br>• *They can be...* This is correct but they are all disadvantages. | Hard engineering like wooden groynes can be used to build up beaches to protect coasts.<br><br>Some people think they are ugly and stop you moving along the beach easily.<br><br>Offshore reefs reduce wave energy and this reduces erosion. They cannot be seen at high tide.<br><br>They are expensive and might be a hazard to boats.<br><br>Sea walls are made of concrete and are expensive at over £5,000 per metre.<br><br>Sea walls protect the base of the cliff from erosion. |

**Overall comment:** There is some good material in this answer, but there is not enough balance between advantages and disadvantages. In addition, the second example is not hard engineering.

---

**Higher Question:** Explain why some cliffs erode more rapidly than others. (6 marks)

| Student answer (achieving Level 2) | Feedback comments | Build a better answer (achieving Level 3) |
|---|---|---|
| In Holderness cliffs are made of soft boulder clay, which is easily eroded.<br><br>Harder rocks like chalk at Flamborough head are not eroded as easily.<br><br>Some cliffs contain weaknesses like joints and faults.<br><br>They get attacked by hydraulic action and lots of weaknesses means lots of erosion.<br><br>Big storms cause lots of erosion. | • *In Holderness cliffs...* This contains a good example of a rock type and location.<br><br>• *Harder rocks like...* This is another good example, but using geographical terms such as abrasion as well as erosion would improve the answer<br><br>• *Some cliffs contain...* This is another good reason.<br><br>• *They get attacked...* In this sentence hydraulic action is a good term to use.<br><br>• *Big storms cause...* This point is correct but needs to be explained. | In Holderness cliffs are made of soft boulder clay, which is easily eroded.<br><br>Harder rocks like chalk at Flamborough head are not eroded as easily by abrasion.<br><br>Some cliffs contain weaknesses like joints and faults.<br><br>They get attacked by hydraulic action and lots of weaknesses means lots of erosion.<br><br>Big storms cause erosion, as some coasts are exposed to a long fetch and large waves. Human actions, e.g. groynes at Mappleton, have made erosion faster elsewhere. |

**Overall comment:** This is a good answer with some examples, but the student needed to add further explanation and remember to use specific geographical terms such as abrasion.

# Chapter 6 River processes and pressures

## How do river systems develop?

### Drainage basin terms

A **drainage basin** is the area of land drained by a river and its tributaries. When it rains, water eventually finds it way into rivers, either by moving across the surface or by going underground and moving through the soil or the rock beneath. Throughout the drainage basin there are a series of changing landforms along the profile.

There are a number of important technical terms associated with drainage basins. These are shown in Figure 1, and their meanings are given here.

Watershed – the boundary of a drainage basin. It separates one drainage basin from another and is usually high land, such as hills and ridges.
Confluence – a point where two streams or rivers meet.
Tributary – a stream or small river that joins a larger stream or river.
Source – the starting point of a stream or river, often a spring or a lake.
Mouth – the point where a river leaves its drainage basin as it flows into the sea.

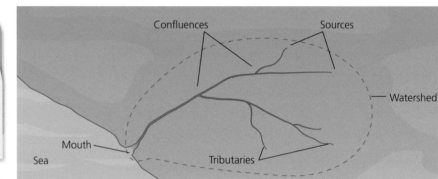

*Figure 1: Drainage basin terms*

### Change in characteristics from source to mouth

A river changes with increasing distance downstream from its source towards its mouth, as it moves from its **upper course**, to its **mid-course** and finally into its **lower course**.

The key characteristics of a river are:
- width – the distance from one bank to the other
- depth – the distance from the surface of the water to the bed
- velocity – how fast the water is flowing
- discharge – the volume of water moving through the river channel in a given time
- gradient – the steepness of the river bed
- load – the material (sediments, debris) carried by the water.

The changes in these characteristics are summarised in the table on the left.

| Characteristic | Change from source to mouth |
|---|---|
| Width | increases |
| Depth | increases |
| Channel shape | broader and shallower |
| Velocity | increases |
| Discharge | increases |
| Gradient | decreases |
| Load | smaller and rounder |

## Afon Nant Peris, Wales

The river Nant Peris rises at 360 m in the Llanberis Pass in North Wales. The river appears as a trickle of water next to the A4086 road. This area receives 2500 mm of rain each year and 80% of it flows out along the river. Figure 2 shows the **long profile** of the Afon Nant Peris, that is the view of the river from its source in the LLanberis Pass to its mouth at Caernarvon. Like most rivers, the Afon Nant Peris has an uneven long profile – steep in some places and much less so in others.

As it flows down the pass, the channel of the river widens and deepens. The river erodes the valley floor, transporting the debris and, later, depositing it elsewhere. In its upper course, the sides of the valley are steep and the river's course is not straight. It flows around the interlocking spurs of higher land. The river is eroding downwards rapidly, and transporting lots of rocks, sand and mud down river.

In its middle course the river flows across a broad area of flat land called a **flood plain**, which it has created. Often the river flows in large curves, called **meanders**. There is both **erosion** and **deposition** in this section of the river, but the deposition is temporary. Sooner or later the **sediments** deposited will be eroded by the river and moved. The gradient here is lower than in the upper course but velocity (speed) is greater because the river is deeper so there is less friction with the banks and bed of the river.

In its lower course the river is known as the Afon Seiont and meanders across a broad flood plain. The river is close to sea level so there is little downward erosion. However velocity and discharge are now at their maximum. The river forms **ox-bow lakes** which may eventually fill with sediment and dry up.

**Top Tip**

These changes in characteristics from source to mouth will not necessarily be gradual. Sometimes there will be sudden and quite dramatic changes, for example when a tributary joins and the discharge increases significantly. High-quality answers to questions asking for these changes to be described might mention this.

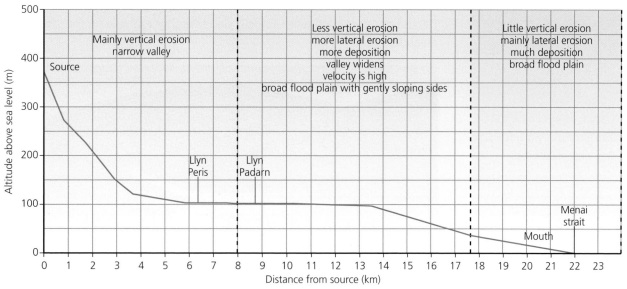

Source: Flint, Fline and Punnett 'Geography and Change' Hodder Staughton 1996 ISBN 0340 647 868 Page 71

*Figure 2: The long profile of the Afon Nant Peris*

*Figure 3: A meander bend*

**Figure 3 shows a meander bend. Explain the changes which might happen to this meander over a period of time. You may use a diagram or diagrams in your answer.**
**(5 marks, June 2007)**

**How students answered**

Most students answered this question poorly. They often described the changes, stating that the bend would get bigger and perhaps become an ox-bow lake. But they did not explain why this happened.

60% (0–1 marks)

Many students gave some explanation of the changes by mentioning erosion and/or deposition and stating where they were happening.

31% (2–3 marks)

Some students answered this question well. They not only mentioned erosion and deposition, but also gave detail of specific process mechanisms such as abrasion. They also established a clear sequence in the changes happening.

9% (4–5 marks)

## River channel processes

### *Erosion*

River erosion involves several different methods, including:

- **Hydraulic action** results from the force of the water hitting the river bed and banks and wearing them away. This is particularly important during high-velocity flow.

- **Abrasion** is caused by the river picking up stones and rubbing them against the bed and banks of the channel in the flow. This wears the bed and banks away.

- **Attrition** is the process by which material carried in the river becomes rounder and smaller over time as it collides with other particles and the sharp edges get knocked off.

- **Corrosion** is dissolving of rocks and minerals by water flowing over them.

### *Transport processes*

A river uses its energy to transport material by:

- Suspension – fine particles travel suspended in the water.

- Saltation – coarse sand and small rocks lifted by the current in times of flood are moved down the river course, before falling back to the river bed.

- Traction – large boulders slide or roll along the river bed in times of flood when the river has a lot of energy.

- Solution – soluble minerals like calcium and sodium travel dissolved in the water.

Deposition takes place when velocity decreases, for example on the inside of a meander.

Interlocking spurs

*Figure 4: Interlocking spurs*

## Landform formation

Rivers produce a wide range of landforms, many of which help make river landscapes attractive places for people to visit and enjoy.

### Interlocking spurs

Near their source, rivers do not have a lot of power as they are very small. They tend to flow around valley side slopes, called spurs, rather than being able to erode them. The spurs are left **interlocking**, with those from one side of the valley interlocking with the spurs from the other side (Figure 4).

### Waterfalls

Where a river flows over bands of rocks of differing resistance, the weaker rock is eroded more quickly and a step may develop in the river's bed. The increased velocity gained by the water as it falls over the step further increases the rate of erosion of the weaker, downstream band of rock. Abrasion and hydraulic action at the base of this step causes undercutting and the formation of a plunge pool. Eventually the overhanging, more resistant rock collapses due to a lack of support, making the **waterfall** steeper. The waterfall in Figure 5 is almost vertical. Over time, repetition of this process means that the position of the waterfall retreats in an upstream direction.

### Meanders

Meanders are large bends in a river's course, found well below the source, often in its lower course as the river approaches its mouth. They are common landforms found on a river's flood plain. The flow of the river swings from side to side, directing the line of maximum velocity and the force of the water towards one of the banks. This results in erosion by undercutting on that side and an outer, steep bank is formed, called a **river cliff**. Deposition takes place in the slower moving water on the inside of the bend, leading to the formation of a gently sloping bank, known as a **slip-off slope**. The cross-section of a meander is, therefore, asymmetrical – steep on the outside, gentle on the inside.

### Ox-bow lakes

As meander bends grow and develop, the necks of land between them become narrower. Eventually the river may erode right through a neck, especially during a flood. Water then flows through the new, straight channel

Resistant rock

Weak rock that has been undercut

Plunge pool

Figure 5: Bridal Veil Falls, Waikato, New Zealand

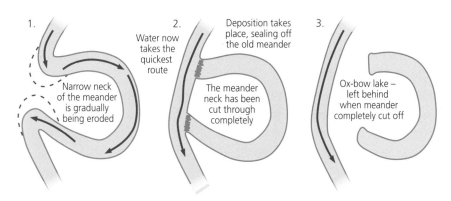

1. Narrow neck of the meander is gradually being eroded

2. Water now takes the quickest route

Deposition takes place, sealing off the old meander

The meander neck has been cut through completely

3. Ox-bow lake – left behind when meander completely cut off

Figure 6: Formation of an ox-bow lake

and the old bend is abandoned by the river. The bend gradually dries up and, helped by deposition at the neck by the river, becomes sealed off as a horseshoe-shaped lake, as Figure 6 shows.

## Flood plains

A flood plain is a wide, flat area of land either side of a river in its lower course. The flood plain is formed by both erosion and deposition. Lateral (sideways) erosion is caused by meanders eroding on the outside of their bends. This makes the valley floor wide and flat. When the river floods, the flood water spreads out on the valley floor, slows down and deposits the sediments it was carrying. This can be seen in Figure 7.

## Levees

**Levees** are natural embankments of sediment along the banks of a river. They are formed along rivers that carry a large load and occasionally flood. In times of flood, water and sediment come out of the channel as the river overflows its banks. The water immediately loses velocity and energy as it leaves the channel and so the largest sediment is deposited first, on the banks. Repeated flooding causes these banks to get higher, forming levees. You can see these alongside the river in Figure 7. Sometimes natural levees are built up further and reinforced by humans to protect settlements, industry and transport links on the flood plain from future floods.

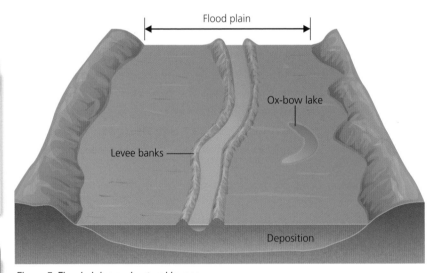

Figure 7: Flood plains and natural levees

## Activity 1

Draw up a table to show all of the landforms that are produced by river processes. Use three headings – 'Erosion', 'Deposition' and 'Erosion and Deposition' – putting each landform under the correct heading.

## Activity 2

Consider these four river landforms – waterfall, flood plain, meander, interlocking spurs.

(a) Which two are most likely to be found in the upper course of a river near the source?

(b) Describe the appearance and position of a flood plain.

(c) Explain how a waterfall is formed.

## The impact of weathering, mass movement and geology on river valley shape and sediment load

**Weathering** is the breakdown of rocks and minerals by physical and chemical processes. The rocks and minerals can then be moved by the processes of erosion. The main agents of erosion are wind, water, ice and gravity. Erosion involves the movement of material, whereas weathering takes place with little or no movement. In addition, on the sides of river valleys the processes of mass movement take place – transferring broken rock, minerals, soil and vegetation down the slope as a result of gravity.

The weathering and mass movement processes act on the valley sides, usually making them less steep over time, as material is moved from the top of the slope to the bottom. However, the river itself erodes its own channel, by wearing away the bed and/or banks, as well as possibly eroding the base of the valley side, making it steeper – and perhaps leading to mass movement.

Weathering can happen in many ways. Some common ones affecting river valleys are:

● Freeze-thaw – this happens when rainwater enters cracks or gaps in a rock and then freezes if temperatures drop below zero degrees. The water expands as it turns to ice and this exerts pressure on the rock, causing it to break down into smaller pieces.

● Rainfall – this happens because rain is slightly acidic. If the air is polluted by factories and vehicles, it can become very acidic. When rain falls on rocks, the acid in it can react with weak minerals causing them to dissolve, and the rock to decay.

● Biological weathering – the roots of plants, especially trees, can grow into cracks in a rock and split the rock apart as you can see in Figure 8.

There are several different processes of mass movement. In river valleys, two of the main forms of mass movement are:

● Soil creep – individual particles of soil slowly move down the slope under gravity and collect at the bottom of the valley sides. The river may then erode this material.

● Slumping – happens when the bottom of a valley side is eroded by the river. This makes the slope steeper and the valley side material can slide downwards in a rotational manner, often triggered by saturation due to rain, which both 'lubricates' the rock and makes it much heavier.

**Geology** influences river valley shape and sediment load in a number of ways: Where a river flows over bands of resistant rocks the valley sides tend to be steep, and the sediment load is small, because erosion is slow. Where a river flows over bands of less resistant rocks the valley sides tend to be gentle and the sediment load is large because erosion is rapid.

*Figure 8: Roots of trees can grow into cracks in rocks and split them apart*

**Top Tip**

Good quality examination answers usually show a clear understanding of processes. Don't just name the process; explain how it works.

## Objectives

- Know the major factors that cause flooding.

- Be able to describe the effects of flooding on people and the environment.

- Understand how the effects of flooding can be reduced.

# Why do rivers flood and how can flooding be managed?

## Causes of flooding

Rivers flood naturally. A river flood is when a river overflows its banks and water spreads out on to the land alongside the river channel. Rivers usually flood as a result of a combination of causes. These are often a mixture of physical and human factors which, together, can result in large amounts of water getting into the river very quickly – so quickly that it fills up and overflows.

### Flooding in Tewkesbury in July 2007

Figure 9 shows the rainfall in the Tewkesbury area in July 2007, together with the discharge of the river Severn from 17 to 27 July. This diagram is called a **hydrograph**. It shows the relationship between rainfall and river flow. The rising limb on the hydrograph shows the rapid increase in water in the river after the heavy rain of 19 and 20 July. The lag time is the difference between the time of the heaviest rainfall and the maximum level of the river. The falling limb is when there is still some rainwater reaching the river but in smaller and smaller amounts.

Point A on Figure 9 shows where some rainwater falls straight into the river. Point B is when a lot of water reaches the river by run-off on the surface. Point C is when some rainfall reaches the river more slowly because it has flowed through the soil and rocks. The hydrograph and Figure 10 show how the heavy rain combined with other factors to flood the area around Tewkesbury. This was a particularly severe flood which cut off the town from the surrounding area. Large numbers of people were forced to leave their homes and move to shelters above the flood waters. When the waters subsided, homes were found to be badly damaged. Carpets, kitchen units, cookers, fires, fridges and central heating boilers all had to be scrapped.

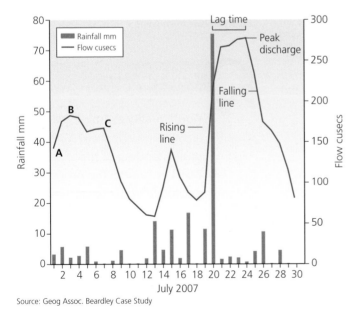

Source: Geog Assoc. Beardley Case Study

*Figure 9: A flood hydrograph of the River Severn in July 2007*

*Figure 10: Flooding in Tewkesbury in July 2007*

## Time lags in hydrographs

Not all hydrographs are the same. Figure 11 shows the hydrographs for the River Wye and the River Severn after a storm which affected them both. The steep upward curve of the hydrograph for the Wye shows a short lag time – the river's maximum flow was soon after the storm. In contrast, the Severn, with its more gently sloping graph, shows a longer lag time. One important reason for the difference is that the Wye flows mainly through moorland, where the vegetation is sparser, so the run-off from the rainfall is rapid and the time lag is short. The Severn, however, flows mainly through coniferous woodland, where the vegetation is thicker and absorbs more of the rain that falls and then releases it quite slowly, so that there is a longer time lag before the water reaches the river.

Other factors which affect these hydrographs are:

- The drainage basin of the Wye is narrower and smaller than that of the Severn, so the water gets to the main river more quickly.

- There are steeper gradients along the Wye than along the Severn, so the water runs off faster.

- There are more and larger urban areas in the drainage basin of the Wye which means that the many impermeable surfaces encourage faster run-off, resulting in a steeper hydrograph.

## Physical factors that contribute to flooding

- Intense rainfall – When rain falls too fast to fully allow its infiltration into the ground, it flows quickly across the surface and into the channel. This happened in Figure 11.

- Snow melt – In some places lots of snow falls during the winter months, but when temperatures rise above zero in the spring all the snow that has built up over months suddenly melts.

- **Impermeable** rocks – Some rock types, like granite, do not let water enter the ground and so rainwater runs off across the surface into the channel.

## Human factors that contribute to flooding

- **Deforestation** – Vegetation collects, stores and uses water from a drainage basin. The less vegetation there is, the more water can reach the channel.

- **Urbanisation** – In towns and cities, rainwater will not infiltrate the hard, man-made surfaces like concrete and tarmac and so it runs off into the channel.

- Global climate change – Increasing global temperatures, partly due to human activities such as burning fossil fuels, can cause more melting of ice in glaciers and, in places, more rainfall and more frequent storms.

ResultsPlus
**Exam Question Report**

**REAL EXAM QUESTION** 95

**Name and explain two physical factors that cause floods.**
**(4 marks, June 2007)**

**How students answered**

Most students answered this question poorly. They often only gave answers that related to human factors, such as urbanisation, rather than the physical factors asked for in the question.

47% (0–1 marks)

Many students gave two causes, such as snow melt or impermeable rock, but did not explain how they cause flooding.

33% (2 marks)

Some students answered this question well. They named two valid physical factors and explained how they caused flooding. This was best done by explaining why water runs quickly across the surface into a river, causing it to rise rapidly and overflow its banks.

8% (3–4 marks)

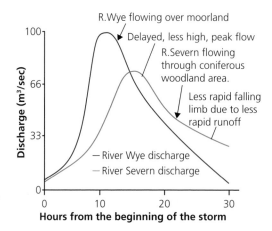

*Figure 11: Hydrograph of the Rivers Severn and Wye*

## Decision-making skills

Create an environmental impact table to rank the effects on the environment in the order that you think are most severe. Give reasons for the order you have chosen.

## Activity 3

Following serious floods in Wales in 2008 the Environment Agency produced a report which showed that :

● One in six properties in Wales was at risk of flooding.

● Over 220,000 properties were at risk of flooding from rivers or the sea in Wales, of which 64,000 were at significant risk.

● 97,000 of these were also vulnerable to surface water flooding, with a further 137,000 properties susceptible to surface water flooding alone.

● 57% of people living in flood risk areas knew they were at risk and, of these three out of five people had taken some action to prepare for flooding.

As well as providing more detailed information on where the greatest risks are, the report also showed that a sizeable part of the nation's important infrastructure and public services were in flood risk areas. For example, over 80% of water and sewage pumping stations/treatment works were in flood risk areas.

*Environment Agency 2009*

(a) How many properties were at significant risk of flooding in Wales?

(b) What percentage of people had taken action to prepare for flooding?

(c) Why is the threat to sewage pumping stations so important?

## Effects of flooding on people and the environment

There are many different and wide-ranging effects that floods can have on people and the environment (see Figure 12). Some of these are very immediate, while others can happen much later – and last much longer.

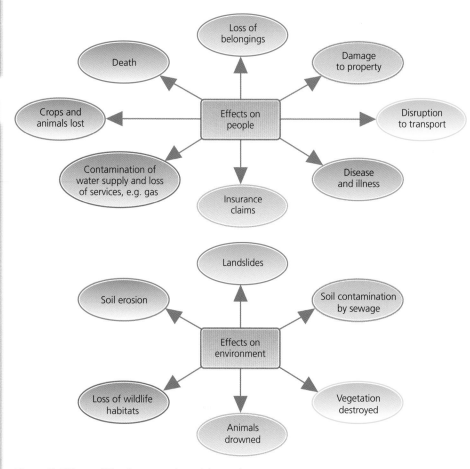

*Figure 12: Effects of floods on people and the environment*

The flooding that affected parts of northern England and the Midlands in the summer of 2007 caused the loss of 13 lives and over £4 billion worth of damage. In Lincolnshire, 40% of the pea crop was damaged, and in north Wales 30 tonnes of debris and earth blocked the only road out of Barland near Presteigne on 21 July.

Floods in Mozambique in Africa during February 2007 led to 121,000 people being made homeless. In Sudan, at least 12,000 livestock and 16,000 chickens were lost, while outbreaks of acute diarrhoea killed dozens of people as water supplies became contaminated.

## Prediction and prevention of the effects of river flooding

### Forecasting

Many rivers, especially those that have a history of flooding, are monitored by the Environment Agency. If river levels rise to potentially dangerous levels, they are able to warn people in areas of risk and, if necessary, evacuate people from their homes to a safer place. They have also produced maps, which are available online, that show areas at different levels of risk.

Computer simulation models are now being used to help predict flooding. In Texas, for example, in response to a prediction from Rice University, the Texas Medical Center was evacuated before a large flood occurred in 2001.

### Building design

There are things that people can do to their property that will make it easier and cheaper to clean up after a flood.

The Environment Agency advice includes:

- Lay ceramic tiles on your ground floor and use rugs instead of fitted carpets.

- Raise the height of electrical sockets to 1.5 metres above ground floor level.

- Fit stainless steel or plastic kitchens instead of chipboard ones.

- Position any main parts of a heating or ventilation system, like a boiler, upstairs.

- Fit non-return valves to all drains and water inlet pipes.

- Replace wooden window frames and doors with synthetic ones.

In many low-income countries, wooden buildings are built on stilts so that the living areas are above the **flood risk** level (see Figure 13).

### Planning

The local government is responsible for giving planning permission for new buildings. In recent years, many new buildings have been built on the flood plains of rivers, putting these properties at significant risk. Now, it is widely accepted that this should only be done if protection measures are put in place. Land use zoning is often used. This means flood risk areas are only used for activities that would not suffer too much from a flood. These include parks and playing fields.

Figure 14 shows how the town of Nome in Alaska has planned its land uses so that most of the area at risk of an extreme, '1 in 100 year' flood event is left as open space or is used for leisure and recreation.

### Education

Local governments or government agencies need to let people know about flood risk and what to do in a flood. Otherwise, emergency plans and strategies are not effective. This is where education is really important. Governments try to educate people by:

- Sending leaflets through the post

- Advertising in newspapers and on television

- Posting information on websites

- Offering helpline telephone numbers

- Having drills and training exercises.

**Top Tip**

Not every flood will have significant effects on all the aspects mentioned. However, is likely that there will be a range of different effects and that both people and the environment will be affected. Some examination questions will ask for reference to both.

97

*Figure 13: House on stilts in Bangladesh*

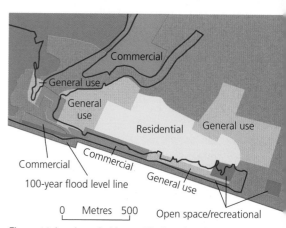

*Figure 14: Land use in Nome, Alaska, showing how it has been planned to cope with possible future flooding*

## Activity 4

Produce a leaflet that could be distributed to homes in a town in the UK at risk of flooding, explaining what they should do before, during and after a flood.

## Activity 5

Write a letter to a newspaper arguing either for or against the use of hard engineering methods in flood management.

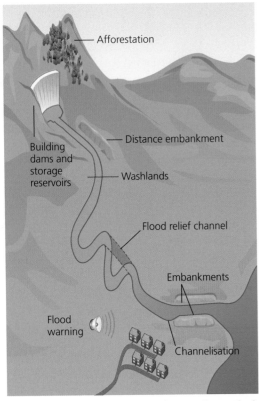

*Figure 15: Flood prevention methods (Harcourt and Warren 2001)*

## Case study: Flooding and flood defences at Bewdley, on the River Severn

Bewdley is in Worcestershire, on the banks of the River Severn. Flooding has been an issue for many years in this historic town, with properties close to the river suffering regular floods. The floods result from heavy rain in Wales where the river rises, and these surges of water work their way downstream through Shrewsbury and then on to Bewdley. In November 2000 there were particularly serious floods which affected 140 homes, causing damage to carpets and furniture, as well as to plaster and even walls. People had to move out of their homes and traffic was excluded from some streets.

The Environment Agency consulted local people about the nature of the flood defences. Local people did not want unsightly flood defences which might spoil the look of their ancient town. In the end, a demountable aluminium barrier system was introduced. The barrier works with a wall below ground level which acts as a cut-off to underground flow. About 450 m of barrier were built, with 150 m of permanent flood defence wall. In total the defences cost £7 million and were completed in two parts – one in the north part of the town and then one in the south. Serious floods along the Severn in 2007 and again in 2012 did not penetrate the flood barriers, and the homes in Bewdley stayed dry.

## Flood management – from hard engineering to integrated and sustainable approaches

There is a wide range of different engineering methods that can be used to try to control rivers in the UK. In the past, the traditional approach was to use **hard engineering** methods to prevent the floods. This tended to ignore the impact of these 'solutions' on all the other features of the river, such as the wildlife, or its use for leisure and tourism. So now the approach is more an **integrated river management** one, looking at all aspects of the river – at both its physical geography and its human geography. This new approach also emphasises the importance of trying to find sustainable solutions to the problems of river floods – solutions which do not damage the environment. Some of these solutions involve hard engineering, but most involve soft engineering, which is thought to be more environmentally friendly. Each method has advantages and disadvantages, as shown on page 99.

## Hard engineering methods

| | Advantages | Disadvantages |
|---|---|---|
| **Embankments (levees)** – these are high banks built on or near riverbanks. | They stop water from spreading into areas where it could cause problems, such as housing. They can be earth and grass banks, which blend in with the environment. | Floodwater may go over the top. They can burst under pressure, possibly causing even greater damage. |
| **Channelisation** – this involves deepening and/ or straightening the river. | This allows more water to run through the channel more quickly, taking it away from places at risk. | More water is taken further downstream, where another town or place at risk might lie. They do not look natural. |
| **Dams** – these can be built upstream to regulate the flow of water. | Water is held back during times of heavy rain or snowmelt. They can also be used for water supply, recreation and HEP. | They can be an eyesore and are very expensive to build. If they burst the damage would be devastating. |
| **Flood relief channels** – extra channels can be built next to rivers or leading from them. | The relief channels can accommodate the surplus water from the river so that it won't overflow its banks. | They can be unsightly and may not be needed very often. Costs can be high. |

## Soft engineering methods

| | Advantages | Disadvantages |
|---|---|---|
| **Washlands** – these are areas on the flood plain that are allowed to flood. | This gives a safe place for floodwater to go. Inexpensive and leaves the natural environment unspoilt. | Flood plain cannot be used for other things. |
| **Afforestation** – trees are planted in the drainage basin. | Trees intercept rainfall and take water out of the soil. This reduces the amount reaching rivers. Wooded areas look attractive and provide wildlife habitats. | The land cannot be used for other activities, such as farming. |
| **Land use zoning** – governments allocate areas of land to different uses, according to their level of flood risk. | Major building projects are allocated to low risk areas. Open space for leisure and recreation is placed in high-risk areas because flooding would be less costly for them. | These may not be the best places for the different activities in terms of public accessibility. |
| **Flood warning systems** – rivers are carefully watched and if the levels are rising, places downstream can be warned. | People living in towns or villages downstream have a chance to prepare or to evacuate. | Sometimes it is not possible to give very much warning and so it is hard to save possessions. |

## Decision-making skills

What factors are used to decide the type of engineering method that is used?

## Skills Builder 1

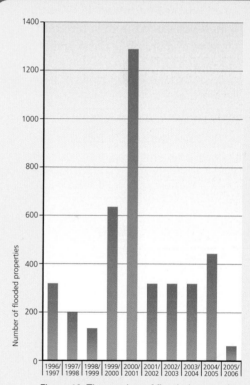

*Figure 16: The number of flooded properties in the south-west of England between 1996/1997 and 2005/2006*

Study Figure 16.

(a) In which years were the (i) highest and (ii) lowest number of properties flooded?

(b) Suggest two physical reasons why an area may have severe floods.

(c) Explain how any two hard engineering methods can be used to protect properties from being flooded.

## Hard engineering and flooding in York

York is a city which is prone to flooding by the Rivers Ouse and Foss which meet in the city. Over the past 20 years hard engineering works have been put in place to protect homes and families. Most of the works consist of raised embankments, some of which are made of earth and some of concrete supported by steel sheeting. In addition there is a natural flood plain upstream of York, called Clifton Ings, which can store 2.3 million cubic litres of water. This lowers the peak flood level in the city by 150 mm. The banks around the Ings were raised in 1982 and sluice controls built for letting flood water in and out of the Ings. All this hard engineering has been very expensive and it has had a significant impact on the environment – it has not benefited wildlife, for example. Further, despite all these works flood water does still overflow the embankments, causing damage to homes, offices and factories. This happened in 2000, leaving the city with a clean-up bill of £1.3 million, not including the cost of damage to people's homes and businesses.

A new system of flood defences was introduced for a trial in 2008. These are temporary hollow plastic barriers, called 'Aquabarriers'. They have holes in one side which allow flood water in and so increase their stability during the flood. Once the flood recedes the water simply drains away out of the holes. These are much cheaper than traditional hard engineering solutions to the problem and can be easily moved.

*Figure 17: Aquabarriers – a new system of flood defence*

### Alternative approaches on the River Skerne near Darlington

A different approach to flood prevention has been developed on the River Skerne in north-east England (Figure 18). In the past, the river had meandered across a flood plain, but over the 200 years before 1998 the river had been straightened and deepened. The flood plain had then been used as a tip for industrial waste. The river was simply a straight urban drain used to carry flood waters through a built-up area. However, from 1998 the river was restored to its natural meandering state by using earthmoving equipment to create large bends in the river. This enables the river to flow from side to side and for water to run off on to the flood plain during flood times.

In addition, a wide range of wetland habitats was created alongside the river. The banks of the meanders were protected with coconut fibre rolls and willow stakes to prevent excessive erosion until the tree roots had become established. Over 20,000 m³ of industrial spoil were removed from the flood plain in order to allow the river to overflow its banks but still protect surrounding housing. The spoil was also used to raise and reshape the valley sides. Over 20,000 trees and shrubs were planted, as well as wildflowers and bulbs. As a result, water quality has improved in the river, flooding has been reduced, and the wildlife – such as swans, fish, dragonfly and even water voles – has increased in the area. This scheme is a much cheaper and more environmentally friendly way of protecting the houses and buildings in the area than traditional hard engineering methods.

## Activity 6

Use the table below to summarise the advantages and disadvantages of the hard engineering flood defences of York and the soft engineering at the River Skerne.

|  | Hard engineering in York | Soft engineering in the Skerne |
|---|---|---|
| Cost |  |  |
| Environmental impact |  |  |
| Degree of success in stopping floods |  |  |
| Impact on wildlife |  |  |
| Other factors |  |  |

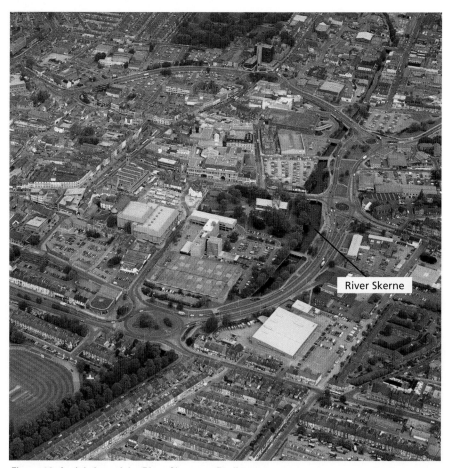

Figure 18: Aerial view of the River Skerne at Darlington

Rivers are vital sources of water, but they can bring destruction and suffering when they flood. Managing rivers so that humans and ecosystems can enjoy the benefits rivers bring, but not the costs, is a difficult challenge.

## You should know...

- ☐ Key terms linked to river systems, such as 'drainage basin'
- ☐ What the long profile of a river is
- ☐ How river characteristics change along the long profile
- ☐ How weathering, erosion, transport and deposition cause river landforms
- ☐ How to describe landforms of upper, mid, and lower course rivers
- ☐ How the landforms are formed
- ☐ How sediment changes as it is moved along a river
- ☐ How flooding is caused
- ☐ How human and physical factors increase flood risk
- ☐ To describe and explain hydrograph shape
- ☐ How climate change could increase flood risk
- ☐ The impacts of flooding, for example on Tewkesbury in 2007
- ☐ How hard engineering schemes might prevent flooding
- ☐ How integrated approaches might reduce flood risk
- ☐ How to evaluate different types of flood management scheme

## Key terms

Deforestation
Deposition
Drainage basin
Erosion
Flood plain
Flood risk
Geology
Hard engineering
Hydrograph

Impermeable
Integrated river management
Interlocking spurs
Levee
Long profile
Lower course
Meander
Mid course

Ox-bow lake
River cliff
Sediment
Slip-off slope
Upper course
Urbanisation
Waterfall
Weathering

### Which key terms match the following definitions?

**A** The relatively flat area forming the valley floor on either side of a river channel, which is sometimes flooded

**B** Not allowing water to pass through

**C** A natural embankment of sediment along the banks of a river

**D** The bend formed in a river as it winds across the landscape

**E** An arc-shaped lake which has been cut off from a meandering river

**F** The dropping of material that was being carried by a moving force

**G** The development and growth of towns or cities

**H** Areas of high land which stick out into a steep-sided valleys

To check your answers, look at the glossary on page 321.

**Foundation Question:** Using a named example, describe the impacts of river flooding. (6 marks)

| Student answer (achieving Level 2) | Feedback comments | Build a better answer (achieving Level 3) |
|---|---|---|
| Flooding happened in Tewkesbury in summer 2007.

There was an unusual amount of rainfall for June and rivers burst their banks.

Lots of homes were flooded.

Tewkesbury was cut off for a while.

Insurance claims were high and insurance will rise in the future. | • *Flooding happened in...* This is a good, specific example.

• *There was an...* Although this is correct, the question is about impacts not causes.

• *Lots of homes...* This mentions an impact, but does not state how many homes were affected.

• *Tewkesbury was cut...* This outlines an impact but could be more specific.

• *Insurance claims were...* This is a good, extended point. | Flooding happened in Tewkesbury in summer 2007.

There were over 5 inches of rain and the rivers Severn and Avon burst their banks.

1,800 homes were flooded and 10,000 had their water supply cut off.

The 4 major roads into Tewkesbury were flooded for a day.

Insurance claims were high and insurance will rise in the future. One year later, 800 people had not moved back into their homes. |

**Overall comment:** The student achieved Level 2 because although the answer was correct, some details were a bit vague. Try to include facts and figures in your answers.

---

**Higher Question:** Explain how human and physical factors can increase flood risk. (6 marks)

| Student answer (achieving Level 2) | Feedback comments | Build a better answer (achieving Level 3) |
|---|---|---|
| Flood risks are basically caused by too much rain, like the 5 inches that fell on Tewkesbury in 2007.

If the ground's already saturated, run-off happens rather than infiltration, making floods worse.

Steep slopes like at Boscastle also increase run-off and flood risk.

Humans can make flooding worse by urban areas and impermeable surfaces.

Deforestation is a natural factor.

It reduces run-off and infiltration. | • *Flood risks are...* This is a good example with details on rainfall.

• *If the ground...* This includes correct geographical terminology.

• *Steep slopes like...* Another good example and physical factor are provided here.

• *Humans can make...* A human factor is detailed here, which is the second part of the question.

• *Deforestation is a...* This is incorrect as deforestation is a human factor.

• *It reduces run-off...* This is unclear as there is confusion over which process is reduced. | Flood risks are basically caused by too much rain, like the 5 inches that fell on Tewkesbury in 2007.

If the ground is already saturated, run-off happens rather than infiltration, making floods worse.

Steep slopes like at Boscastle also increase run-off and flood risk.

Humans can make flooding worse by urban areas and impermeable surfaces.

Deforestation is a human factor that reduces infiltration.

It leads to more rapid run-off and a much steeper hydrograph. |

**Overall comment:** The student uses good geographical terminology, has some balance of human and physical factors, and uses examples which is good. The answer achieved Level 2, but only a few minor changes would be needed to score Level 3.

# Chapter 7 Oceans on the edge

104

## How and why are some ecosystems threatened with destruction?

The term ecosystem describes a grouping of plants and animals that is linked with its local physical environment – through use of soil nutrients, for example. The oceans, covering two-thirds of our planet, are home to distinctive **marine ecosystem** communities composed of fish, aquatic plants and sea birds – as well as tiny but very important organisms such as krill and plankton.

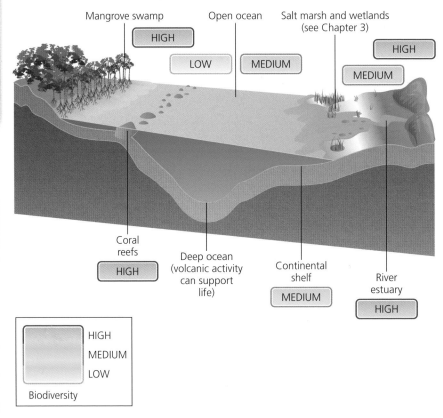

*Figure 1: Biodiversity in the oceans' ecosystems*

**Coral reef** ecosystems have an incredible level of **biodiversity** (their variety of plant and animal species). Although they cover less than 1% of the Earth's surface, they are home to 25% of all marine fish species (see Figure 12, page 114). Biodiversity is also high in waters close to the edges of the world land masses – above the **continental shelf**. There, species enjoy shallow warm water, enriched with silt nutrients from river **estuaries**. Even in the deep ocean, where light cannot penetrate, unique ecosystems are found. Underwater volcanic activity can create densely populated sites of plant and animal life. Here, often at great depth, life has evolved that can survive truly extreme conditions of heat and pressure.

## The global pattern of mangrove swamps

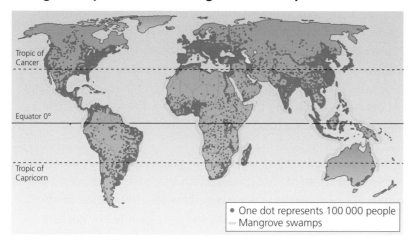

Figure 2: The global pattern of mangrove swamp and populated areas

Mangroves are areas of swampy forest, originally found in estuaries and along marine shorelines in around 120 tropical and subtropical countries. Many are densely populated, especially in coastal Asia (Figure 2). Mangrove plants have evolved to tolerate daily tidal flooding and high salinity. Long twisting roots anchor the trees against a constantly ebbing and flowing tide (Figure 3). The roots trap mud, making a **habitat** for lobster and prawn.

Figure 3: Mangrove forest

### Global threats to mangrove swamps

These species used to be fished in **sustainable** ways that preserved the **mangrove swamp**. However, prawns can be more intensively harvested from ponds dug in mud on cleared land. Global demand for prawns has caused widespread removal of mangrove vegetation. Mangrove swamp naturally covered 200,000 km² of the Earth's surface but only half that area now remains, often because of this prawn **aquaculture**. Vietnam's Mekong Delta is a typical site where mangrove trees have been removed, leaving a flat muddy plain studded with blue plastic-lined ponds. Other major areas of loss include Thailand (half lost since 1960), the Philippines and Ecuador.

Mangrove removal creates problems of coastal erosion and loss of nursery grounds for fish. Crocodiles, snakes, tigers, deer, otters, dolphins and birds all lose an important habitat. Carbon dioxide stored over centuries in the rich mud beneath the swamp is released when the trees are removed.

Mangrove swamp is also nature's defence against tsunamis. The enormous ocean wave that struck the coast of south-east Asia in 2004 killed 230,000 people. Where they were still in existence, mangrove trees shielded lives and property. In Thailand, recent conversion of mangrove habitat into prawn farms and tourist resorts close to Phuket contributed significantly to the catastrophic losses experienced there.

Prawn aquaculture can also cause serious **pollution** problems. Antibiotics and pesticides used in the prawn pools frequently leak into the delicate ecosystem of neighbouring areas.

Quick notes (Mangrove swamp):
- Because of commercial pressures, **unsustainable** exploitation now takes place, such as removal to create space for aquaculture.
- Of 200,000 km² originally, only 100,000 km² now remain.

## Skills Builder 1

Study Figure 3

(a) Name one adaptation shown that helps the trees survive in a wet tidal environment.

(b) Describe two ways in which undisturbed mangrove forest can support human activities.

## Unsustainable use and disruption of marine ecosystems

Ecosystems are made up of different animal and plant populations, all of whom are dependent on one another in some way. A natural balance exists among these populations. For example, the number of top carnivores such as sharks is always relatively low compared with fish they feed on, such as tuna. This is because big hunting sharks use a lot of energy chasing their prey and each must eat large numbers of tuna to stay healthy. These relationships are shown in the illustration of a **food web** for an open ocean (Figure 4).

Ecosystems are balanced in other ways too. The term **nutrient cycle** describes the movement and re-use of important substances – such as nitrogen – within an ecosystem. For example, fish take in nitrates when they eat submerged plants and algae. Waste products from the fish are converted by bacteria into ammonia and eventually into nitrates. The nitrate is absorbed by algae, restoring the original balance. The entire cycle can then begin again.

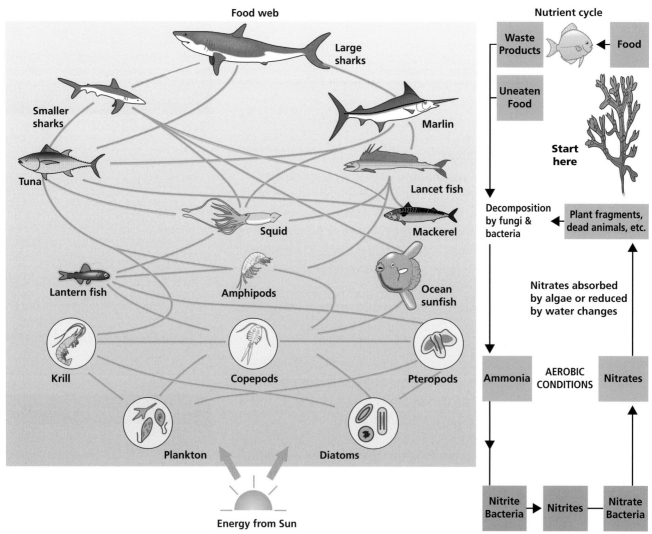

Figure 4: Marine food web and nutrient cycle for an area of open ocean

## Food web disruption

**Overfishing** by humans interferes with the natural balance – the equilibrium – of ocean animal populations. Excessive hunting of a particular fish species such as tuna triggers a series of impacts. First, humans directly reduce the number of tuna. Next, this action impacts indirectly upon sharks. These top carnivores can no longer find enough tuna to meet their own needs and numbers of sharks fall too. In contrast, numbers of organisms lower down the food web might actually increase at first – mackerel numbers, for example, could boom with fewer tuna preying on them.

If fishing ceases, the original balance may later be restored. In some cases, however, overfishing has been known to cause permanent species loss.

Worldwide, scientists say 5 million ocean species have yet to be discovered – and we are wiping them out before we have found them.

| | |
|---|---|
| **Cod fishing** | North America's cod population has never recovered from overfishing in Newfoundland during the 1970s. |
| **Baiji white dolphin** | Some large marine species, including the Baiji white dolphin, were hunted to **extinction** during the twentieth century. |
| **Whales and sharks** | Humpback whale and great white shark populations are unlikely to ever regain their original sizes. |

Another human impact that brings dire results to estuaries and oceans is **eutrophication**. This process occurs when excessive nutrients are added to a body of water. Nitrate fertilisers often get carried by rain **run-off** from farmland into rivers or over cliff edges into coastal water. In contrast to toxic pollutants that kill, the result here is one of 'over-feeding'. The nutrient-enriched waters initially experience a sudden growth in marine life. Tiny organisms flourish, creating an explosion of life called an algal bloom.

For a variety of reasons, the presence of so much algae uses up most of the water's oxygen. Fish and crustacean (crab and prawn) species suffocate in the de-oxygenated water. Eutrophic conditions can be found all over the world, with Japan and the Gulf of Mexico particularly badly affected (Figure 5 on page 108). The North Sea is another 'nitrate hotspot' where lobster populations have been lost due to a lack of oxygen.

**ResultsPlus**
**Build Better Answers**

**EXAM-STYLE QUESTION**

**Study Figure 4. Describe how an increased demand for squid in restaurants would impact on this marine ecosystem. (4 marks)**

■ **Basic answers** (0–1 marks)
Simply state there would be fewer squid left in the ecosystem.

● **Good answers** (2 marks)
Also identify linked effects, such as fewer tuna or a growth in lantern fish.

▲ **Excellent answers** (3–4 marks)
Also make explicit mention of the distinction between direct and indirect impacts.

## Activity 1

Why are there so few big fierce animals?

This deceptively simple question is a useful way to start discussing the workings of food webs. Think about how there are losses of energy at each level of the food chain (kinetic energy from movement, as well as heat losses). Also consider the physical losses of matter that take place due to urine and excretion.

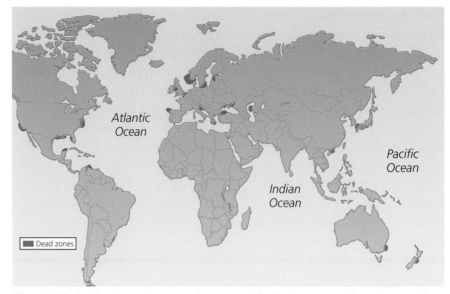

*Figure 5: 'Marine dead zones' – areas where eutrophication has occurred on a large scale*

Closely linked with eutrophication is the problem of **siltation**. Run-off from the land also washes soil into estuaries and coastal waters. This is common in parts of south-east Asia that have been badly deforested. After heavy rain, underwater plant communities become buried in silt. Seagrasses are deprived of the sunlight they need for photosynthesis. These plants are the starting point for the entire marine food web and if they die then so too do the fish.

## Climate change and the oceans

**Climate change** has started to create a warmer world. However, we should not expect to see a uniform increase in sea temperatures as a result. The world's oceans are dynamic systems where significant flows of warm and cool water can be identified (Figure 6). These flows could be disrupted or change direction because of climate change. For example, one impact of a major global temperature rise could be wide-scale melting of ice in Antarctica and Greenland. Billions of gallons of ice-cold water would be dumped into the world's oceans, leaving some seas cooler than before.

Complex physical factors such as ocean currents make it hard to predict exactly how marine life will be affected by global warming. However, a number of early impacts are already visible. Scientists have identified some stresses that climate change seems already to be bringing to oceans and their ecosystems, including **bleaching** (see table on page 109).

### Activity 2

Study Figure 6.

1. Figure 6 shows the UK enjoying a warm current. What is this current's name?

2. Study the cold ocean current flowing south from Greenland. It is fed by cold meltwater running off Greenland's ice sheets, especially in summer. What do you think will happen to the temperature of water surrounding the UK if the Earth gets significantly warmer?

3. How could this impact on life in the UK?

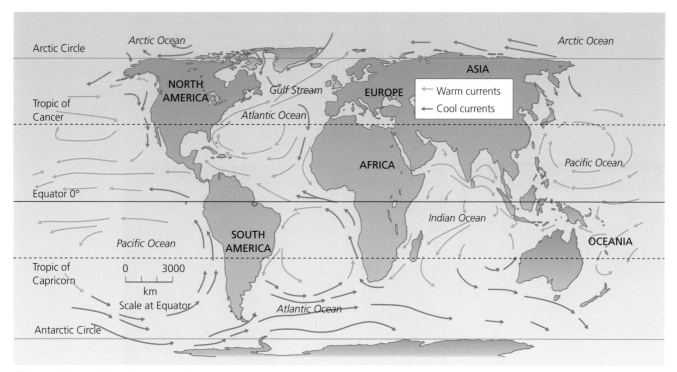

*Figure 6: The world's major warm and cool ocean currents*

| | |
|---|---|
| **Warmer water** | Some computer models suggest that fish in tropical seas will face a famine if their water gets too warm. Microscopic plankton that fish feed on become stressed and less productive in warmer waters. This is because warmed-up surface water tends to mix less well with the colder nutrient-rich water below. Fewer nutrients coming to the surface means less plankton growing. Food shortages then develop throughout entire food webs, causing reductions in population size at every level of the food chain. Warmer water may also have the effect of triggering species migration into cooler areas (see page 110). |
| **More acidic water and bleaching** | The majority of the world's coral reefs are in danger of being killed off by climate change. This is because the oceans absorb some of the extra carbon dioxide that humans are putting into the atmosphere. This helps slow down the actual rate of global warming, but is bad news for marine life. Carbon dioxide dissolves in water to form carbonic acid. Small rises in acidity can seriously damage coral reefs, which then appear bleached after losing their vibrant colour. Australia's Great Barrier Reef may lose 95% of its living coral by 2050 if current trends continue. |
| **Higher sea levels** | Melting ice sheets on land masses will bring **sea-level rises**. Coastal marine ecosystems such as mangrove swamp or UK salt marshes would become permanently submerged. Unique maritime communities might disappear altogether in some locations if a significant global sea level rise of 10 cm or more takes place. Recent events in Antarctica already suggest that ice sheets there are becoming less stable as a result of warmer temperatures. The collapse of the Larsen B ice shelf in 2002 worried many scientists who believe land ice is now moving into the sea at an accelerated rate. (See Figure 7 on page 110). |

Biodiversity changes are already being seen in British waters because of climate change. Water in parts of the North Sea is between 1 °C and 2 °C warmer than 20 years ago. During this time, the sandeel, a cold-water species of fish, has become less common. Scientists think this is the reason for falling populations of guillemots, puffins and other birds that have sandeels in their diet. However, there have been increased sightings of tuna, stingray and other warm water fish species.

## Skills Builder 2

Study Figure 7.

(a) Estimate the size of the area of polar sea ice that broke apart.

(b) Explain how human activities may have contributed to this event.

(c) Describe the possible impacts of melting polar land ice on two named marine ecosystems.

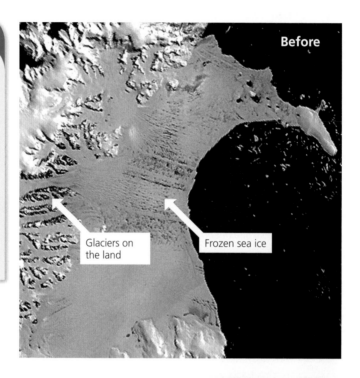

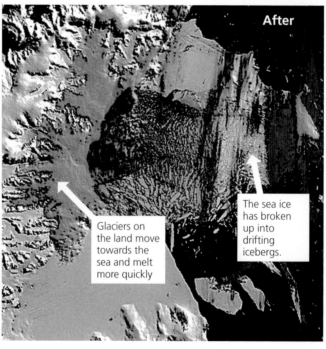

*Figure 7: A major break-up of the Larsen B ice shelf in the Antarctic, over just five weeks in 2002*

# How should ecosystems be managed sustainably?

## Pressure in the Firth of Clyde

The Firth of Clyde is a 60 km stretch of sea water along Scotland's west coast (Figure 8). The waters of the Clyde are home to 40,000 animal and microbe species. Nature-lovers regularly catch glimpses of seals, harbour porpoises and large basking sharks. Leatherback turtles and killer whales are also sometimes seen. However, the water is used in a number of ways that put pressure on this marine wildlife:

- Fishing – This is still an important source of local income, along with kelp (seaweed) harvesting. Thanks to a warm ocean current called the Gulf Stream, commercial fish and crustacean population sizes are very high. However, over-fishing has caused numbers of species – like cod – to crash. Some fishing takes place using 'active' methods, where ships drive in circles, closing their nets around shoals of fish. There are also 'passive' methods, where pots and traps are laid along the seabed.

- Tourism and leisure – Falling incomes from fishing and farming have led local businesses to try and make more of tourism and the leisure opportunities offered by the water and coastline. The Firth of Clyde is now the UK's second-largest yachting centre. Snorkelling and kayaking are also popular activities. However, such activities disturb wildlife.

- Sewage disposal – In the past, on-land sewage treatment facilities were limited. Waste from toilets often flowed straight into the sea, damaging sea life. Thankfully, this is less of a problem now due to tougher new laws.

- Military testing – The last Ice Age left the Firth of Clyde with deeply eroded seafloor valleys. This makes a perfect testing ground for the Royal Navy's nuclear submarines. A serious accident would devastate the ecosystem.

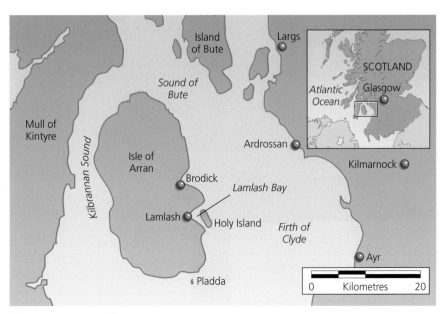

*Figure 8: The Firth of Clyde*

## Objectives

- Learn about the threats faced by marine wildlife in Scotland's Firth of Clyde.

- Explain why attempts to manage marine ecosystems often cause conflict and debate.

- Make an assessment of local and global attempts to manage oceans more sustainably.

## Decision-making skills

Read the bullet list, left. Rank the pressures in the Firth of Clyde in order of their severity. Use as your criteria: scale, immediacy and difficulty to manage.

## Activity 3

Look at Figure 8.

(a) Identify which major ocean the Firth of Clyde is joined to.

(b) Describe the physical factors that attract people to this stretch of water.

(c) Describe how pressures on the Firth of Clyde ecosystem have changed over time.

## Scallop fishing in Lamlash Bay

Much of the Firth of Clyde has been completely over-fished. Lamlash Bay has been especially badly affected. Scallops used to thrive here. In restaurants, these shellfish are an expensive seafood delicacy. They live on the seafloor and fishermen harvest them, using heavy dredging machinery made up of metal chains and rollers. This equipment scours the seabed and in so doing destroys the maerl – a pink-coloured cousin of coral. Maerl is an important nursery habitat for many fish species such as cod, plaice and haddock. With the maerl mostly gone, these fish species have all but vanished too.

The first people to notice change were Lamlash residents, some of whom enjoy scuba diving. They saw places where the seafloor now looked like the surface of the moon! A dead wasteland replaced the previously colourful and species-rich environment. Along with other concerned locals, the divers formed an organisation called COAST (Community of Arran Seabed Trust). The group started campaigning for a no-take zone to be established in Lamlash Bay as part of a sustainability plan. No-take zones have helped regenerate fish stocks in other parts of the world by offering species a protected spawning ground. COAST's wider aims were to:

- Improve the marine environment, and reverse the maerl and fish decline

- Sustain the livelihood of people dependent on tourism, as well as fishing

- Increase the popularity of the area as a diving and tourist destination.

In 2008, the Scottish government made part of Lamlash Bay a no-take zone. All fishing within the specified area has been banned while scientific trials are carried out. Tests will establish whether the seabed can regenerate naturally. The remainder of the bay is now a management area where scallops can still be fished, but only in less destructive ways.

## Activity 4

1. The Lamlash scheme is currently unique in Scotland. If it is such a good idea, then why have more no-take zones not been established in other places?

2. Do you think it should be easy to get different stakeholders such as conservationists and fishermen to talk to one another?

3. Do you think it will be a relatively simple matter to reach similar compromises in other threatened places?

## Decision-making skills

Create a table based on the opinions shown in Figure 9 to show arguments *for* and *against* the no-take zone.

'My family have been fishing for generations and it's the only work I know. If I can't fish, how am I going to pay for Christmas this year? Plus the sea is there for us to use – I don't see the point in keeping people like me away from doing honest work.' *Clyde scallop fisherman*

'I can't fish for fun much anymore – there's little left in the bay for me to catch. We should give the fish time to recover and leave them alone for a while.' *Amateur angler, Lamlash*

'It's a challenge managing places like this. We need to keep jobs, otherwise all the young people move out. That can leave us with a lot of independent elderly and no one paying the taxes we need to give them care. We know fishing jobs will be lost. But maybe some of the fishermen can refit their boats to take tourists on trips.' *Council Officer, Ayr*

'I retired here for the peace and the views. I feel sorry for the fishermen, but I approve of any attempt to try and limit industrial-type activities. Mind you, I hope this doesn't attract too many visitors – we don't want the roads getting too busy.' *Pensioner, Lamlash*

'It's a disgrace the way the fishermen have been allowed to ruin the sea life over in Lamlash Bay. The no-fish zone is the best news we've had in years. Finally our wildlife will be restored.' *School teacher, Ayr*

'The no-fish zone is good news for me. Our firm is going to invest in some new glass-bottomed boats. Lamlash is getting a lot of publicity and plenty of visitors will pay to see the wildlife. We've got porpoises and seals, and even basking sharks sometimes swim in these waters.' *Boat tour operator, Arran*

*Figure 9: A view of Lamlash Bay, now a no-take zone*

## Local views on managing the Firth of Clyde

Many outsiders, both working and retired, have moved to places bordering the Firth of Clyde so that they can enjoy the sea views. They want to see the water treated well and some would like it to be kept completely free from commercial exploitation. In contrast, other people – such as local fishermen – rely on the Firth of Clyde for their livelihood. Some see laws like the no-fish zone in Lamlash Bay as a step taken in the wrong direction.

### *Future plans for the Firth of Clyde*

The recent establishment of a no-fish zone in Lamlash Bay has shown that it is possible to make major adjustments to how local waters are managed. However, even bigger changes could soon be under way in the Firth of Clyde.

First, it is one of ten stretches of Scottish water that may soon be designated as a Coastal and Marine Park (CMP). The aim would be to ensure that coastal and marine-based activities are managed in sustainable ways to bring long-term economic benefits to people, while protecting the environment. The CMP, if established, will be run along the lines of Britain's existing National Parks.

This would mean much closer monitoring of commercial activities like fishing and sea-bed drilling for oil, gas or other minerals (which are all very important to the Scottish economy).

Secondly, the Scottish Marine Act is a new set of laws to help manage future conflicts in Scottish waters. For example:

- Scotland's government wants 31% of electricity to come from renewable sources. Major tidal and offshore wind resources can be exploited to assist with this. However, big new off shore developments could interfere with navigation for ships. Critics say that offshore wind farms with unsightly turbines – although designed to help the environment – will ruin the look of local landscapes (see Figure 10).

- Habitats of species like dolphins and other marine mammals need protecting from pollution. However, rising energy prices mean that companies want to exploit the remaining oil and gas resources found under the seabed – activities that can pollute.

Can compromises be reached that will keep everyone happy? Only time will tell.

*Figure 10: Offshore wind turbines*

**Results Plus**
**Build Better Answers**

### EXAM-STYLE QUESTION

**Study Figure 10. Explain one advantage this site brings for building wind power turbines. (2 marks)**

■ **Basic answers** (0 marks)
State that they are in the sea.

● **Good answers** (1 mark)
State that the advantage is 'they are at sea and not on land'. This response only gains one mark, as it does not explain why this is a good thing.

▲ **Excellent answers** (2 marks)
Explain that a marine site will not be too close to anyone's home, unlike a land site. This response gains full marks.

## Activity 5

1. Should access to coral reefs be restricted?

2. If tour operators increased their prices, then visitor numbers would fall but overall revenues might stay high. However, tourists with less money would no longer be able to visit. Is this fair? What rules would you make?

*Figure 11: A tourist diver visits a coral reef*

## Sustainable management at the local scale

### Sustainable management of the Coral Triangle

Coral reefs are alive. Although they have the appearance of rocks, reefs are made of living animals. Each piece of coral contains a polyp. These marine animals (ranging in size from 1 mm to 20 cm) live alongside others, forming large colonies. The coral is the polyp's external skeleton made of calcium carbonate. These skeletons make up the reefs we find in clear, warm, sunlit seas.

As well as being home to one quarter of the world's fish species, coral reefs form natural barriers that protect nearby shorelines from the erosive power of the sea. They offer essential protection to coastal dwellings and beaches in areas like Florida. Coral reefs are also good for tourism because they attract scuba divers and swimmers (Figure 11). However, coral reefs are under threat. Local pollution from tourist vessels and from agricultural run-off damages coral communities. The global threat of acidified sea water accelerates the destruction. Estimates suggest that 70% of the world's coral reefs will be destroyed by 2050. Good ecologically based management is needed to help restore reef ecosystems.

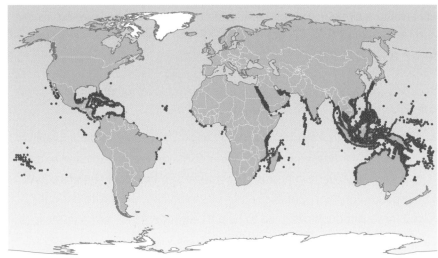

*Figure 12: Global distribution of coral reefs*

The Coral Triangle is a 6 million km² area of the south Pacific that contains three-quarters of all known coral species. 120 million people gain their food and income here. As a result of this population pressure, over-use problems are acute. The six governments of the Coral Triangle – Indonesia, Malaysia, Papua New Guinea, the Philippines, the Solomon Islands and Timor Leste – want to discourage non-sustainable activities. In 2007, they established a partnership to conserve the coral reefs and the fisheries they support. A range of strategies are being introduced, including:

● Marine-protected areas where reef populations can remain unharvested

● New sustainable fisheries management rules (including quotas if necessary)

● Protection measures for threatened species

● Establishing climate change adaptation plans.

### Shetland Islands aquaculture

Commercial fishing of waters like the North Sea has become unsustainable. Too many fish are being caught before they have had time to reproduce and breed the next generation. This is mostly due to improvements in technology:

- Increased longlining (some ships lay a 150 km length of baited hooks on the seabed).

- Sonar is now used to detect shoals of fish that might otherwise have been missed.

- Huge factory ships now have freezers, allowing them to stay out at sea longer.

Levels of success are so high that future generations of fishermen may soon have nothing left to catch. So overfishing is an economically – as well as environmentally – unsustainable practice.

One solution is to stop fishing altogether, as in Lamlash Bay. However, this leaves consumers without fish to eat and fishermen with no source of income. Another solution is to instead introduce aquaculture. Intensively farmed salmon and cod are now raised in caged enclosures along many northern European coastlines. Shetland Aquaculture is an association made up of nearly 50 producers based in the UK's northernmost islands. It was established in 1984 to provide an alternative to traditional unsustainable fishing methods.

Since that date, the industry has grown from strength to strength: 1,200 residents of the Shetland Islands now work in the aquaculture sector, and production has expanded from just 50 tonnes of fish in 1984 to more than 50,000 tonnes today. Initially, the focus was on salmon but cod, sea trout, haddock, halibut and mussels are now reared. Help has come from researchers at the North Atlantic Fisheries College in the Shetland town of Scalloway. The college has its own marine hatchery, and supplies both brood fish and eggs to help commercial firms become established.

Around half of all fish eaten in the UK has been farmed in places like the Shetland Islands. Globally, aquaculture has grown at around 8% per annum since the 1950s (Figure 13). But fish farming can bring new dangers to marine ecosystems. Outbreaks of parasites and disease are common among caged fish who live in cramped conditions, and hundreds of thousands of salmon escape from North Atlantic farms each year, allowing these serious problems to spread to wild populations.

Clearly there are always tensions when trying to achieve economic *and* environmental sustainability. Attempts to provide sustainable incomes for coastal fishing communities in Shetland have continued to threaten the wider environment. In contrast, total protection for the environment was achieved in Lamlash Bay (page 112) when fishing was entirely prohibited. But this left fishermen there with no source of income – and local consumers without seafood.

## Skills Builder 3

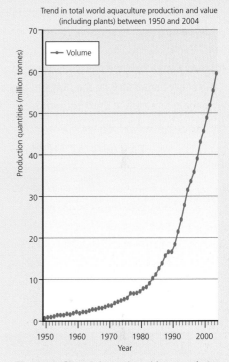

Figure 13: Growth in total world aquaculture production since 1950

Study Figure 13.

(a) Describe how aquaculture production has grown since 1950.

(b) Describe the advantages and disadvantages of this growth for wild fish populations.

## Activity 6

(a) Should people living in the Faroe Islands be allowed to keep killing Pilot whales if it is an important part of their cultural tradition?

(b) How do we balance the global need to preserve species with the rights of local people to embrace their heritage?

### TAKING ACTION

| IWC |
| --- |
| The International Whaling Convention (IWC) was established in 1946 to oversee the management of the whaling industry worldwide. In 1986 it issued an indefinite ban on commercial whale hunting. |

| UNCLOS |
| --- |
| The United Nations Convention on the Law of the Sea (UNCLOS) requires that the 156 nations who have signed it must follow the IWC guidelines. |

| CITES |
| --- |
| The Convention on International Trade in Endangered Species of Wild Fauna and Flora (CITES) gives global protection to all of the great whales. |

## Global actions to maintain ocean health

Global actions are needed to tackle pollution and to save threatened species from overfishing and extinction. International organisations already play an important role in helping countries to reach agreements. For example, the United Nations Food and Agriculture Organisation (FAO) regulates the management of deep sea fisheries. In addition, individuals can do their bit by making shopping choices that do not support unsustainable fishing. For example, many shoppers avoid buying tinned tuna that has been caught in nets that also may have trapped dolphins. Since the 1980s, many tuna products have carried the label 'dolphin-friendly', to show that no dolphins were harmed in the process of catching the fish.

### Protecting endangered marine species

International laws exist to protect endangered species. Many species of whale were hunted almost to extinction during the twentieth century, bringing a public outcry in many countries. Now whale populations are being helped to recover by three global agreements (see table on left).

It is not just whales that have been helped by new international laws. CITES has helped protect other species too. In 2007, sturgeon fish were added to its endangered list and can no longer be hunted legally in the wild (sturgeon eggs are used to make caviar, an expensive delicacy).

However, international agreements are not always fully effective:

- Japan has a long history of defying international whaling laws and continues to slaughter whales.

- Norway has objected to proposals for the south Pacific to be made into a whale sanctuary.

- Each year in the Faroe Islands, around a thousand pilot whales are massacred after they run aground in shallow water. Local people see the hunt as an important part of their culture.

### Tackling pollution

Shipping is an enormous industry, with over 90% of all trade between countries involving sea travel. However, sub-standard ships and illegal activities cause marine pollution – especially along busy shipping lanes. Vessels are expected to follow international rules set out by UNCLOS. One success has been the retirement of single-hulled oil tankers after the single-hulled *Prestige* went down off the coast of Spain in 2002, causing great damage to sea species. A global phase-out of these old tankers is under way, which should be an enormous step taken towards maintaining ocean health.

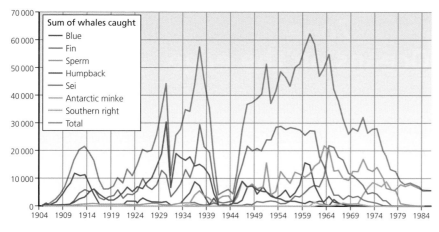

*Figure 14: Twentieth-century whale harvests in the Antarctic Ocean*

It is also now illegal for ships that have recently delivered oil to use seawater to wash out their tanks. Empty tanks are sure to still have plenty of oil stuck to the sides. Flushing tanks clean causes significant oil pollution (and brings in marine species to become 'stowaways', moving around the world inside the ships). Flushing is therefore banned, although monitoring what actually happens on ships is hard once they are out at sea.

A far more difficult problem to tackle on a global scale is the growth of the 'Pacific Garbage Patches'. These are enormous rubbish-strewn regions of the north Pacific. Researchers in 1999 counted one million pieces of floating plastic per square mile, most of it in the form of tiny fragments. These fragments are the remains of plastic bottles and bags that have been broken down by **attrition** and other erosion processes. They are captured and kept in place by a circulatory ocean current called the North Pacific Gyre. The current's flow creates giant pools of 'rubbish soup' (Figure 15). This rubbish has been flushed into the ocean by run-off and sewer discharge from thousands of different cities all over the world. This is a tough global problem to tackle, but more countries (such as Wales) are now passing laws that attempt to limit the use of 'throwaway' plastic bags.

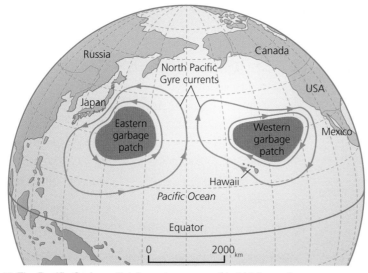

*Figure 15: The Pacific Garbage Patches – two areas of 'rubbish soup'*

## Skills Builder 4

Study Figure 14.

(a) In what year were most Blue whales caught?

(b) What happened between 1939 and 1945?

(c) Describe changes in the numbers of Sei whales caught each year.

(d) Explain why fewer whales of any species were caught after the 1970s.

**Build Better Answers**

**EXAM-STYLE QUESTION**

**Study Figure 15. Describe the size and location of these areas of 'rubbish soup'. (3 marks)**

■ **Basic answers** (0–1 marks)
Ignore part of the question and only describe either the size or location of the garbage patches.

● **Good answers** (2 marks)
State that two large patches are present, one on each side of the Pacific.

▲ **Excellent answers** (3 marks)
Also use the scale to estimate the size of the patches – or are far more precise about the location (e.g. proximity to the USA).

# examzone
## Know Zone
## Oceans on the edge

The oceans are a very important resource. Millions of people depend on oceans for food and income. Human actions are rapidly degrading some parts of the oceans, increasing the threat of extinction and resource collapse. Climate change is a long-term threat to the health of the oceans.

## You should know...

☐ What different types of marine ecosystem there are

☐ The distribution of mangrove swamps or coral reefs globally, and their changing pattern

☐ Which human activities, like overfishing, are degrading marine ecosystems

☐ How these human activities can lead to damage and destruction

☐ What nutrient cycles and food webs are

☐ How humans can damage marine food webs and nutrient cycles

☐ How climate change might damage marine ecosystems

☐ That marine resources are under increasing pressure

☐ How this pressure affects ecosystems locally, such as the Firth of Clyde

☐ Why, locally, people's views on their ecosystems can lead to conflict

☐ What sustainable management means

☐ How sustainable management can protect ecosystems like the Firth of Clyde and Coral Triangle

☐ How global actions such as CITIES and the IWC can help improve ocean health

## Key terms

| | |
|---|---|
| Aquaculture | Habitat |
| Attrition | Mangrove swamp |
| Biodiversity | Marine ecosystem |
| Bleaching | Nutrient cycle |
| Climate change | Overfishing |
| Continental shelf | Pollution |
| Coral reef | Run-off |
| Estuaries | Sea level rise |
| Eutrophication | Siltation |
| Extinction | Sustainable |
| Food web | Unsustainable |

### Which key terms match the following definitions?

A An illustration of the grouping of animals and plants found in an ecosystem, showing the sources of food for each organism

B The deposition of silt (sediment) in rivers and harbours

C The number and variety of living species found in a specific area

D A tidal swamp dominated by mangrove trees and shrubs that can survive in the salty and muddy conditions found along tropical coastlines

E The loss of oxygen in water after too much nutrient enrichment has taken place

F Hard stony ridge, just above or below the surface of the sea, formed by the external skeletons of millions of tiny creatures called polyps

G Long-term changes in global atmospheric conditions

H Water that flows directly over the land towards rivers or the sea after heavy rainfall

To check your answers, look at the glossary on page 321.

**Foundation Question:** Using examples, describe the threats facing marine ecosystems. (6 marks)

| Student answer (achieving Level 2) | Feedback comments | Build a better answer (achieving Level 3) |
|---|---|---|
| Marine ecosystems, like coral reefs, can be overfished. | • **Marine ecosystems like...** The student names an ecosystem and a threat. | Marine ecosystems, like coral reefs, can be overfished. |
| They might suffer from bad fishing types, like dynamite fishing, which destroys coral. | • **They might suffer...** This is an extension of the first point and provides a detailed type of fishing. | They might suffer from bad fishing types, like dynamite fishing, which destroys coral. |
| Divers break off bits of coral. | • **Divers break off...** is another correct threat and gains a mark, but is a bit vague on what divers do. | Divers break off bits of coral to collect as souvenirs. |
| This is happening a lot in the coral triangle area due to lots of tourists. | • **This is happening...** The student uses a named example, linked to a threat, which is good. | In the Mekong delta area of Thailand they are expanding aquaculture. |
| | | Mangroves are being cut down and replaced with ponds used to farm shrimps. |

**Overall comment:** The student would have scored additional marks if they had been more specific about the threat from divers, and remembered that a range of examples are needed for a longer answer question.

- - - - - - - - - - - - - - - - - - - - - - - - - - - - - - - - - - - - - - - - - - - - - - - - - -

**Higher Question:** Using named examples, explain the short- and long-term threats facing marine ecosystems. (6 marks)

| Student answer (achieving Level 2) | Feedback comments | Build a better answer (achieving Level 3) |
|---|---|---|
| Climate change is a short-term threat to marine ecosytems as climate changes quickly from day to day. | • **Climate change is...** This part of the answer is incorrect because climate change is a long-term threat. | Climate change is a long-term threat, as water temperatures gradually rise due to global warming. |
| Ecosystems like coral reef can be affected by rising water temperatures and this causes coral reef to become bleached. | • **Ecosystems like coral...** This is good because it explains why bleaching happens, but there is no example given. | Ecosystems like coral reef can be affected by rising water temperatures and this causes coral reef to become bleached. |
| Another short-term threat is from overfishing, such as what happened in the Firth of Clyde. | • **Another short-term threat...** This is good because a threat and an example are given. | A short-term threat is from overfishing, such as what happened in the Firth of Clyde. |
| Fisherman caught too many fish so fish stocks fell and the catch fell too. | • **Fishermen caught too many**... This part of the answer uses good terminology and extends the point above by adding detail. | Fisherman caught too many fish so fish stocks fell and the catch fell too. |
| This is not sustainable. | • **This is not sustainable.** This is correct, but it could be linked more carefully to the example above. | Fishing like this is not sustainable as long-term fish stocks could disappear. |
| A long-term threat would be like tourism. | • **A long-term threat...** The threat of tourism is not explained here. | Tourists often break coral off for souvenirs when diving, but coral might recover so this is a short-term threat. |

**Overall comment:** The student used some good terminology and some examples of threats and places. However, they did not really cover both short-term and long-term threats, and made some errors.

# Chapter 8 Extreme climates

## Objectives

- Describe the characteristics of fragile polar areas and hot arid areas.

- Explain how ecosystems and people have adapted to life in these environments.

- Understand that people living there make a unique and valuable contribution to world culture.

## What are the challenges of extreme climates?

The most important **extreme climates** are those where it is extremely cold (**polar**) or extremely hot and dry (**hot arid**). These two types can be subdivided further, as in the table below.

| Polar | **Glacial** – ice-covered places, e.g. Greenland |
| | **Tundra** – places with frozen soils, e.g. Alaska |
| Hot arid | **Deserts** – truly arid places with less than 250 mm rain a year, e.g. Sahara |
| | **Drylands** – semi-arid places with 250–500 mm rain a year, e.g. Sahel |

Places with extreme climates experience temperature or rainfall conditions that limit **flora** (plant) growth and **fauna** (animal) numbers. Without the aid of technology, human populations living in these places are limited too. Where extremely high or low temperatures are found, people's day-to-day existence is threatened. In contrast to life in a more **temperate climate** like that of the UK, special measures must constantly be taken just to stay alive. Places with an extreme climate have a low **carrying capacity**. This means that the number of people they can support without the aid of technology is lower than in other places where temperatures are more moderate and rainfall supplies are larger and more regular.

Of course, once money and technology becomes available, anything is possible. For example, 2 million people live in Las Vegas, in the Nevada desert – just 300 km from Death Valley, one of the driest (and deadliest) places on Earth. Anyone left without water in Death Valley would die from dehydration within a day, if not hours. It is the water piped from the River Colorado that keeps life going in Las Vegas (see Chapter 4).

### Deserts and drylands

A truly arid climate is one that receives less than 250 mm rainfall each year. Arid regions are also known as deserts. Large areas of the Earth's surface are arid, including the Sahara and the Australian, Arabian and Kalahari deserts. Most are located in the Tropics, for reasons explained in Chapter 3. Key extreme desert facts include:

**Hottest place on Earth**
A temperature of 58 °C was once recorded in the North Sahara at a place called El Azizia in Libya. Death Valley in Nevada (57 °C) comes a close second.

**Temperature range**
Night-time temperatures in Death Valley are 20–30 °C cooler than daytime temperatures (as low as 4 °C).

**Rainfall variability**
There are parts of Chile's Atacama desert that have not seen rain for 400 years. In many places there, average rainfall is just 1 mm per year. When rain does occur in desert areas, flash flooding may result.

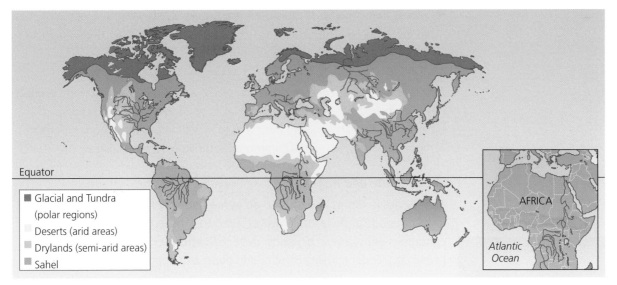

*Figure 2: Deserts, drylands, and polar regions*

On the borders of deserts there are semi-arid areas – also called drylands – where annual rainfall ranges from 250 mm to 500 mm. Higher rainfall areas show **seasonality** in their annual rainfall pattern. In California, for example, rains fall between November and March.

The Sahel is a long strip of drylands that borders the south of the Sahara (Figure 2). Compared with the Sahara, conditions in the Sahel are less testing for humans. Natural water supplies are sufficient for around 50 million people to live here – provided that their lives are well adapted to the conditions. Permanent, settled lifestyles exist side by side with nomadic (migratory) behaviour. Along the Sahel's northern edge, the nomadic Tuareg people live a migratory lifestyle. They live in small tribes consisting of 30–100 family members, who keep camels, goats, cattle and chickens which graze the land.

### Fragile desert and dryland flora and fauna

Deserts and drylands are **fragile** environments, meaning they are easily threatened by natural or human-induced changes. Ecological communities in arid areas have evolved in highly specialised ways to deal with challenging conditions. They cannot tolerate environmental changes and, as a result, many species will die out.

In the hottest desert regions, few animals can survive apart from tough scorpions and small reptiles. In dryland areas, with a greater supply of water, biodiversity rises as tough grasses, shrubs, cacti and hardy trees begin to form the basis of a larger food web. The plants and animals have evolved ingenious **adaptations** to help them deal with a shortage of water:

- Grasses with root systems that can reach as far as 50 metres underground.

- Trees that store water, e.g. acacia trees, with their short, fat trunks.

- Seeds that only germinate after heavy rain. Plants immediately produce brightly coloured flowers to attract insects.

## Skills Builder 1

Study Figure 2.

(a) Describe the distribution of deserts.

(b) Using examples, describe how plants and animal communities have adapted to survive the extreme conditions found in deserts and drylands.

(c) Explain one change that these communities are vulnerable to.

● Insects that can collect moisture from the air. By standing still, some beetles let early morning mist condense as water droplets on their bodies, which then runs into their mouths.

Deserts often experience an explosion of life after heavy rainfall. If the ground has been baked hard then temporary lakes will form. Dormant seeds and hibernating animals awaken. Birds and animals are attracted from far afield. Temporary plant and animal communities thrive for a while (Figure 3).

### Vulnerability and fragility

The species in desert and dryland areas are naturally resilient to seasonal or longer periods of drought. However, even they have their limits. When dry conditions get worse in drylands, **desertification** sometimes takes place, revealing the vulnerability of semi-arid plant and animal communities.

● Human causes of desertification include over-grazing by goats and cattle. This has been made worse over time by restrictions placed on the movement of nomadic herds by settled farmers.

● Physical changes include the occurrence of long-term and cyclical droughts (see Chapter 4, page 60).

## Extreme cold

Cold polar climates are found in areas of high **latitude** – especially inland areas, far from the warming influence of the sea (Figure 4). Places where the land is partly or completely covered with ice are called **glacial** regions. 98% of Antarctica is covered with snow and ice. The coldest ice-covered glacial regions such as central Antarctica do not support any life at all. In the northern hemisphere, lying south of the glacial regions are the **tundra** regions. They are not directly beneath ice but do experience very cold weather for most of the year, leaving the ground beneath permanently frozen. This gives them a distinctive appearance and environment, as you will see. Around 20% of the Earth's land surface is tundra. Key extreme cold facts include:

Figure 3: A desert blooms

| Coldest temperature ever recorded | Antarctica has a recorded low of -89 °C. The coldest temperatures for populated continents are -81 °C in Yukon, Canada and -55 °C in Siberia, Russia. |
|---|---|
| Seasonality | Alert Bay, on the north-east coast of Ellesmere Island, Canada is 82.5° north. It receives no direct sunlight for 50 days of each year. |
| Variability | Temperatures are highly variable through the year in some tundra regions. Summer in Anchorage, Alaska can be hot (26 °C) while winter is very cold (−15 °C). |

Figure 4: A view of the Alaskan tundra

In the high latitudes of the northern hemisphere, conditions are cold and icy but less extreme than in Antarctica. Daytime temperatures at the surface of Alaskan glaciers can reach as high as 12 °C in summer. Daytime temperatures keep skiers warm and the reflected sunlight even brings sunburn.

The tundra landscape (Figure 4) is found south of the glacial regions in the northern hemisphere. Much of Alaska – a separate US state to the west of Canada – is tundra. In populated regions, such as the northern town of Barrow, the Sun never rises above the horizon for up to one month of the year and temperatures may fall to –50 °C. Sunlight here is so weak that the growing season is generally under three months. Native people hunt extensively during the most brightly lit months, building up meat stores for the year, or working alongside settlers to mine and extract resources, including oil.

### Fragile glacial and tundra flora and fauna

Away from the deepest ice sheets, small specialised and tough alpine plants grow on the exposed rock surfaces that can be found poking through the snow in glacial areas. Along with lichen and moss, they form the basis for simple food webs. Plants survive strong winds by growing close to the ground in crevices and cracks. Spiders and tiny mice can also scratch out a living on glacial mountain sides. Like desert ecosystems, these organisms tolerate environmental change poorly.

Tundra food webs are far more complex, with around 1,700 species of vegetation and animals. A range of adaptations help plants and animals to survive here:

| | |
|---|---|
| **Water-loving plants can thrive** | In the flat low-lying tundra regions, surface layers of soil melt in summer, leaving many areas waterlogged and creating perfect conditions for sedges and moss (Figure 4). |
| **White fur as camouflage** | Plants form the basis for a food chain with many plant-eaters and meat-eaters present. Some, like snowshoe rabbits and snowy owls, have evolved white fur and feathers so that they cannot easily be seen against winter snow. |
| **Caribou have two layers of fur** | The caribou is a type of deer that has adapted to survive the bitter tundra cold. They have two layers of fur to keep the heat in and large hooves to help them travel over soggy ground. |

### Vulnerability and fragility

In both glacial and tundra regions, we find that flora and fauna can be vulnerable to sudden environmental changes:

- Bursting lakes – Catastrophic changes occur when a glacial meltwater lake bursts, which is likely to become more common due to climate change and warmer temperatures in polar regions. Meltwater running off a glacier sometimes builds up behind a natural barrier, such as a ridge or bank of soil. Failure of the barrier can release a surge of meltwater, destroying local ecosystems.

- **Solifluction** – Tundra regions are vulnerable to seasonal soil melting. This can trigger a process called solifluction. Rather like a landslide, whole sections of a slope start to move under gravity. The surface vegetation gets rolled beneath the moving mass of soil, like the tracks on a tank. Again, a warming atmosphere increases the likelihood of this happening. Heat from buildings can also cause melting.

## Skills Builder 2

Study Figures 2 and 4.

(a) Describe the distribution of glacial and tundra regions.

(b) Using examples, describe how glacial or tundra plants and animal communities have adapted to survive the extreme conditions found in these places.

(c) State and explain one change that a named polar ecosystem is vulnerable to.

## The adaptations people make

### Living with extreme cold

Many of the Earth's coldest regions show evidence of long-term settlement, despite the challenging environmental conditions found there. In Chapter 2, we learned about the ancient historical activities of some groups of people living in the world's far north. During the last Ice Age, people were migrating out of Asia and into North America across ice and land bridges, spreading their culture through regions that remain cold today. More recently, Scandinavian Vikings were exploring Greenland during the Medieval Warm Period, around the tenth and eleventh centuries.

Today, people can be found living in all the cold regions of the Arctic Circle. In Russia, we find nomadic reindeer herders called the Khanty people. In Alaska, native Inupiat and Yup'ik tribes have been joined by more recent settlers. The adaptations to cold conditions seen in Alaskan settlements are thus a mixture of ancient and modern technologies:

| | |
|---|---|
| **Buildings** | High-pitched steep roofs allow snow to slide off, while triple-glazed windows keep the heat indoors. In areas of **permafrost**, houses are raised on stilts to prevent their heat from melting the frozen ground beneath (Figure 5). Ice loses volume when it melts, causing the land to sink and subside, which would damage the building. Even worse, melting the ice might trigger solifluction flows. |
| **Transport** | Roads can also suffer from permafrost melting, causing cracks to develop. In order to prevent this, roads are built on 1–2 metre thick gravel pads that stop heat transferring from vehicles to the frozen soil beneath. |
| **Farming methods** | Native Alaskans have always relied on hunting and fishing for food rather than farming the land, because of the permafrost and very short growing season for crops. |
| **Clothing** | Traditionally, Inupiat and Yup'ik people favoured coats made of caribou skin (with the double lining of fur) and sealskin boots. Goose down was used as a lining. Now they also wear modern man-made textiles, like Gore-tex. |
| **Energy use** | Energy use is high for people living in cold environments, especially at high latitudes where sunlight is extremely limited for part of the year. In parts of Alaska, geothermal heat provides energy and hot water, a beneficial side-effect of volcanic activity along this part of America's Pacific coast. |

### Exploring polar cultures

Many of Alaska's native people are part of a larger ethnic group called the Inuit who are spread throughout the Arctic Circle – in Canada, Alaska, Russia and Scandinavia. In all these places, unique cultures have survived into the modern age.

*Figure 5: An Alaskan house built on stilts to avoid melting the permafrost*

| Yup'ik masks | The Yup'ik community in south-western Alaska are well known for their carved masks, which are a unique feature of their storytelling culture. |
|---|---|
| Inupiat art | You may be familiar with the art of totem poles. Tlingit village in Alaska attracts tourists to see an impressive collection of poles that have been collected there. |
| Inupiat whaling ceremony | Some aspects of polar culture are controversial. In Alaska, the Inupiat are still allowed by US law to hunt and kill bowhead whales. Communal sharing of freshly cut whale meat is an important ritual that the Inupiat people believe helps bind their culture together. |

Other cold places such as Iceland and Greenland have their own highly distinctive cultures. Life in a challenging environment has evolved over time so that practices born in hardship have become treasured traditions. For example, in Iceland, people still love to eat 'rotten shark'.

### Iceland's 'rotten shark'

Until recently, Iceland was regarded as one of the least developed countries of the northern hemisphere. It has a harsh glacial, stormy and volcanic environment. However, thanks to investment in banking and hi-tech research, the Iceland economy has made enormous progress over the past century.

Despite this economic advancement, Icelandic people still like to share 'rotten shark' at parties. The meat of the Greenland Shark is naturally poisonous because the shark contains fluids that allow it to live in cold water without freezing. To make the meat safe, Icelanders used to bury chunks of the meat for months at a time before eating. During this time, the fluids drain from the shark, making it safe.

The decomposition that takes places also gives the meat a very strong ammonia smell that many people find disgusting. Although modern Icelanders no longer need to rely on it for survival, many have learned to enjoy the taste of 'rotten shark'. They view it as part of their heritage. You can watch a short film about this at: http://video.nationalgeographic.com/video/player/places/culture-places/food/iceland_rottensharkmeat.html

Polar cultures are valued round the world. Many people alive today in Europe and North America share cultural roots with Icelanders, being also descended from Scandinavian Vikings. The Vikings were seafaring people from Sweden and Norway who travelled widely throughout the northern hemisphere around one thousand years ago (see page 30). Also, UNESCO (the United Nations Educational, Scientific and Cultural Organization) has pledged to safeguard Inuit cultures, which, due to their unique character, are viewed as being an important part of global cultural heritage.

Figure 6: Many people in Iceland enjoy eating 'rotten shark'

Quick notes
(Iceland's 'rotten shark'):

- Extreme climates gave rise in the past to unique and interesting cultures.
- Originally, many cultural practices developed through necessity.
- These cultures live on today, through tradition and familiarity – and sometimes to attract tourists.

## Living with aridity

The word 'arid' might at first bring to mind inhospitable places where culture is unlikely to blossom. But remember that the first world's first great ancient civilisations – the Egyptian pharaohs and the Arabian and Persian empires – allflourished in places we classify as arid and semi-arid regions.

Desert and dryland cultures flourish today, especially when newer technologies have been introduced alongside older ones. In the Middle East, desert cities like Dubai and Abu Dhabi have air-conditioned buildings and enjoy desalinised water supplies paid for with oil money (Figure 7). Whether in the Middle East, or the Sahel, common adaptations to the arid climate include:

| Buildings | Flat roofs collect water, while walls are painted white to reflect sunlight and keep buildings cool. In poorer areas, windows are small to reduce light and keep temperatures low. Where there is money, there are bigger windows – and air-conditioning. |
|---|---|
| Transport | In the early twentieth century, camels were still the dominant transport for Middle Eastern and African desert tribes. Today, modern roads allow motorised transport for those with money. Rich people in the Gulf states drive powerful off-road vehicles in areas where sand frequently covers over the roads. |
| Farming methods | Water is incredibly precious in deserts and drylands. The ancient Egyptians relied on irrigation from the Nile to help their farming. The Nile is now dammed at Aswan, giving Egypt reliable water supplies for its crops. Where irrigation is not available, traditional **nomadic pastoralism** is still practised, with goats and sheep being moved to new places when they begin to put too much stress on local vegetation. |
| Clothing | Middle Eastern tribes of Bedouin people favour loose-fitting clothing, often white to reflect sunlight. To avoid sunburn and protect the wearer from wind and sand, heads and faces are often covered with head-scarves called *kufiya*. Clothes were originally woven from sheep or goat wool. |
| Energy use | For traditional societies, energy use has often been low, thanks to the high temperatures. Wood fires keep nomads warm inside their tents at night. Where oil has been found in the Middle East – and where wealth is now high – energy use has, of course, soared, in response to rising demand for air-conditioning and refrigeration in permanent settlements. |

*Figure 7: The Dubai skyline*

## Exploring arid cultures

Some of the world's desert and dryland cultures have given us truly amazing cultural and archaeological artefacts:

- **Egypt's Pyramids at Giza** – Perhaps the greatest wonder of the ancient world, Egypt's pyramids lie in the desert, an important reminder that harsh climates have never stopped human progress.

- **Ethiopia's Hadar area** – The earliest known human ancestor, a 3.5-million-year-old skeleton, was found here. This area, now a UNESCO World Heritage Site, has enormous significance for global culture.

- **South America's Nazca Lines** – In the desert of Peru, giant shapes have been 'drawn' in the desert. They were constructed around 1,500 years ago, but they can only be seen properly from high above – by aeroplane – making them one of the world's greatest archaeological mysteries.

The Sahel region that you have been learning about in this chapter is also home to many rich cultures and traditions:

- The use of instruments such as the Hausa flute and gourd rattle has produced a distinctive musical culture not found anywhere else (Figure 8). For instance, Mali's Dogon people use music and dance in a festival, called a 'Dama', to honour the dead.

- The city of Timbuktu in Mali was an ancient place of learning and is said to have been home to 25,000 scholars in the sixteenth century.

- In nearby Niger, the Tuareg tribe are excellent craftsmen, renowned for their indigo cloth, gold and silver jewellery and carved wooden masks.

However, culture clashes are also found in the Sahel. In Sudan's Darfur region, for example, the recent wars between tribes have resulted in 300,000 people being killed and around 2 million displaced since 2003. Other conflicts are currently underway in Chad, Niger, Mali and Senegal. Too often, the same problem lies at the root of the conflicts – different groups of nomadic cattle-herders and settled farmers are fighting over limited grass, water and soil.

**ResultsPlus**
**Build Better Answers**

**EXAM-STYLE QUESTION**

**Describe how building styles are adapted to the extreme climate you have studied. (3 marks)**

■ **Basic answers** (0–1 marks)
State that buildings are white in arid places.

● **Good answers** (2 marks)
Also say something about why this is the case (because white reflects sunlight, keeping temperatures lower inside).

▲ **Excellent answers** (3 marks)
Not only describe a traditional type of adaptation, but also say that modern houses have air-conditioning (especially where oil money is found).

*Figure 8: Musicians of Mali's Bobo tribe playing traditional instruments*

## Objectives

- Describe the human and physical threats faced by one extreme environment.

◉ Explain how climate change could further threaten this environment and its people.

◉ Understand that a range of local and global management strategies can help protect this environment.

# How can extreme environments be managed and protected from the threats they face?

## Human threats to extreme climates

Extreme climates give rise to physical environments that are sensitive to change. Human actions often threaten to change local conditions and disturb the delicate balance found in a number of ways. The tables below and opposite show examples of the impacts in the Sahel and in Alaska.

**Pollution** occurs because of resource **exploitation**, often in previously pristine wilderness environments. One reason for this is that valuable oil underlies many of the world's arid and polar regions. **Land degradation** sometimes results from poor management, accelerating desertification in drylands and permafrost melting in some polar areas.

It is not just the physical geography that is affected. Tourism and settlement by in-migrants can result in **cultural dilution**, because outside influences cause a culture to lose its unique characteristics. Ancient languages become lost – and replaced by English or Spanish for instance.

*Figure 9: Striking a pose for the tourist gaze*

|  | Impacts in the Sahel |
|---|---|
| Pollution | One theory suggests that air pollution in North America and Europe may have made drought and desertification worse in the Sahel region since the 1960s. Clouds of sulphur particles emitted from power stations and factories may have affected patterns of cloud formation in ways that have left the Sahel without its normal level of rainfall. |
| Land degradation | Desertification is an enormous problem. Traditional nomadic herders find they cannot migrate with their cattle as easily as in the past, for various reasons. International food-growing companies have seized the best land. National boundaries established in the colonial era interfere with age-old migration routes. Civil war and political instability today drive people into marginal lands. The result is often overgrazing, removing the marginal vegetation and allowing soil erosion. |
| Cultural dilution | The Sahel countries were colonised by European powers in previous centuries. Many have experienced disruption and civil war since independence. As such, it is not surprising to find that some native languages and art have been lost and forgotten. Today, foreign tourists are a vital source of income, and tourism continues the process of cultural dilution and loss. In the case of the Dogon people of Mali, they sometimes perform their sacred Dama funeral ritual for the amusement of tourists. Some people think that cultures like the Dogon should not be exploited like this (Figure 9). |
| Limited economic opportunities and out-migration | Young people have migrated away in search of work. Most head south to coastal cities such as Lagos, where export-oriented industries are based, and the high population density provides more opportunity to work in formal and informal service industries. Some head north to the Mediterranean and cities like Cairo – or a passage to Europe. |

How can extreme environments be managed and protected from the threats they face?

129

| | Impacts in Alaska |
|---|---|
| **Pollution** | Alaska has suffered badly from the extraction of oil in the North Slope area. In 1989, an oil tanker, *Exxon Valdez*, ran aground on the Alaskan coast while transporting oil to market. Only 15% of the 11 million spilled gallons was ever recovered, and 5,000 sea otters and many seals and eagles were killed. More recently, a broken pipeline in 2006 spilled 200,000 gallons of oil in the fragile North Slope region. |
| **Land degradation** | Millions of square kilometres of permafrost have been damaged because people didn't take account of the sensitive soil conditions. Wide-scale melting can be found around warm urban areas such as Fairbanks and Barrow. However, the greatest human threat comes from global warming. Temperatures are rising in Alaska, perhaps by as much as 5 °C over the past century. Scientists are also reporting an increased frequency of landslides in the permafrost. |
| **Cultural dilution** | In the past, 20 native languages were widely spoken in Alaska. As elsewhere in North America, European languages such as English have been adopted by the youngest generations of tribes (Figure 10). In-migration by new settlers since the 1700s has brought new languages. Native names were replaced by names like Peter and John after missionaries visited. In the 1970s, American schooling insisted on classroom use of English. Now, some languages, such as Eyak, have lost their last speaker, while others are on the verge of dying out. |
| **Limited economic opportunities and out-migration** | Many young native Alaskans have left the region in search of better paid work in the contiguous United States – the 48 states south of Canada. The Alaskan climate and environment limits opportunities in agriculture and tourism. Low population density means limited service sector employment too. |

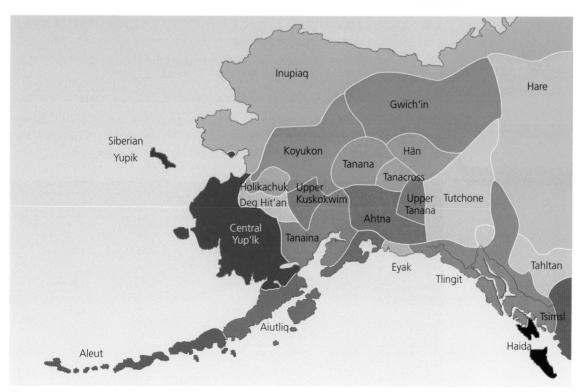

*Figure 10: Threatened languages of Alaska*

### EXAM-STYLE QUESTION

**Choose one extreme environment. Describe one threat that it faces on account of human activity.**
**(2 marks)**

■ **Basic answers** (0 marks)
State a natural problem affecting humans, instead of a human-caused problem (a few mention earthquakes).

● **Good answers** (1 mark)
Name a human-produced problem such as 'permafrost is melting in Alaska' or 'desertification is caused by farmers in the Sahel'.

▲ **Excellent answers** (2 marks)
Also fully describe how this would impact on the landscape or ecosystems. 'In Alaska, vegetation is threatened by landslides and pools of water.' 'In the Sahel, grasses die and soil erosion takes place.'

## Climate change and extreme environments

Climates and environments everywhere face change and challenge because of global warming, as we learned in Chapter 2. Events under way in arid and polar regions are especially concerning. The additional stress in arid and dryland regions endangers some of the poorest and most vulnerable people on the planet. Impacts in polar regions also threaten people and ecosystems living there – while melting polar ice, in turn, brings a sea-level rise that threatens societies on a worldwide scale.

### Climate change and desertification

Some predicted impacts of climate change for the Sahel drylands include:

| | |
|---|---|
| **Increased aridity** | Drylands where rainfall is already low could turn into fully arid desert as the twenty-first century progresses. Most predicted changes of weather show the Sahel rain belt being pulled either further north or south, away from populated areas. Dramatic drying is likely. |
| **Nomads losing their livelihood** | Nomadic herders in the northern Sahel will be unable to find enough vegetation for their cattle to graze. |
| **Sand dune threats** | Old inactive sand dune systems in the Sahel could re-activate and begin to move again, threatening crops and grazing land. |
| **Climate change refugees** | Millions of people, unable to find enough water and land to grow food or feed their cattle, may be forced to migrate elsewhere. Will already over-populated coastal cities in Africa be able to cope with an even greater influx of migrants? |

Reports suggest that people are already feeling the effects of climate change. In some parts of the Sahel, women now walk as far as 25 km a day to fetch water for their families. According to the World Bank, Niger's village of Feteye is on the front line against climate change. Millet crops have failed and sand dunes are advancing. People living there face drought and famine.

However, evidence for the Sahel is still inconclusive, and some scientists think there is a possibility of rain returning as a result of global warming. This would lead instead to the 'greening' of the Sahel. The climatic system is complex and there remains much uncertainty over what will happen in the future.

*Figure 11: The creeping desert*

How can extreme environments be managed and protected from the threats they face?

131

## Climate change and polar melting

Polar regions face their own climate change threats, including:

| | |
|---|---|
| **Changes in sea ice cover** | Arctic and Antarctic sea ice are melting. Scientists are recording record low levels in summer. This is changing sea temperatures and threatens the decline of species like polar bears that hunt seals on sea ice. |
| **Land glacier retreat** | Most of the world's glaciers are now shrinking in size and scientists blame a warmer climate. All over the world, the pattern seems to be the same. Tourism in icy regions such as Canada is threatened. |
| **Permafrost melting** | In Alaska and Siberia, up to 40% of permafrost is expected to thaw if climate change predictions are right. As we have already learned, this poses many threats to people and the tundra ecosystem. Temperatures in parts of Alaska have risen by 5 °C over the past century. Permanent, rather than seasonal, melting of permafrost is already triggering long-term flooding of some parts of the tundra. |
| **Species migration** | One possible beneficial effect is that the treeline could move further north as temperatures get warmer. Areas of tundra might be colonised by coniferous forest, bringing increased animal and bird life. |

In south-west Alaska, the Yup'ik people believe they are already suffering the effects of climate change. Described by one newspaper as 'America's first community of global warming refugees', the Yup'ik no longer take their dog teams safely out on to Bering Sea pack ice in search of fish. The ice has lost half of its thickness, making it unsafe. In the village of Newtok, all the buildings are sinking and tilting as the permafrost beneath melts. Whereas once the Yup'ik drove their building piles 3 metres deep into the ground, now it must be 4 metres. It is clearly becoming warmer. According to scientists at NASA, temperatures in Newtok have risen by 4 °C since the 1960s – and by as much as 10 °C in winter months.

For complex reasons, global warming is felt most by people living in polar regions. The loss of snow cover also sets off a vicious circle for the Yup'ik. With less bright white snow to reflect the sunlight, the ground warms and melts even more.

### Activity 1

(a) Will the world see millions of climate change refugees in your lifetime?

(b) Where will they go?

(c) Should countries like the UK offer them homes?

(d) This may turn out to be one of the biggest issues of the twenty-first century. What are your views?

### Decision-making skills

Which do you consider to be the major threats facing either polar or arid areas? To help you with your answer, use the tables on page 128 or 129 and 130 or 131.

*Figure 12: A retreating glacier – the Rhone Glacier in 1996 and 2000*

## Sustainable management of extreme environments

Extreme environments are under threat due to a combination of local and global pressures:

- Global warming is changing world temperatures, rainfall patterns and global sea level. Temperature changes are especially high near the poles, where warming of several degrees has been recorded since the 1950s.

- Local population growth, land-use changes (often linked to the actions of global companies) and political problems (such as the civil war in Sudan's Darfur region) threaten environments and ecosystems.

How can we ensure that polar and arid places have a sustainable future? How can we ensure that future generations of communities living there today can still enjoy and use these special environments? How can we make sure that biodiversity levels remain the same and species are not lost?

### Sustainable management of arid regions

In Chapter 4 you read about the use of **intermediate technology** to help aid water supplies in drought-stressed parts of Tanzania. You should now re-read that section (pages 68–69). The WaterAid website contains details of many more projects, plenty of which are located in the Sahel region, especially Burkina Faso and Mali (www.wateraid.org).

*Figure 13: Rows of solar panels quietly produce energy in America's Mojave Desert*

One of the most exciting developments for arid areas is the move towards greater use of solar power. In America, Australia and the Middle East, enormous areas of desert are already being used to collect sunlight in photocells (Figure 13). In the near future, countries with the greatest amounts of intense sunlight may come to be regarded as the world's most resource-rich places – rather than those with the most oil, as is the case today.

### Sustainable management of polar regions

Iceland is an excellent example of a cold region where every effort has been made to use sustainable energy supplies. The island is located along the mid-Atlantic ridge, a zone of volcanic activity. Thanks to the up-welling of magma beneath Iceland, groundwater becomes heated and sometimes pressurised to form steam. This steam can be harnessed to drive turbines that generate electricity (Figure 14, page 133) while hot water can be used directly to heat buildings. It is a renewable energy resource that can be used to heat and light greenhouses, allowing Icelanders to grow fresh fruit and vegetables and heat water for fish farms all through the year.

In Alaska, similar technology has been installed in many places located close to the coastal tectonic plate boundary that forms part of the Pacific 'Ring of Fire'. Unlike Iceland, this is a plate boundary where dangerous, unpredictable and explosive activity sometimes takes place. However, it still produces beneficial resources such as hot springs and renewable energy. The first Alaskan volcano to be tapped for geothermal power is Mount Spurr, a tall, snow-capped and steep-sided volcano in the Aleutian Volcanic Arc. A tourist resort situated at Chena Hot Springs near Fairbanks is entirely powered by geothermal power.

How can extreme environments be managed and protected from the threats they face?

133

| Local action in Greenland | Great effort has been made to secure sustainable hunting for communities who see it as an important part of their cultural heritage. Scientific recommendations are closely followed in relation to how many seals can be hunted, for instance. The World Heritage Site at Ilulissat Icefjord is carefully managed to minimise threats from tourism, such as the removal of cultural artefacts. |
|---|---|
| Local action in Canada | Canada is home to approximately 16,000 of the estimated 20,000–25,000 polar bears in the Arctic regions. For the Canadian Inuit, polar bears are especially significant culturally, spiritually and economically. Small, sustainable hunting quotas are therefore allowed for the Inuit at a subsistence level. |

Another example of sustainable living, which we have already seen, is the construction of houses on stilts (Figure 5, page 124) – an adaptation which helps preserve the permafrost in its pristine state for future generations.

### Global actions to protect extreme environments

Polar and arid regions are widely regarded as special places. International efforts have been made to try and ensure they have a sustainable future. The best known of these is the 1961 Antarctic Treaty. It has become one of the most successful international agreements of all time, restricting commercial exploitation of the Antarctic continent. Following on from this, the 1998 Protocol on Environmental Protection to the Antarctic Treaty is one of the toughest sets of rules for any environment in the world. Under the agreement, no new activities are allowed in Antarctica until their potential impacts on the environment have been properly assessed and minimised. Tourist boat operators taking visitors there have to follow incredibly strict guidelines.

As climate change begins to damage the Arctic, we are starting to see global efforts focused on this polar environment as well. In 2008, the US placed Arctic polar bears on its endangered species list. If it really wants to help the bears, however, the US still needs to do far more to curb its own $CO_2$ emissions. Only global action on climate change will stop more Arctic ice from being lost, depriving the bears of their hunting ground.

The threat posed to the world by ice melting makes global action to protect polar environments absolutely essential. Loss of arid environments is perhaps seen as a less urgent concern by many. However, efforts are still made to unite the international community and help to halt desertification and 2006 was the 'International Year of the Deserts and Desertification'. According to the UN, the aim was 'to get the message across that desertification is a major threat to humanity, compounded by both climate change and loss of biological diversity. **Land degradation** affects one-third of the planet's land surface and around one billion people in over a hundred countries.'

*Figure 14: Geothermal energy generation in Iceland*

### Skills Builder 3

Study either Figure 13 or Figure 14.

(a) Describe how energy is being generated in the extreme environment shown.

(b) Explain why this is a type of renewable energy.

### Decision-making skills

Suggest what is meant by a sustainable future for extreme environments and explain how renewable energy can assist with this.

For either polar or arid environments, identify the features of what a sustainable future may look like. Explain how renewable energy can assist with this.

# examzone

## Know Zone
## Extreme climates

Some places are just very difficult to live in. Extreme heat and intense cold make survival a major challenge for plants, animals and humans. Extreme climate regions are often untouched, but they are facing growing threats to their very survival.

## You should know...

You must study polar regions and hot arid areas. You need to know:

- ☐ What is meant by an extreme climate
- ☐ Facts and figures for extreme climates
- ☐ Where extreme climates can be found
- ☐ How plants and animals have adapted to live there
- ☐ How people adapt and cope with life in extreme climates
- ☐ The value and uniqueness of the peoples who live there
- ☐ How climate change could alter extreme climate zones
- ☐ How extreme climate zones could be affected by other threats
- ☐ How local people are adapting to the threats they face
- ☐ How global actions might protect extreme environments

## Key terms

Adaptation
Carrying capacity
Cultural dilution
Desertification
Dryland
Exploitation
Extreme climate
Fauna
Flora
Fragile environments
Glacial regions
Hot arid regions

Intermediate technology
Land degradation
Latitude
Nomadic pastoralism
Permafrost
Polar region
Pollution
Solifluction
Temperate climate
Tundra

### Which key terms match the following definitions?

**A** The position of a place north or south of the Equator, expressed in degrees

**B** Plants

**C** The maximum number of people that can be supported by the resources and technology of a given area

**D** A climate that is not extreme (in terms of heat, cold, dryness or wetness)

**E** Permanently frozen ground, found in polar (glacial and tundra) regions

**F** The spread of desert conditions into what were semi-arid areas, caused by natural climate change or by human activities, such as overgrazing and deforestation

**G** Animals

**H** The movement downhill of soggy soil when the ground layer beneath is frozen

To check your answers, look at the glossary on page 321.

**Foundation question:** Describe how plants and animals are adapted to either polar or hot arid extreme environments. (3 marks)

| Student answer (achieving 2 marks) | Feedback comments | Build a better answer (achieving 3 marks) |
|---|---|---|
| *In polar places like the Arctic plants have very small leaves.* | • *In polar places...* scores 1 mark. It is a good example of an adaptation. | In polar places like the Arctic plants have very small leaves. |
| *They grow close to the ground.* | • *They grow close...* This also scores a mark as it is another clear adaptation. | They grow close to the ground. |
| *Maybe only a few inches high.* | • *Maybe only a...* As the question asks for plants and animals, another point on plants is not needed and does not score a mark. | Animals usually have thick, white fur, such as the Arctic Fox. |

**Overall comment:** The candidates should have spotted that the question asked for plants and animals. Read questions carefully and underline or highlight key words.

---

**Higher question:** Explain how plants are adapted to either polar or hot arid extreme environments. (3 marks)

| Student answer (achieving 2 marks) | Feedback comments | Build a better answer (achieving 3 marks) |
|---|---|---|
| *Plants need to have shallow roots.* | • *Plants need to...* This is a correct point but is a description rather than an explanation. | Plants need shallow roots to avoid the permafrost. |
| *Some plants grow close to the ground.* | • *Some plants grow...* This is another correct point but, like the first, it describes rather than explains. Together with the first point, it scores 1 mark. | Plants grow close to the ground to avoid wind damage. |
| *Polar regions are cold and dry. Plants have small leaves to limit water loss.* | • *Polar regions are...* This is a correct answer and it explains why, so scores 1 mark. | Polar regions are cold and dry. Plants have small leaves to limit water loss. |

**Overall comment:** The student tended to describe rather than explain. The last point was linked to a reason, but the first two were not.

# Unit 2 People and the planet

## Your course

This unit focuses on human geography and the topics link together to build an overall understanding of how humans interact with the planet. You will learn about how populations grow and change, where people live and work, and how they exploit and use resources. There are three sections:

**Section A** topics are **compulsory** and introduce you to the main aspects of how people live on our planet. You will study **all** topics:

- Topic 1 (Chapter 9): Population dynamics
- Topic 2 (Chapter 10): Consuming resources
- Topic 3 (Chapter 11): Globalisation
- Topic 4 (Chapter 12): Development dilemmas

**Section B** will cover how people interact with different aspects of our planet on a small scale and you will study **one** topic:

- Topic 5 (Chapter 13): The changing economy of the UK
- Topic 6 (Chapter 14): Changing settlements in the UK

**Section C** will cover how people interact with different aspects of our planet on a large scale and you will study **one** topic:

- Topic 7 (Chapter 15): The challenges of an urban world
- Topic 8 (Chapter 16): The challenges of a rural world

## Your assessment

- You will sit a 1-hour 15-minute written exam worth a total of 78 marks. Up to 6 marks are available for spelling, punctuation and grammar.

- There will be a variety of question types: short answer, graphical and extended answer, which you will practice throughout the chapters that you study. You will answer **all** the questions in Section A, **one** question from Section B and **one** question from Section C.

- **Section A** contains questions on the compulsory topics.

- **Section B** contains questions on the two small-scale topics.

- **Section C** contains questions on the two large-scale topics.

**Remember** to answer the questions for the topics that you have studied in class!

---

Study the photograph of Las Vegas.

(a) Explain the term 'sustainable city'.

(b) Describe two ways of making cities more sustainable.

# Chapter 9 Population dynamics

## How and why is population changing in different parts of the world?

### Global population growth and future projections

During 2010 at least another 79 million people were added to the world's population. This annual increase was less than during the 1980s and 1990s. Population growth rates have fallen from 2.1% per year to around 1.95%. Besides showing the upward curve of global population since 1800, Figure 1 shows how long it has taken for the world's population to increase by 1 billion. It took 118 years from 1804 to double from 1 to 2 billion. Since then, the length of time has shortened considerably to a mere 12 years between 1987 and 1999.

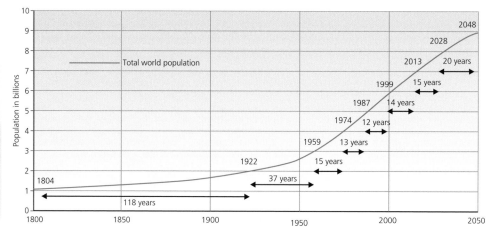

Figure 1: Global population growth, 1800–2050

What the graph also shows is that the length of time it will take to add further billions will start to increase, from 12 to 14 to 20 years. In short, the rate of population growth is predicted to slow down. The interesting question is – given this slowing down – when exactly will the point of **zero population growth** be reached? There is much disagreement about the likely date. Some have suggested it may be as early as the 2020s. Others say it will not be before 2060. There are still others who are much less optimistic. They fear that the global total will continue to rise and reach 10 billion by the year 2100.

No matter where you are in the world, population change (growth and decline) is produced by two processes – **natural change** and **migration** change. Figure 2 shows that natural change depends on the **birth rate** and the **death rate**. If there are more births than deaths, population will increase. If there are more deaths than births, population will decrease.

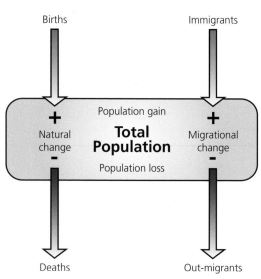

Figure 2: Elements of population change

How and why is population changing in different parts of the world?

139

Migration is the movement of people into and out of an area or country. If **immigrants** (incomers) exceed **emigrants** (outgoers), there will be a gain in population. If the situation is reversed, there will be a loss of population. But, of course, any change in the total size of the global population results only from natural change. Migration simply helps change the distribution of that global population.

So the overall rate of global population growth depends on the difference between the birth and death rates. The more the birth rate exceeds the death rate, the higher the rate of **natural increase**. Remember too that a rise in the rate of natural increase has a multiplier effect and adds even more population growth. This is shown in Figure 3.

The most important factor responsible for the rising rate of population growth has been the fall in the death rate, as shown in the table on the right. (You will also see that the trend in the birth rate has been downward too, working in the opposite direction.) A whole range of factors has contributed to the fall in mortality. They include:

- The development of modern medicines

- The introduction of vaccination and immunisation programmes

- Better healthcare – more doctors, nurses and hospitals

- More hygienic housing

- Cleaner drinking water and better sewage disposal

- Better diet.

A reduction in the **infant mortality rate** has also contributed to this lowering of the overall death rate.

Forecasting future population figures is a difficult business, because there are often surprise changes. Who really knows what is likely to happen to the global birth and death rates? Will the spread of birth control and worries about population pressure on the world's resources continue to lower the birth rate? What about the HIV/AIDS pandemic? Will this cause the death rate to rise significantly? Is there some new highly infectious and lethal disease just around the corner? What will happen if there is an outbreak of nuclear war? Only time will tell.

**Changes in the global birth, death and infant mortality rates, 1800–2050**

| Year | Birth rate per 1,000 people | Death rate per 1,000 people | Infant mortality rate per 1,000 live births |
|------|------|------|------|
| 1800 | 40 | 35 | no data |
| 1850 | 40 | 34 | no data |
| 1900 | 37 | 28 | no data |
| 1950 | 37 | 20 | 126 |
| 2000 | 23 | 9 | 57 |
| 2050 | 14 | 10 | 10 |

## Activity 1

Study the table above.

(a) What changes do you think have helped to lower the global infant mortality rate?

(b) How might the infant mortality rate differ between developing and developed countries?

(c) Calculate the natural increase rates for 1950 and 2050. What does the difference between the rates tell us about future population change?

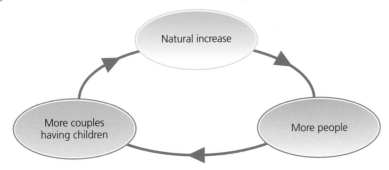

Figure 3: The population multiplier effect

# National variations in population change and structure

## Population change and the demographic transition model

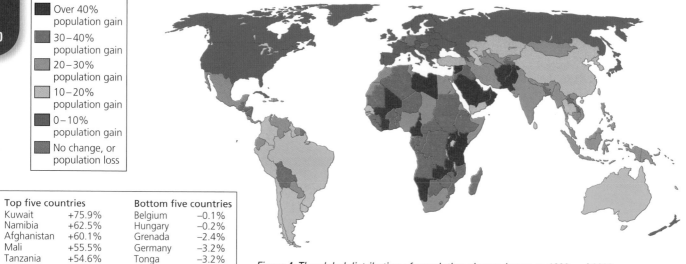

| Over 40% population gain |
| 30–40% population gain |
| 20–30% population gain |
| 10–20% population gain |
| 0–10% population gain |
| No change, or population loss |

| Top five countries | | Bottom five countries | |
| --- | --- | --- | --- |
| Kuwait | +75.9% | Belgium | −0.1% |
| Namibia | +62.5% | Hungary | −0.2% |
| Afghanistan | +60.1% | Grenada | −2.4% |
| Mali | +55.5% | Germany | −3.2% |
| Tanzania | +54.6% | Tonga | −3.2% |

Figure 4: The global distribution of population change between 1990 and 2000

The global population continues to increase, but its distribution across the face of the Earth is changing. Figure 4 gives us a snapshot of that change between 1990 and 2000. It shows some strong contrasts; for example between the high rates of growth in Africa and the Middle East and the little or no gain in North America, Europe and Russia.

Bearing in mind Figure 2 and the mechanisms of population change, let us look at Figure 1 again (see page 138 for both figures). What factors might explain the global pattern of population growth? The low rates of growth over much of the northern hemisphere might tempt us to think that the temperate environment has in some way played a part. Equally, the high rates of growth in Africa might cause us to think that they have something to do with the tropical environment. But neither explanation really holds true. We would be better advised to remember Figure 3 (page 139) and think in terms of culture and even more importantly, **development**.

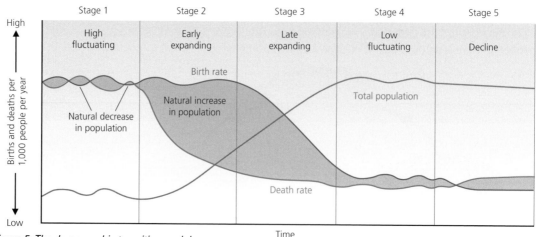

Figure 5: The demographic transition model

How and why is population changing in different parts of the world?

141

It has been observed that as countries develop, their birth and death rates change and, as a result, so too does the rate of natural change. These changes underlie a generalisation known as the 'demographic transition model', which is shown in Figure 5. The model suggests that countries pass through five different stages:

**Stage 1: High fluctuating** – A period of high birth and death rates, both of which fluctuate. Natural change hovers between increase and decrease. Reasons for the high birth rate include:

⦿ Little or no birth control

⦿ High infant mortality rate, which encourages couples to have more children

⦿ Children are seen as an asset and status symbol.

Reasons for the high death rate include:

⦿ High infant mortality

⦿ High incidence of disease

⦿ Poor nutrition and famine

⦿ Poor housing and hygiene

⦿ Little or no healthcare.

**Stage 2: Early expanding** – A period of high birth rates, but falling death rates. The population begins to increase rapidly. Reasons for the falling death rate include:

⦿ Lower infant mortality

⦿ Improved healthcare and hygiene

⦿ Better nutrition

⦿ Safer water and better waste disposal.

**Stage 3: Late expanding** – A period of falling birth rates and death rates. The rate of population growth slows down as the rate of natural increase lessens. Reasons for the falling birth rates include:

⦿ Widespread birth control

⦿ Preference for smaller families

⦿ Expense of bringing up children

⦿ Low infant mortality rate.

## Skills Builder 1

(a) Complete the birth and death rates table for 2007 by calculating the rates of natural change.

(b) Explain why Cambodia and Chile have lower death rates than the UK and Germany?

| Country | Stage | Birth rate Births/ 1,000 | Death rate Deaths/ 1,000 | Rate of natural change |
|---------|-------|--------------------------|--------------------------|------------------------|
| Swaziland | 1 | 27.0 | 30.4 | |
| Cambodia | 2 | 25.5 | 8.2 | |
| Chile | 3 | 15.0 | 5.9 | |
| UK | 4 | 10.7 | 10.1 | |
| Germany | 5 | 8.2 | 10.7 | |

**Results Plus**
**Exam Question Report**

**REAL EXAM QUESTION**

**Choose a developed country that you studied. Describe the changes and the problems that have been caused by population change. (4 marks)**

**How students answered**

Many students did not choose a developed country. They stated the problems of population change, but gave little indication about how these problems might be caused.

28% (0–1 marks)

Most students identified only one problem, usually an ageing population, and explained this in terms of improvements in life expectancy. Changes in birth rate were mentioned only rarely.

61% (2 marks)

Some students identified two problems, usually an ageing population and falling numbers of young people, and explained these in terms of both death rate and birth rate changes. Some also mentioned migration patterns.

11% (3–4 marks)

## Skills Builder 2

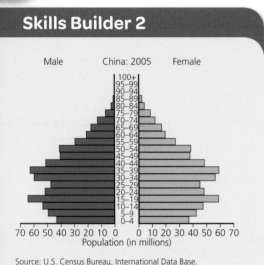

Male     China: 2005     Female

Source: U.S. Census Bureau, International Data Base.

*Figure 6: Population pyramid for China, 2005*

Study Figure 6.

(a) Describe the distinctive features of China's population pyramid.

(b) At what stage in the demographic transition would you place China?

**Stage 4: Low fluctuating** – A period of low birth and death rates. Natural change hovers between increase and decrease. The population as a whole 'greys' – it becomes older. The death rate kept low by improving healthcare. The birth rate kept low by:

- Effective birth control

- More working women delaying the age at which they start having a family.

**Stage 5: Decline** – A period during which the death rate slightly exceeds the birth rate. The result is natural decrease and a decline in population. The population becomes even 'greyer' because modern medicine is keeping elderly people alive longer. Fewer people in the reproductive age range (15–50) means a lower birth rate. This stage has only recently been reached – by some European countries. It raises some interesting questions. Do populations continue to decline to the point where they disappear altogether? Or will immigration keep up the numbers?

Finally, a word or two of warning about the demographic transition model:

- It is a generalisation.

- Not all countries will follow the same pathway.

- Countries that do appear to follow the transition will do so at different speeds – some much faster than others. The important factor is the speed of development.

### Population structure and change

Two of the most important characteristics of a population – age and gender – can be shown in a single diagram known as a **population pyramid.** The male population is shown on one side of the pyramid and the female on the other side (Figure 7). The vertical axis of the pyramid is divided up into age-groups, usually of five years. The youngest age group, 0 to 4 years, is at the bottom and the oldest age group, over 90 years, at the top. The number of males or females in a particular age group is shown by a horizontal bar. The bar is drawn proportional in length to either the number (as in Figure 7) or the percentage of all males or females in that age group.

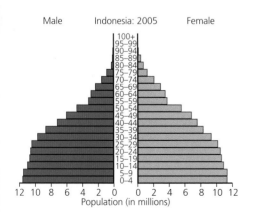

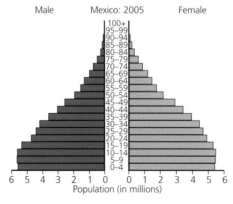

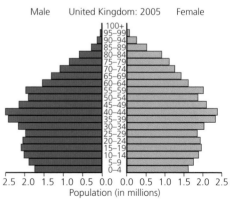

Source: U.S. Census Bureau, International Data Base.

*Figure 7: Population pyramids for Indonesia, Mexico and the UK, 2005. Note the different horizontal scales used.*

How and why is population changing in different parts of the world?

143

The overall shape of the pyramid tells us about the present balances between the different age groups and between males and females. The diagram can be very helpful when making forecasts about future population totals and population growth rates.

A broad-based and rather squat pyramid shape, like that for Indonesia (a developing country), shows what is called a **youthful population**. There are plenty of young adults in the population and they are responsible for a high birth rate and many young children. But because the death rate is high and **life expectancy** low, there are not many people aged over 55. The 'young' part of the population (under 19) is much greater than the 'elderly' part (over 60).

The population pyramid for Mexico (a newly-industrialised country) is more even-sided and taller than that for Indonesia. This means that the death rate is lower and life expectancy is greater. The 'young' part of the population is still larger than the 'elderly' but less so than in Indonesia.

The population pyramid for the UK (a developed country) has almost lost its 'pyramid' shape because it bulges in the middle. The base of the pyramid, the young population, has been 'eroded' away. Now the 'elderly' are equal in number to the 'young'. Clearly, the birth rate has declined and so too has the death rate. More people are living longer. The pyramid tells us that we have an **ageing** or **greying population**. (Factors such as war, emigration for work and HIV/AIDS can also 'erode' the young population.)

From these three population pyramids, we can now understand that their shape is controlled by:

◉ The birth rate – the higher it is, the broader the base of the pyramid.

◉ The death rate – the lower it is, the taller the pyramid.

◉ The balance between the two rates – whether births exceed deaths or vice versa.

We can also relate these three pyramids to the demographic transition model (see Figure 5 on page 140). The population pyramid for Indonesia is typical of a country at Stage 2; that for Mexico typical of Stage 3, and that for the UK typical of Stage 4 (possibly 5). Clearly, as a country develops, its population pyramid changes.

Age and gender are two physical qualities by which all of us can be pigeon-holed or classified. They are also, as we have just seen, two aspects of **population structure**. Admittedly, they are the most important, but there are others – ethnicity, religion, occupation and class, for example. Each of these is a different way of viewing and analysing the structure of a population. Again, just as the age structure of a population changes over time, so too can any of these other structural elements.

**ResultsPlus**
**Build Better Answers**

**EXAM-STYLE QUESTION**

**Explain two problems faced by countries with youthful populations. (4 marks)**

■ **Basic answers** (0–1 marks)
Describe one problem, usually too many people to feed.

● **Good answers** (2 marks)
Link at least one described problem with an explanation. They also include some detail, such as young people do not contribute (dependents) and need support, placing strain on the economy.

▲ **Excellent answers** (3–4 marks)
Offer two descriptions with explanations, perhaps adding strain on educational services, such as the need for teachers. Some identify potential for future growth of high numbers of young people in the population as a problem.

## Activity 2

1. Find out what different groupings are recognised in the UK under each of the four structural headings: ethnicity; religion, occupation and class.

2. Carry out a brief survey of ethnicity and religion in your class. What are your results?

## Activity 3

1. Name two countries suffering from overpopulation. What is your evidence for this?

2. Can you identify two countries that might be suffering from underpopulation? What is your evidence for this?

## Decision-making skills

Some countries are underpopulated. Assess the likely success of the following ideas to increase population:

- Tax credits
- A 'birth bounty' – a payment for each child born
- Opportunities for parental leave for both parents for up to a year after the birth of a child
- Free nursery care for all up to the age of three
- Guaranteed employment at the same level on return to work following maternity/paternity leave

Also think about how migration policies could help to solve underpopulation.

# How far can population change and migration be managed sustainably?

In order to answer the question – How far can population change and migration be managed sustainably? – we first need to ask another question: Why is it necessary for governments to manage populations, particularly their numbers and movements?

## Managing populations

One of the duties of government is to monitor what is happening to a country's population. Is it growing in number or declining? Is it changing? And, most importantly, is it changing in a way that is likely to lead to problems? The key to answering this last question lies in the balance between the resources of a country and its population. As shown in Figure 8, if population outweighs resources, then all sorts of problems, known collectively as **overpopulation**, are likely to arise. Feeding all the people is an obvious challenge. Can enough food be produced to ensure that people do not suffer from malnutrition or starvation? Unemployment is another challenge – can sufficient work be found so that people are able to support their families? Can sufficient decent housing be provided so that people do not have to live in slums or shanty towns?

Conversely, but much less commonly, the balance may tip in favour of resources. This is known as **underpopulation**. Generally speaking, its problems and challenges are much less demanding than those of overpopulation. Providing services and exploiting resources in underpopulated areas are two of the challenges.

An optimum population exists when resources and population are equally balanced. Achieving this sustainable situation is probably the main aim of most governments.

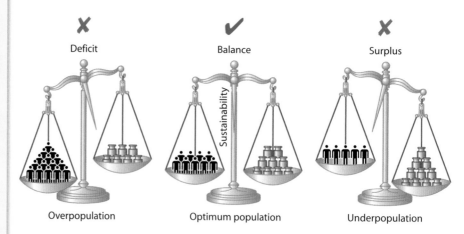

Figure 8: Three different relationships between population and resources

There are many examples of governments deciding that they need to manage their populations. In the majority of cases, the reason has been the need to control numbers – to stop them passing the **tipping point** that leads to overpopulation. So what does a government do to control population numbers? The target of any action is most often the birth rate. If this is lowered, then in time the whole population will become less. The birth rate is usually lowered by encouraging birth control (Figure 9) and making it expensive to have too many children. However, it is important to understand that there are some religions that are strongly against such policies.

When the aim of government action is to increase population, again the target is likely to be the birth rate. Couples may be offered money or other benefits to have more children. But there is another possible action. That is to encourage migrants to come to the country, particularly young adults, who are likely to want to have children eventually.

We will now take a look at two very different case studies – one about dealing with population growth, the other about population decline. They are also different in terms of size. China is the most populous country in the world – and one of the largest, while Singapore is one of the smallest. But they do have one thing in common – most of their people are of the same ethnic origin – Chinese. China covers an area of 9.6 million km² and has a population of 1,350 million. Singapore occupies a small island of about 700 km² and has a population of 5.3 million.

ResultsPlus
**Build Better Answers**

**EXAM-STYLE QUESTION**

145

**Explain how governments can encourage a rise in birth rate. (4 marks)**

■ **Basic answers** (0–1 marks)
State that some countries need to do this and might identify a country. Some confuse pro-natal policies with anti-natal policies.

● **Good answers** (2 marks)
Describe at least one policy with links to named location (e.g. Singapore).

▲ **Excellent answers** (3–4 marks)
Two policies or more are described and their effect is explained. Improved maternity leave therefore encouraging women to take time off to have children; cheap nurseries and access to the best schools also feature frequently in answers.

## Decision-making skills

Argue the case *for* and *against* population control. You should consider it from an environmental, economic, political, social and cultural viewpoint.

*Figure 9: Birth control devices*

# Case study: China's one-child policy

Faced with a high rate of growth in its already huge population, the Chinese government introduced 'voluntary' schemes to control birth rate as early as the 1970s. With the birth rate already falling fast, it introduced its 'one-child policy' in 1979 (Figure 10). For nearly twenty years after that, no couple was supposed to have more than one child, and those who did were penalised in various ways. All couples were closely monitored by female health workers who were trusted members of the Communist Party. Couples with only one child were given a 'one-child certificate' entitling them to such benefits as cash bonuses, longer maternity leave, better childcare, and preferential access to housing. In return for the certificate, couples would have to pledge that they would not have more children. Unmarried young people were persuaded to postpone marriage. Couples without children were advised to 'wait their turn'. Women with 'unauthorised' pregnancies were pressured to have abortions. Those who already had a child were urged to use contraception or undergo sterilisation. Couples with more than one child were virtually forced to be sterilised.

Since 1996 this **anti-natalist** policy has been relaxed a little, particularly in rural areas. The birth rate fell from 34 per 1,000 in 1970 to 13 per 1,000 in 2008, and the annual population growth rate fell from 2.4% to 0.6%. Even so, the total population has grown from 996 million in 1980 to 1,350 million today. The brake has certainly been put on population growth, and as the cutback in children works its way up the population pyramid, so its effect will become stronger.

One thing seems very clear. The policy has been much more effective in urban areas than the countryside. In cities, finding enough living space for a family of three is difficult. Raising a child there is much more expensive. In rural areas, however, there is always the need for an extra pair hands to help on the family farm. In short, there are two very different attitudes towards children.

China's one-child policy remains very controversial. Population growth fell very rapidly before it was introduced in 1979 as a result of changes in Chinese society, land reform and, no doubt, a 'voluntary' policy that may not have been entirely voluntary in practice.

The one-child policy had a number of unwanted consequences. The Chinese tradition is to prefer sons. So as couples are limited to having only one child, there has been widespread sex-selective abortion. If you look closely at the age bars in Figure 6 on page 142, you will see there are more males than females below the age of 45 years. There are now 120 males to every 100 females. This is having consequences:

- Parents 'spoil' their 'one-boy' child and as a result he tends to be obese, demanding and delinquent. They are referred to as 'little emperors'!
- Because of the increasing shortage of women of marrying age, bartering for brides and 'bride kidnapping' have become common in rural areas and prostitution has increased in the cities.
- There may well be a future shortage of labour, particularly if life expectancy remains low and illness rates remain high.

*Figure 10: Promoting the happy image of a 'one child' policy*

**Case study quick notes:**
- Population management needs tough government.
- Population management can also be tough on people.

How far can population change and migration be managed sustainably?

147

## Case study: Singapore's 'Have three or more' policy

Since the mid-1960s, the Singapore government has controlled the size of its population. First, it wanted to reduce the rate of population growth, because it was worried that the small island would soon become overpopulated. This policy was so successful that in the mid-1980s the government was forced to completely reverse the policy. The old family planning slogan of 'Stop at two' was replaced by 'Have three or more – if you can afford it'. It now adopts a **pro-natalist** policy. Instead of penalising couples for having more than two children, they have now introduced a whole new set of incentives to encourage them to do just that. These include:

- Tax rebates for the third child and subsequent children
- Cheap nurseries
- Preferential access to the best schools
- Spacious apartments.

Pregnant women are offered special counselling to discourage 'abortions of convenience' or sterilisation after the birth of one or two children.

**Case study quick notes:**
- Governments are able to control population numbers in a variety of ways.
- Control is usually achieved by a 'stick and carrot' approach.

*Figure 11: Singapore's 'Have three or more – if you can afford it' policy offers incentives to have lots of children*

## Managing migration

Governments are particularly interested in migration. Is the number of people entering the country (immigrants) exceeding the number of people leaving (emigrants)? If it is, then the situation (net in-migration) will be adding to any growth in population coming from natural increase. This situation might then begin to ring alarm bells – about possible overpopulation, unemployment and housing shortages. So, in this situation, some sort of action will need to be taken to check the inflow of migrants. If the situation is the complete opposite, then a government might take action to stop it, particularly if the loss of people involves a 'brain drain' of its best workers.

So whether or not a government decides to manage or control migration depends on what its impacts are likely to be. Those impacts can be both positive and negative. Government policies can also change in response to changing situations. This is well illustrated by what has happened in the UK since the middle of the twentieth century.

### Activity 4

Read the case study on page 146 and look back at Figure 6 on page 142. Can you pick out any evidence of the one-child policy in China?

### Activity 5

Try and find out if the population of the UK has ever been managed. Is it being managed today?

*Figure 12: Immigrants from the West Indies arriving in the UK in 1956*

## Decision-making skills

Draw a quadrant (as shown) which looks at the positives and negatives of migration into the UK for both the host country and the source country.

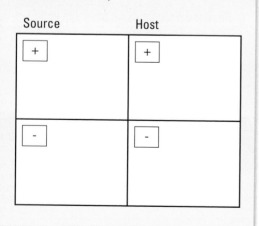

### Open-door policy

The UK's post-war immigrants came mainly from colonies in the Caribbean, and from what had been the Indian Empire (India, Pakistan and Bangladesh). Immigration was encouraged by an Act of Parliament which gave all Commonwealth (ex-colonial) citizens free entry into the UK. The first ship to bring in immigrants from Jamaica docked at Tilbury (Essex) in June 1948.

It is estimated that as a result of its open-door policy during the 1950s and 1960s over a quarter of a million immigrants came from the Caribbean. Roughly the same number came from what had been the Indian Empire. By 1971, there were over 1 million immigrants from Commonwealth countries. The new settlers took up a variety of jobs. Many found work in textile factories and steelworks. Many drove buses or worked on the railways. Later arrivals, particularly from India, opened corner shops and restaurants or ran Post Offices.

### Closing the door

By the 1970s, the UK had more than enough labour, and controls were introduced to reduce the migrant arrivals. Figure 13 shows the number of UK residents who were born abroad. Despite the controls on immigration, clearly the number has been steadily rising. So too has the percentage of the UK population that they represent.

Enough time has passed for us to see and understand the impacts of this post-war immigration. Its positive economic impacts were:

◉ It met the shortage of unskilled and semi-skilled labour.

◉ It played an important part in the post-war reconstruction of the country.

On the negative side:

◉ Public money had to be spent on meeting the everyday needs of the immigrants and their families – housing, schools, healthcare, etc.

◉ When the economy went into recession in the 1970s, these immigrants added to the burden of unemployment.

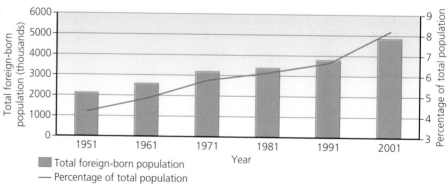

*Figure 13: UK residents born abroad, 1951–2001*

How far can population change and migration be managed sustainably?

149

As for the social impacts, it is clear that the immigration created great tensions (Figure 14). The UK was not used to having sizeable ethnic groups in its population. There was hostility towards the immigrants, and they were discriminated against – and abused. As a consequence, they tended to settle in particular areas, for personal security reasons. Better to live with people from the same ethnic group than run the risk of being victimised by white neighbours. Most often, immigrants became segregated, to form 'ghettos' in areas of rundown housing in the inner areas of towns and cities. Clearly, there were some serious negative social impacts.

Slowly, the social situation has changed. Discrimination has been made illegal. UK law now states that all citizens, regardless of ethnicity, should enjoy equal opportunities. Slowly, most white people have come to realise that they are not threatened by immigrants. They have come to realise that there are positives. Ethnic groups add to the country's skill base and culture. The offspring of the original immigrants have made their way in the UK, and many now occupy well-paid and responsible jobs. They have moved into areas of better housing. They represent the country in a range of sports. They have seats in Parliament. They are now truly UK citizens. The situation is still not one of complete harmony, and many still live in poverty. But the situation is much better than it was 40 years ago.

### Re-opening the door

Since the 1970s the immigration of New Commonwealth citizens has been subject to some form of government control. In the late 1990s the economic situation in the UK began to change. A period of economic boom saw the country short of labour once again. This time, the search for willing workers turned to Eastern Europe.

In 2004 the East European states of the Czech Republic, Estonia, Hungary, Latvia, Lithuania, Poland, Slovakia and Slovenia joined the EU. Since then, many of their citizens have come to work in the UK. Figure 15 shows the push and pull factors. In most cases, these **economic migrants** intend to stay only until they feel they have made enough money to take home.

*Figure 14: Inner city race riots*

### Skills Builder 3

Take a look at a town or city you know. Do members of different ethnic groups live in particular areas? Draw a simple sketch map to show these areas.

Quick notes (the UK opens its doors): The UK's need for labour led to it becoming a multi-ethnic society.

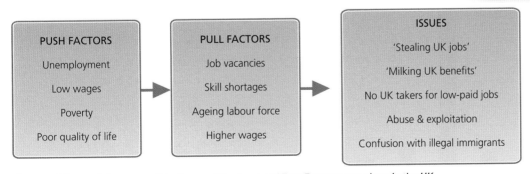

*Figure 15: The push factors, pull factors and the issues of East European workers in the UK*

## Activity 6

Explain in no more than 150 words why economic migrants from Eastern Europe have been attracted to the UK.

Figure 16 shows that over half of the East European migrants come from Poland, the largest of the new member states. The vast majority of migrants are young and single, with over 80% of them aged between 18 and 34.

Some UK newspapers (and citizens) take a very negative attitude towards these economic migrants. They are accused of depriving UK workers of jobs and taking advantage of our state benefits system. Figure 15 (page 149) shows these as two of a number of issues relating to these economic migrants. But the critics choose to ignore four important facts:

- The migrants contribute to the UK's economy by the taxes they pay.

- The jobs that many of them take up are mainly low-paid. Such jobs are often avoided by UK workers.

- The migrants have a strong work ethic, which can directly benefit employers. They are efficient and polite. Sadly, there are employers who unfairly exploit these qualities.

- Less than 5% of them receive any sort of state benefit.

Many of these workers are now returning home – persuaded by the economic recession and better employment prospects back home. Also important is the changing exchange rate. With the pound sterling falling in value, workers are not able to send or take home so much money.

One of the advantages of belonging to the European Union is that workers are free to move between member countries. All they need is a passport or national identity card. The UK is unable to control this movement of economic migrants.

While it is becoming easier to migrate within the EU, it is becoming more difficult for migrants to enter it from other parts of the world (Figure 17). In order to enter the UK from outside the EU, for example, you need a visa. There are various types – visitor, business and working holiday – and they are usually valid for less than a year.

### Quick notes
### (Eastern European workers):
The 'boom and bust' cycles that most countries go through mean alternating shortages and surpluses of labour. The economic migrant can play a useful part in smoothing out the differences.

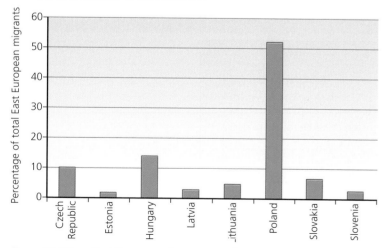

*Figure 16: Where East European migrants come from*

*Figure 17: UK border control*

How far can population change and migration be managed sustainably?

151

If you wish to come to the UK to work and settle down, then you have to go through the points-based system, which was introduced in 2008. It allows British businesses to recruit the skills they need from abroad. Only migrants with those skills will be able to come and live in the UK.

As national boundaries are tightened, the volume of illegal immigration increases. There are two 'porous' frontiers through which most illegal immigrants enter the EU – in the Mediterranean countries (see Figure 18) and along the eastern border with Belarus, Russia and the Ukraine.

Stopping illegal immigration is a particular challenge facing the USA along its 2000 km border with Mexico. Hundreds of thousands of people attempt to cross this border each year, desperate to enjoy what the USA has to offer – well-paid jobs, better schools and medical facilities. The USA is forced to spend huge sums of money each year on constantly guarding and patrolling the border, as well as on prisons to hold those illegal immigrants who are caught (and subsequently deported). It is not only these huge costs, but the USA regards these illegal migrants as causing employment, racial and cultural tensions in cities. Nonetheless, those illegal migrants who manage to evade the authorities play an important part in the US labour market, especially in California and other border states.

There are times, however, when governments are unable to control migration. This is well illustrated by what happened during the so-called Arab Spring (2011–12) when huge numbers of refugees moved out of Tunisia, Libya, Egypt and Syria into neighbouring countries to escape the violence and bloodshed of civil war.

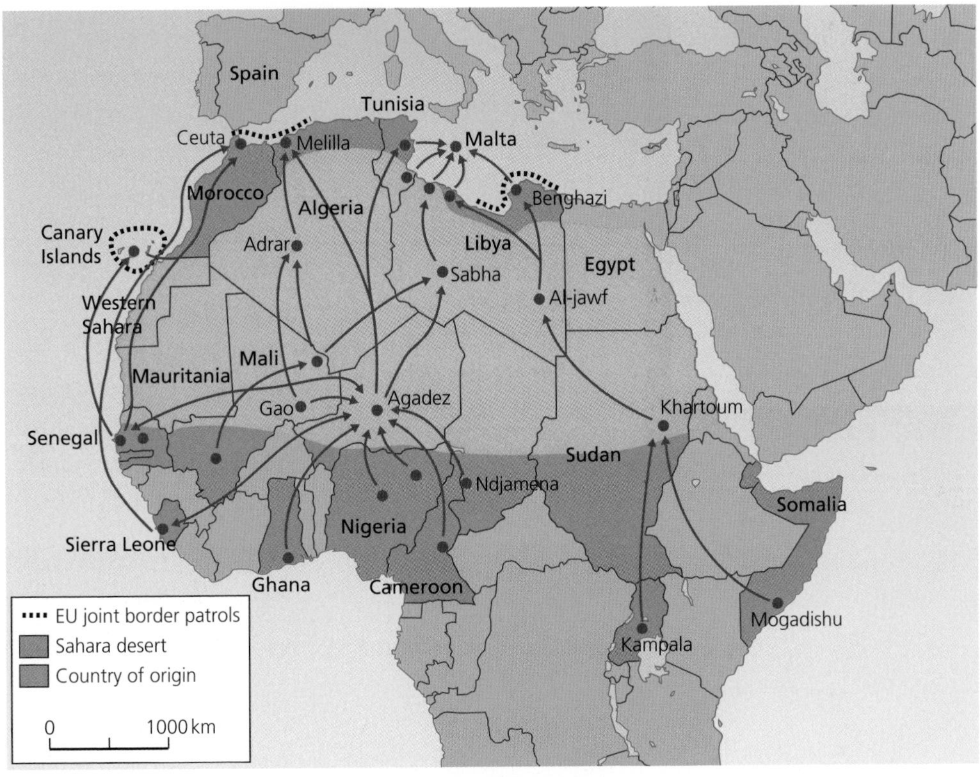

> ## Skills Builder 3
>
> Study Figure 18.
>
> Describe the routes being taken by illegal immigrants into the European Union.

Figure 18: Illegal entry routes in to the European Union from Africa

# Know Zone
# Population dynamics

Population is always changing. You can study the effects of both natural change from births and deaths, and migration. This change has important implications so countries try to manage both population and migration.

## You should know...

- ☐ How the world's population is still growing at a decreasing rate
- ☐ What causes population to grow or decline
- ☐ Why some countries have more rapid growth than others
- ☐ How birth rates and death rates can change population
- ☐ Why these rates vary around the world
- ☐ How immigration and emigration can affect population growth
- ☐ Why infant mortality rate varies across countries
- ☐ How you can fit countries pattern of population change into the Demographic Transition Model
- ☐ How you can draw population pyramids to show population structure
- ☐ What is meant by sustainable or optimum population growth
- ☐ How countries try to manage growth to avoid overpopulation and underpopulation
- ☐ What issues force people to migrate
- ☐ How and why countries manage migration
- ☐ How migration can have both benefits and costs for source and host
- ☐ Why economic migration is becoming so widespread

## Key terms

| | | |
|---|---|---|
| Ageing population | Infant mortality rate | Population structure |
| Birth rate | Life expectancy | Tipping point |
| Death rate | Migration | Underpopulation |
| Development | Natural change | Youthful population |
| Economic migrant | Natural increase | |
| Emigrants | Overpopulation | Zero population growth |
| Immigrants | Population pyramid | |

### Which key terms match the following definitions?

**A** Economic and social progress that leads to an improvement in the quality of life for an increasing proportion of the population

**B** A person leaving a country or region to live somewhere else (for at least a year)

**C** The change (an increase or a decrease) in population numbers resulting from the difference between the birth and death rates over one year

**D** The number of births per 1,000 people in a year

**E** A person arriving in a country or region to live (for at least a year)

**F** The point at which the momentum of a change becomes unstoppable

**G** Diagrammatic way of showing the age and sex structure of a population

**H** The average number of years a person might be expected to live

To check your answers, look at the glossary on page 321.

**Foundation Question:** Describe the key features of Stage 1 of the Demographic Transition Model. (3 marks)

| Student answer (achieving 2 marks) | Feedback comments | Build a better answer (achieving 3 marks) |
|---|---|---|
| This stage has a very high birth rate.<br><br>It also has a very high death rate.<br><br>This is because of the high amount of disease. | • **This stage has...** This is correct and scores 1 mark.<br><br>• **It also has...** This is another correct statement and scores 1 mark.<br><br>• **This is because...** This is an explanation rather than a description. As the question asks for a description, this part of the answer does not score any marks. | This stage has a very high birth rate.<br><br>It also has a very high death rate.<br><br>Both the birth rates and the death rates fluctuate from year to year. |

**Overall comment:** The student should have circled the command word – *describe* – to make sure that they responded to the question being asked.

**Higher Question:** Explain why Stage 1 of the Demographic Transition Model is called the 'High Fluctuating' phase. (3 marks)

| Student answer (achieving 2 marks) | Feedback comments | Build a better answer (achieving 3 marks) |
|---|---|---|
| The very high birth rates go up and down because of infant mortality varying.<br><br>The death rate is high and it fluctuates because of famines. | • **The very high...** Although this is a true statement, it is not really an explanation as it only implies fluctuation.<br><br>• **The death rate...** This is a valid point as it explains why the death rate fluctuates. | The infant mortality fluctuates because of AIDS and poor hygiene, which leads to disease. The birth rate is high because there is no contraception.<br><br>The death rate is high and it fluctuates because of famines. |

**Overall comment:** The student attempts to explain in their answer, but the explanations need to be developed.

# Chapter 10 Consuming resources

154

## Objectives

- Be able to define and give examples of different types of resource.

- Be able to explain why some countries have a higher demand for resources than others.

- Understand that global economic growth is putting pressure on the supply of energy.

*Figure 1: The tar-sand reserves of Alberta, Canada*

## How and why does resource consumption vary in different parts of the world?

### Classifying resources

A resource is anything that can be used to meet a human need. Resources are divided into two types:

- **natural** – parts of the environment, such as minerals, climate and soils

- **human** – properties of a population, such as its technology, capital and skills.

This chapter is about natural resources and they, in turn, are widely recognised as being of two types:

- **non-renewable resources** – energy and mineral resources like coal, oil and diamonds – cannot be 'remade', because it would take millions of years for them to form again. These physical resources exist in a fixed amount that is gradually being used up.

- **renewable resources** – energy resources like solar, water and wind power – renew themselves indefinitely (for ever) and offer continuous flows of supply. They too are physical resources.

Nowadays, a new category of renewable resources is recognised. These are:

- **recyclable resources** – biological resources like wood – which can, by deliberate human intervention, be deliberately renewed. This means that they can be managed so that they are usable now but will last into the future.

Cutting across these three categories, we can begin to recognise at least four different types of resource:

- **Physical** – e.g. water, wind and sunlight

- **Energy** – e.g. coal, oil and gas

- **Mineral** – iron ore, bauxite and uranium

- **Biological** – timber, maize and soya bean.

This is not a very satisfactory classification, as a number of those resources can be of two types. For example, coal is both an energy and a mineral resource; water is both a physical and an energy source.

You will also come across another term, **sustainable resources**. These are natural resources which need to be managed if they are to be available for future generations. Timber is a prime example of a sustainable resource. Thus sustainable resources may be distinguished from renewable resources, such as wind and solar power, by the fact that renewable resources 'renew' themselves, without human intervention.

How and why does resource consumption vary in different parts of the world?

155

These distinctions between different types of resource are important when it comes to using them. For example, some will always be available, whilst others will run out. But availability depends not just over time and on whether a resource is renewable or non-renewable. Availability also depends on where the resource is located. Many renewable resources are widely available around the world – and they always will be. Non-renewable resources tend to be localised in particular places – and they will eventually run out.

Space and time are also important when it comes to the **consumption** of resources. Where are the resources being consumed? In other words, where is the demand located? Demand and consumption change over time. In most cases, they increase and are driven by increases in population and by economic development. Clearly, increasing consumption emphasises the importance of using sustainable resources.

Another way of looking at resources is to consider the benefits that come from their use and compare these with the costs. Remember that resource use is not always plain sailing – there are difficulties, and exploitation can have adverse impacts. With the era of 'cheap oil' coming to an end, we must weigh up the relative costs and benefits of **alternative energy** resources (see table).

## Decision-making skills

Study the table below. Draw a scale from minus three to plus three. Plot each of the three resources listed by their environmental cost and economic benefit for the future.

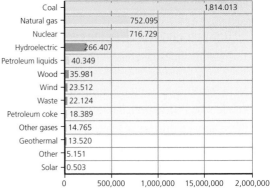

Figure 2: The energy sources used to generate electricity in the USA, in billion kilowatt hours

## The evaluation of three alternative energy resources

| Resource | Benefits | Costs |
|---|---|---|
| **Tar-sand oil** (non-renewable)<br><br>There are enormous reserves of tar-sands, especially in Canada (see Figure 1), and they have attracted more attention since oil prices have risen rapidly. | • There are about 300 billion barrels of oil in the tar-sands (equivalent to Saudi-Arabian oil reserves).<br>• Tar-oil extraction would make profits for the major oil companies.<br>• There would be tax benefits for countries with tar-sands.<br>• Dependence on the dangerous Middle Eastern region would be reduced.<br>• Continued production of oil would avoid the costs of switching to other fuels, such as hydrogen. | • Heavy oils, such as those from tar-sands, produce up to three times more $CO_2$ than 'light' oils.<br>• Tar-oil extraction uses vast quantities of water – up to six barrels for each barrel of oil.<br>• Ancient spruce forests will need to be removed to scrape away the oil-rich sands, and pollution of groundwater and rivers will be inevitable.<br>• 300 billion barrels is about five years' supply, thus it only delays the need to search for alternative technologies. |
| **Biofuels** (recyclable and sustainable)<br><br>There are several types of biofuel, but the best known are ethanol (often extracted from corn) and bio-diesel, extracted from crops such as soya or palm oil. | • $CO_2$ is absorbed when they are grown – but $CO_2$ is released when they are used.<br>• They can be grown in many different environments allowing every country the possibility of producing some of its own fuel.<br>• Internal combustion engines need very little modification to cope with biofuels, so there is no need for costly changes to vehicle designs. | • They are not carbon neutral because the farming methods themselves release $CO_2$.<br>• The amount of land needed to replace conventional fuels would be enormous. All the current food crop regions of Europe and the USA would need to be devoted to biofuels to meet current demand for fuel.<br>• They would be grown as 'monocultures', reducing habitat variety. In some areas of the Tropics palm oil is already replacing tropical rainforest. |
| **Solar energy** (renewable and sustainable) | • Solar energy is unlimited.<br>• It is environmentally friendly, with minimal carbon emissions after initial panel production.<br>• New thin panels are being developed that will be much more efficient than present technology. | • It is intermittent, so back-up systems and new ways of storing electricity are needed.<br>• Current production is tiny (see Figure 2).<br>• It is relatively expensive. Present panels last about twenty years and, for household usage, do not even pay for themselves.<br>• Huge areas of panels are required to produce significant amounts of energy. |

## Activity 1

(a) Identify two forms of 'alternative' energy.

(b) Draw up a table to show the costs and benefits of developing them.

Note: You should use the internet to conduct your research, but remember that this a very controversial topic and many websites are arguing strongly *for* or *against* the development of alternative energy sources.

## Skills Builder 1

Study Figure 2 on page 155.

(a) Identify three renewable sources of energy for electricity production.

(b) Calculate the percentage of electricity produced in the USA using fossil fuels.

## Skills Builder 2

Study Figure 3.

(a) Calculate the percentage of global diamond production that comes from developed countries.

(b) Using an atlas, locate the five largest diamond-producing countries.

## A changing world of 'haves' and 'have nots'

It might seem obvious that countries with large quantities of natural resources would be more developed – and more successful – than those without those resources. After all, if the person sitting next to you has more gold, diamonds and oil than you have, you might suppose that they are wealthier than you are. And, of course, there are examples of countries – such as Saudi Arabia with its oil – where the wealth is a result of an abundant natural resource. But the picture is much more complicated than common sense might suggest. Take a look at Figure 3, which shows the world's major producers of diamonds.

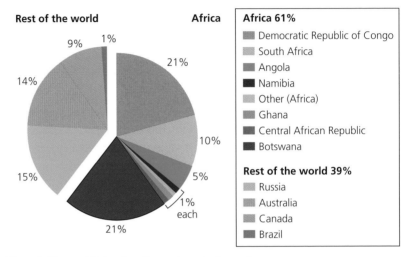

Figure 3: The world's leading diamond-producing nations

It is obvious that although there are a few developed countries that produce diamonds there are a large number of developing countries too, some of which are very poor indeed. The Democratic Republic of Congo (DRC) and Botswana are the world's two largest diamond-producing nations. But the DRC is 177th out of the 179 countries on the United Nations Human Development table. In the DRC, the average life expectancy is 48. Botswana is placed 126th on the list, with an average life expectancy of 53. So the world's two largest diamond-producing countries are poor countries, in one case extraordinarily so. It would be fair to conclude that, in both of these countries, the money being made out of diamonds is not benefiting most of the people. And that is why common sense can mislead you. Countries are made up of different groups of people – different interest groups – and there are often considerable gaps in wealth between the rich and the poor. In the DRC, a very small number of people have done very well indeed as a result of the diamond mines – but most of the country's people have not.

It must be remembered, however, that the resource situation is constantly changing. Non-renewable resources become exhausted, while advances in technology bring new resources into focus. For example, some countries that currently have to import oil and gas may do rather better when it comes to alternative sources of energy. Madagascar, one of the world's poorer countries, currently imports oil. Quite fortuitously, recently discovered reserves of tar-sand oil are just beginning to be exploited, but with a worrying amount of damage to the environment. But the good news does not end there. Madagascar is also well placed with the respect to the other energy resources shown in the table on page 155 – biofuels and solar energy.

How and why does resource consumption vary in different parts of the world?

157

Global wealth – just like resources – is also unevenly distributed. Figure 4 shows the size of each country in terms of the average wealth of its people rather than the true geographic size of the country.

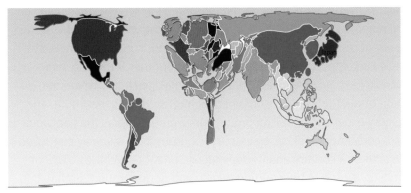

Figure 4: World wealth distribution in 2002, by country

It is obvious that Europe, the USA and Japan are the wealthiest global regions, and equally clear that South America and, above all, Africa are very poor indeed. In fact, the richest 20% of the human race (most of whom live in the three wealthy regions) consume over 86% of all global resources, whilst the poorest 20% consume only 1.3% of global resources. These inequalities have increased dramatically in recent years.

But this world of 'haves' and 'have nots' is not quite a simple as it appears. Over 20 million US citizens live on 'trailer parks', in what we would call 'caravans'. At the same time, there are a significant number of millionaires living in Mumbai, a city better known for its slums.

Figure 5 shows that as societies develop, it is highly likely that their impact on the environment will increase. One of the many ironies of the battle against global warming is that the societies that are least equipped to cope with its effects – Malawi and Mozambique, for example – are not generally responsible for the problem in the first place. Figure 5 shows that the present impact of both China and India is low. But these countries, with their huge populations, are expected to develop into the emerging superpowers, raising concerns that the scale of their future consumption and resource use may have a significant impact on the future of the planet.

## Skills Builder 3

Study Figure 5.

(a) Identify the global 'average' ecological footprint, in hectares per person.

(b) Describe the relationship between human development rank and ecological footprint.

(c) Identify two countries that do not 'fit' the general relationship.

## Watch Out!

Whenever you write 'in developing countries…' or 'in the UK…', remember that there is lot of variation within countries. Try to avoid using a statement such as 'The UK is rich, so…' and replace it with 'Many people in the UK are relatively rich so…'.

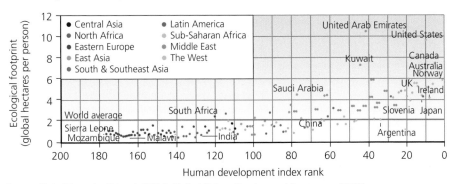

Source: Global Footprint Network (2006); United Nations Development Programme (2006).

Figure 5: The relationship between development and ecological footprints

## Activity 2

1. Using the Internet, define the term 'peak oil'.

   www.peakoil.org

   www.dieoff.org/l1.html

   www.eia.doe.gov

   www.cges.co.uk

   www.iea.org

2. Suggest reasons why some countries face more challenges than others as oil production declines.

## The uneven patterns of oil supply and oil demand

The distribution of oil is determined by nature, but the production of oil is affected by our technical ability to extract oil, some of which is found in very difficult environments. As with the tar-sand oil which we studied earlier, there are potential sources of oil in a number of locations that are not shown on the current reserves map (Figure 6). Current production is dominated by the Middle East, especially Saudi Arabia. Many countries, including some in the Middle East, have reached what is known as 'peak oil', where the production of relatively cheaply obtained oil has reached its maximum, and they are now experiencing falling production. Figure 7 shows the per capita (per head) consumption of oil, by country. Consumption is largely controlled by the wealth of a country and its dependence on motor vehicles. 70% of oil production is used in transporting people and goods around, both within countries and between them. Oil is consumed at the rate of about 1,000 barrels (160,000 litres) a second.

The USA has less than 5% of the world's population but it consumes 25% of the oil. This is not just because the Americans are wealthy – there are other wealthy nations whose people use much less oil. It is the country's high standard of living – relying on air-conditioning, for example – and, more than anything, the dependence on cars that causes the massive consumption. This dependence – which is built into the American way of life – is caused by:

- The poorly developed public transport system – within cities and between cities

- The pattern of low-density urban settlements, requiring long journeys to work, school and the shops

- The long history of very low petrol prices, which is only recently coming to an end.

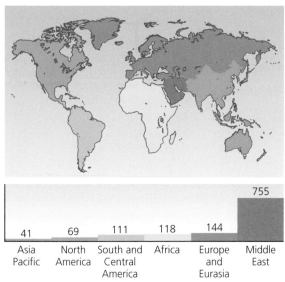

Figure 6: Oil reserves by global region, in billion barrels, 2007

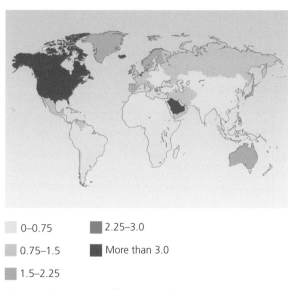

Figure 7: Consumption (in tonnes) of oil per head, by country, 2007

How and why does resource consumption vary in different parts of the world?

159

As a result, there are about 260 million cars in the USA – not far off one for every person in the country.

The USA situation is especially worrying because there are many people in the world who would like to share the 'American Dream' of low-density living with very high energy usage. Unfortunately, economic growth causes increases in the use of resources, which in turn increase the emission of greenhouse gases. The table below shows the scale of the environmental problem if countries like China and India move towards being more like America.

**People and cars: the situation in 2008**

|  | Population | Number of cars per 100 people | Car ownership |
|---|---|---|---|
| USA | 300 million | 87 | 260 million |
| China | 1,300 million | 4 | 59 million |
| India | 1,100 million | 1 | 12 million |

As China and India continue to grow rapidly (see Figure 8) it is unlikely that their citizens will give up dreams of owning a car and enjoying the type of lifestyle that is accepted as 'normal' in the USA and other developed countries.

It is also unlikely that these countries – and others in South and East Asia that are developing fast – will accept too many lectures about the need to slow down their rate of growth from countries that have already reached high levels of consumption. In fact, exactly the opposite is likely to happen as businesses seek to expand their markets in these global regions.

It is certain that the continuing growth of the global economy will put still more pressure on both the supply and consumption of oil. This situation is set to create more tensions at an international level between the oil producers and the oil consumers. The former are very much in the driving seat. Organisations such as **OPEC** (Organisation of the Petroleum Exporting Countries) and **GECF** (Gas Exporting Countries Forum) now wield immense global power. They are able to raise prices and reduce supplies at will. Thus international relations are, and will continue to be conditioned by the agreements that oil-hungry countries manage to seal with individual oil-rich countries. The best that can be hoped for is that such agreements will be honoured.

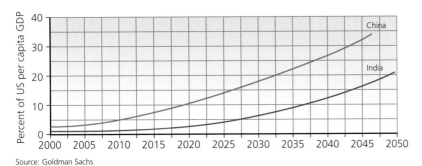

Source: Goldman Sachs

*Figure 8: The growth of the economies in China and India compared with the US*

**Build Better Answers**

**EXAM-STYLE QUESTION**

**Outline how increasing car ownership might cause an increase in the demand for oil. (2 marks)**

■ **Basic answers** (0 marks)
Fail to identify the link between car usage and demand for oil.

● **Good answers** (1 mark)
Recognise that increasing car ownership is likely to increase demand for oil.

▲ **Excellent answers** (2 marks)
Also recognise this link and add some data to support the point.

**Skills Builder 4**

Study Figure 8.

Describe the growth of the Chinese and Indian economies shown in the figure.

## Solar power – a bright future?

Of the sustainable sources of energy, the possible contribution of solar power to meeting future global energy demands has been rather overshadowed by the debate about wind power. Solar power can be harnessed to generate electricity in two ways:

- by using lenses or mirrors to concentrate sunlight and then using it to heat water into steam, which drives turbines

- by using banks of photovoltaic (PV) cells.

The best locations for solar power are the desert regions of the tropics and sub-tropics. Here skies are clear for at least 300 days a year and solar radiation is intense. What this means is that there is much less of the world with potential solar power compared with wind power. Nonetheless, despite having fewer sunshine hours, some European countries (Germany, Italy, Spain and the Ukraine) are already putting solar power to use.

An important point is that we already have the technology needed to convert this renewable resource into energy. At present, it requires huge areas of PV cells to generate significant amounts of electricity. No doubt, banks of PV cells will soon be developed that are both more compact and efficient. However, the space requirements of harnessing solar power in this way are not really a problem, for the areas with the greatest solar power potential are sparsely populated, if at all. It has been estimated that an area the size of Portugal covered with arrays of PV cells (Figure 9) could produce as much electricity as the total output from all of the world's present power stations.

So, solar power seems to offer a future that would be less controversial than wind power but just as sustainable. It certainly can be expected to play a major part in meeting the world's rising demand for electricity and in shifting the world towards a more sustainable energy future. Harnessing the solar power potential of deserts, such as the Sahara, is likely to require international cooperation, which might not be simple. But this would not be a problem for countries such as Australia, China and the USA which happen to have huge deserts of their own.

Because weather is constantly changing, it is almost impossible to maintain a reliable and continuous supply of power from weather-dependent energy sources such as wind and sunlight. However, solar power presents an additional and unique drawback – night time. Clearly, efficient ways of storing solar power, or the electricity it generates, will need to be devised if the potential of solar power is to be fully exploited.

Figure 9: A field of photovoltaic solar panels

**Quick notes:**
Solar power is a sustainable source of energy with a bright future, but not everywhere. It promises to give the world's dry areas a useful role in helping to meet the global energy demand in a sustainable way.

# How sustainable is the current pattern of resource supply and consumption?

## The different theories on how far the world can cope

The world's population is predicted to reach 10–12 billion in the second half of this century. Are global resources going to be sufficient to support that number of people?

Although many experts are optimistic about the world's ability to produce enough food in total, there will still be the problem in the future – as now – of unequal distribution. Today, when the world's population is between 6 billion and 7 billion, there are many people – about 1.5 billion – who are never very far away from famine. About 4,000 babies (under the age of 1) die every day from a lack of fresh water and basic sanitation.

Are there too many people in the world? Is there a limit to the number of people that the world's resources can support? At first sight, these questions seem simple enough to answer but, in fact, they are two of the most controversial of all issues, and have been argued over for centuries. You don't need to take sides in the debate (although you can), but you do need to understand the different points of view.

### *The Malthusian theory*

Thomas Malthus (1766–1834), an English economist, believed that because population increases in a different (and faster) way than food supply, there would inevitably come a time when the world could not cope. He argued that population would increase geometrically (1, 2, 4, 8, 16, 32, 64, 128, etc.) because two people would have four children, and those four would have eight, and so on. Food supply, on the other hand, would only increase arithmetically (1, 2, 3, 4, 5, 6, 7, 8, etc.) because improvements in farming practices could only come gradually. From these calculations, Malthus predicted in 1798 that population growth would outstrip food supply, leading first to tensions in society and ultimately to deaths from famine, war and disease – causing a sudden fall in the population. Figure 10 shows a series of these collapses. After each one, the population rises rapidly again until it hits the 'barrier' of the slowly increasing food supply – and collapses again.

Malthus wrote at a time when the rich owned most of the resources and enjoyed a life of comfort, whereas the poor had very little access to resources, especially land. He thought that it was wrong to give assistance to the poor because he believed that they were less able to understand these risks and thus limit population growth.

Malthus's predictions turned out to be wrong because food production increased rapidly. At the end of the nineteenth century there were more people who were, by and large, better off. If there was a relationship between people and resources, it was obviously not quite as Malthus had argued.

## Objectives

- Know that there are several different ways of explaining the link between population and resources.

- Be able to describe the main aspects of Malthus's and Boserup's theories.

- Explain the ways in which sustainable development is interpreted by different groups.

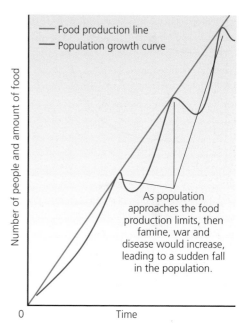

Figure 10: The relationship between food supply and population, according to Malthus

## Skills Builder 6

Describe the growth of the population shown in Figure 10.

The pattern seemed to be repeated in the twentieth century, when population growth was more rapid than at any other time in human history (from about 1 billion to about 6 billion) and living standards and quality of life rose globally more than in any other century.

Those with Malthusian views today – the 'neo-Malthusians' – tend to be building on the 'common sense' idea that poverty is caused by there being too many people in the world. They look at images of famine victims, often in Africa, and assume that famines are the result of too many people rather than, as is often the case, poor distribution of food or too much land devoted to profitable crops for export – much as it was in Malthus's own day.

### The Boserupian theory

Esther Boserup (1910–1999) a Danish economist, had a very different view from Malthus. She suggested that population growth has a positive impact on people that will enable them to cope, because as resources start to run out we are forced to 'invent' or innovate our way out of the problem (see Figure 11). This belief relates to the old saying 'necessity is the mother of invention' – the process of invention is really driven by basic need, rather than by curiosity or individual genius. Some believe that we 'invented' farming because we were starting to run out of hunted food. The same explanation might be suggested for the 'Green Revolution' or even genetic modification.

According to this idea, population growth becomes something positive and, what is more, very important to our development as a species. This rather reassuring and optimistic view has not gone unchallenged. The fact that we have invented our way out of food production problems in the past is no sure indicator that we can do so in the future. So, although for some, the 'more people equals more wealth' idea is confirmed by the past 200 years, for many others this ignores the environmental impact of population growth which, they say, will bring it all to a very abrupt and unpleasant halt.

A few years ago the debate about the growing global population tended to concentrate on the use of mineral resources – not just oil and gas, but also metals. Recently, the debate about the future of the human population has returned to its original focus – food. Food is seen as a resource under increasing pressure.

As with oil, there are two global distributions associated with food – production (i.e. supply) and consumption. There is a mismatch between the two distribution patterns. Two critically different conditions result from this mismatch:

- **food security** exists where people do not go hungry or live in fear of starvation (supply meets or exceeds demand). The people have access to sufficient food. They either produce all they need or they are able to import any shortfall in domestic production.

- **food insecurity** exists where people do not have access to an adequate supply of food (demand exceeds supply). This most frequently occurs when food production fluctuates on a seasonal or annual basis. When production dips, hunger and starvation take hold.

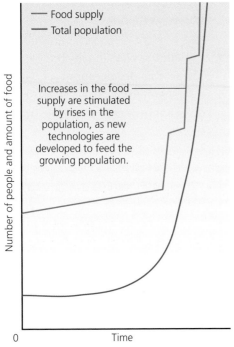

Number of people and amount of food

— Food supply
— Total population

Increases in the food supply are stimulated by rises in the population, as new technologies are developed to feed the growing population.

0      Time

*Figure 11: The relationship between food supply and population according to Boserup*

## Skills Builder 2

Study Figure 11.

(a) Compare the growth of total population and food supply.

(b) Compare the growth rate of food production with that shown on Figure 10.

How sustainable is the current pattern of resource supply and consumption?

163

So what do we make of the present global food situation? What we see is that food production continues to rise. During the first decade of the twenty-first century, FAO statistics suggest that food production grew by an average of around 1.6% each year. So perhaps Boserup was right. The human population seems to be able to cope with the challenge of feeding increasing numbers of people – for the moment at least. It seems to have innovated its way out of the problem. This has been thanks to advances in such matters as plant and livestock breeding, irrigation and fertilisers. But does the world really enjoy food security? How can this be so when the UN states that in 2010 there were 925 million hungry people in the world – that is, over 10% of the global population (Figure 12). What is perhaps also surprising is that 19 million of those hungry people lived in the developed world. This suggests that the neo-Malthusian may also be valid. That people are going hungry because of a poor distribution of food and because increasing amounts of land in developing countries are being taken over to produce food for export.

## The challenges for resource consumption centre on achieving sustainability

Since the 1980s the idea of **sustainable development** has become very well known. In fact, the word 'sustainable' is now widely used – from governments to corporations, from schools to local authorities, from charities to churches. The word is used – and misused – so much that it is easy to lose track of what it really means. The original definition given in 1987 runs as follows:

> *Sustainable development is development that meets the needs of the present without compromising [limiting] the ability of future generations to meet their own needs.*

So the challenge is to limit the consumption of resources today for the benefit of people tomorrow. But how is this to be done? Let us look at three possible ways of moving towards a sustainable use of resources.

**Education** – Changing minds and behaviour is the goal of education. High on the list here is to change the attitudes of key players in the exploitation of resources – governments and businesses. Instead of maximising short-term taxes and profits, the objective should be to adopt a longer-term view. This should focus particularly on:

- rationing the use of non-renewable resources

- using those resources more efficiently

- being careful with those resources whose exploitation has adverse environmental impacts.

Then there is much that we can do as individuals in terms of changing our lifestyles. 'Live simply' campaigns stress the need to change our own habits. To live sustainably means that we should live in a way that conserves resources. The table on page 164 illustrates four simple actions that we might take. But even these are not straightforward; each has a downside. Overall, though, our individual challenge should be to reduce our **carbon footprint**.

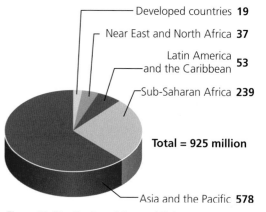

Developed countries **19**
Near East and North Africa **37**
Latin America and the Caribbean **53**
Sub-Saharan Africa **239**

**Total = 925 million**

Asia and the Pacific **578**

*Figure 12: Distribution of the world's hungry people*

## Skills Builder 7

Write an account analysing in some detail what is shown by Figure 12.

**Reducing our use of resources – benefits and problems**

| Ways of reducing our use of resources | Benefits | Costs and problems |
|---|---|---|
| Using local farmers' markets – do we really need imported food? | Supports local farmers and reduces the costs of transport and the 'carbon footprint'. | Buying locally produced greenhouse-grown tomatoes in the spring is more damaging to the environment than buying imported tomatoes grown outside in Spain. It costs more energy to heat the greenhouses than to transport Spanish tomatoes. |
| Reduce the unnecessary luxuries in your life – how many pairs of shoes do you have? | Reduces the waste of resources in their manufacture and their transport, quite apart from the packaging. | What about the foreign workers who make these 'unnecessary' goods? Will they still have jobs if we all cut back on these luxuries? |
| Get on your bike | Cars are great polluters. The average journey by car is just 8.7 miles, and with 33 million cars on the UK roads there is enormous scope here to reduce $CO_2$ emissions. | The old and the infirm cannot jump on to bikes. It is also hard to imagine how the economy would continue to operate as effectively if we cut back on road freight. |
| Recycling and conservation | There are huge savings here as materials are reused and resources saved. | The impact on global warming is quite small, and it certainly does not help to improve the lives of the poorest, especially if our trips to recycling centres are in our 4X4s. |

**164**

## Activity 3

Using the following link: www.carbonfootprint.com/:

(a) Measure your carbon footprint.

(b) Identify three ways in which individuals can reduce their impact on resources.

## Decision-making skills

Study the table above. Carry out a scaled cost-benefit analysis of the four schemes. (Use a minus for costs and a plus for benefits).

## Activity 4

Think of other actions that might be taken to reduce resource consumption. Use the three headings: education, conservation and recycling.

**Conservation** – Its main focus is maintaining the health of the natural world, its habitats and biodiversity. It is about the proper use of resources, their allocation and protection. Allocation of resources means sharing out available resources to consumers in a fair and economic way. It is really resource management and it is clearly a government task. Indeed, it is generally true that responsibility for conservation as a whole rests with governments. Typical conservation actions include designating national parks, setting fishing and timber-felling quotas, and creating programmes that encourage the use of alternative energy sources.

**Recycling** – This is an excellent way of saving resources, especially energy. Saving energy certainly means burning less oil. UK households produce over 30 million tonnes of waste a year, of which slightly less than 20% is collected for recycling. This percentage figure is quite low compared with some other EU countries. There is still a great deal of waste that could be recycled that ends up in landfill sites which are harmful to the environment. Here are some interesting facts about recycling (www.recyclingguide.org.uk/):

- 1 recycled tin can would save enough energy to power a television for 3 hours.

- 1 recycled glass bottle would save enough energy to power a computer for 25 minutes.

- 1 recycled plastic bottle would save enough energy to power a 60-watt light bulb for 3 hours.

- 70% less energy is required to recycle paper compared with making it from raw materials.

- Up to 60% of the rubbish that ends up in the dustbin could be recycled.

- The unreleased energy contained in the average dustbin each year could power a television for 5,000 hours.

- On average, 16% of the money you spend on a product pays for the packaging, which ultimately ends up as rubbish.

- As much as 50% of waste in the average dustbin could be composted.

- Up to 80% of a vehicle can be recycled.

- 9 out of 10 people would recycle more if it were made easier.

The problem here is that whilst governments may be trying to use these ways to move towards a sustainable use of resources, they are at the same time making economic growth a top priority. Economic growth inevitably means more consumption of resources, and achieving the various goals becomes even more difficult.

## Will technology save the day?

There are still a number of people who believe that technology will come to the rescue. Building on the ideas of Boserup and others, they see the increased pressure on food, energy and water resources as likely to produce a positive response. But the history of technological 'fixes' is a very mixed one. The beauty of the food situation is that much can be done straightaway without having to wait for some technological breakthroughs. For example, reducing the wastage during the stages of harvesting and subsequent processing of food would have a positive effect. Feeding the huge mountains of waste food, instead of crops, to livestock would do likewise.

The situation with oil and gas is very different. Whilst we are already exploiting alternative sources of energy which are both renewable and sustainable, there will still be a big shortfall in energy supply when the oil and gas eventually run out. Yes, we do need to look for some technological fix, and the good news is that there appears to be one waiting in the wings – the so-called **hydrogen economy**.

The basic idea of the hydrogen economy is to use hydrogen in a fuel cell to create power. This promises to be a completely clean technology. The only by-product is water. The hydrogen comes from the electrolysis of water. So there are no dangers to the environment like oil spills and no greenhouse gases are emitted into the atmosphere. No longer would there be dependence on the oil states. Another big advantage is that it is easy to split water molecules into hydrogen and oxygen (electrolysis). We already know how to do this splitting. It can be done almost anywhere, even in the home, provided there is a supply of electricity. But here's the snag – the splitting needs a large supply of electricity. So how is this electricity to be produced, if you are not to burn fossil fuels? The answer lies in some of those alternative sources of energy, such as water, wind, solar and tidal power. However, to bring the supply of electricity from these sources up to the demands of a hydrogen economy is an immense challenge in itself.

**ResultsPlus**
**Build Better Answers**

**EXAM-STYLE QUESTION**

**Outline how technology might help solve shortages of resources. (2 marks)**

■ **Basic answers** (0 marks)
Misunderstand the word 'technology' or 'resources'. Include some information about how consumption can be reduced.

● **Good answers** (1 mark)
Suggest that technology might lead to higher production, but do not offer any evidence or illustration.

▲ **Excellent answers** (2 marks)
Show that a particular technology (e.g. hydrogen cells) might replace a non-renewable resource such as oil.

# exam zone
# Know Zone
# Consuming resources

Resource consumption is a major concern globally. There are different views as to how sustainable our consumption is as supplies of many resources, such as oil, are finite. There is also concern over the 'two-speed' world, with developed countries consuming 80% of the resources but only containing 20% of the population.

## You should know...

☐ How resources can be classified by type and availability

☐ How exploiting resources has both benefits and costs

☐ Why the supplies of some resources are running out

☐ How demand for resources varies between countries and is linked to their state of development

☐ How supplies of resources lead to a world of 'haves' and 'have nots' as countries have different levels of resource base

☐ Why there is concern over dwindling supplies and growing demands for oil

☐ Why there are issues surrounding the supply and consumption of energy

☐ That different theories exist about how far the world can cope with the current consumption of resources

☐ How we can become more sustainable in our resource consumption

☐ How technology may help solve shortages of resources

## Key terms

Alternative energy

Boserupian theory

Carbon footprint

Consumption

Food insecurity

Food security

Human resource

Malthusian theory

Natural resource

Non-renewable resource

Recyclable resource

Renewable

resource

Sustainable development

Sustainable resource

### Which key terms match the following definitions?

**A** Development that meets the needs of the present without compromising (limiting) the ability of future generations to meet their own needs

**B** The skills and abilities of the population

**C** A measurement of all the greenhouse gases we individually produce, through burning fossil fuels for electricity, transport, etc., expressed as tonnes (or kg) of carbon dioxide equivalent

**D** Those resources – like coal or oil – that cannot be 'remade', because it would take millions of years for them to form again

**E** Resources – such as wood – that can be renewed if we act to replace them as we use them

**F** Energy sources that provide an alternative to fossil fuels

**G** The using up of something

To check your answers, look at the glossary on page 321.

**ResultsPlus**
**Build Better Answers**

**Foundation Question:** State **three** ways we can use resources more sustainably. (3 marks)

| Student answer (achieving 2 marks) | Feedback comments | Build a better answer (achieving 3 marks) |
|---|---|---|
| We can recycle our waste.<br><br>We can recycle our mobile phones.<br><br>We can use bikes instead of cars. | • *We can recycle our waste* is correct and scores 1 mark.<br><br>• *We can recycle our mobile phones* does not gain any marks as it is an example of the first point, rather than a new point.<br><br>• *We can use...* scores 1 mark. However, it would be better to extend this answer and link it to resources. | We can recycle our waste.<br><br>We can buy local food and cut down on food miles, saving on aeroplane costs and fuel.<br><br>We can use bikes instead of cars so that we use less petrol. |

**Overall comment:** When you are asked to state three ways, make sure that you list three different ways. In this question, the student also needed to link their answer to the idea of sustainability to be sure of the mark.

**Higher Question:** Explain how developing sustainable transport can lead to a reduction in resource use. (3 marks)

| Student answer (achieving 2 marks) | Feedback comments | Build a better answer (achieving 3 marks) |
|---|---|---|
| The best way of not using scarce resources is to walk or cycle.<br><br>It is also useful to cut down on car use by developing good public transport, such as buses.<br><br>Another way is to develop bus lanes. | • *The best way...* is a straightforward and correct explanation that scores 1 mark.<br><br>• *It is also...* This is also correct and is another explanation linked to resource use. It scores 1 mark.<br><br>• *Another way is...* This is potentially a good idea but it is not linked to an explanation so does not score any marks. | The best way of not using scarce resources is to walk or cycle.<br><br>It is also useful to cut down on car use by developing good public transport, such as buses or trains.<br><br>Another way is to develop bus lanes. These lanes make the buses faster and more popular and will save on fuel. |

**Overall comment:** The student demonstrated that they knew what sustainable transport was and linked their ideas to resource use. The only mistake was not to explain bus lanes.

# Chapter 11 Globalisation

## Objectives

- Learn what the global economy is.

- Recognise that the employment sectors of countries change with development.

- Understand that globalisation is having different impacts in different parts of the world.

## How does the economy of a globalised world function in different places?

### Employment sectors and changing balances

The process of **globalisation** is gradually drawing the countries of the world together into a single economic system. This system is known as the **global economy**. The processes involved are a set of policies aimed to encourage that 'drawing together'. The global economy, like the economy of any country, is divided into three or four sectors. Each sector involves a different type of activity and employment:

- The **primary sector** – working natural resources. The main activities are agriculture, fishing, forestry, mining and quarrying.

- The **secondary sector** – making things, either by manufacturing (a TV or a car) or construction (a house, road or new airport).

- The **tertiary sector** – providing services: commercial (shops and banks); professional (solicitors and dentists); social (schools and hospitals); entertainment (restaurants and cinemas) and personal (hairdressers and fitness trainers).

- The **quaternary sector** – a new sector that is mainly found in developed countries. It is about research, information and communications. Sometimes known as the 'knowledge sector' or 'high-tech industry', it also includes nano-technology and biotechnology, producing computer hardware and software, as well as the development of new military equipment (Figure 1).

Figure 1: A typical activity in the quaternary sector

The **employment structure** of a country means how the workforce is divided up between the main employment sectors. The relative importance of these sectors changes over time and from country to country.

### Watch Out!

You may come across the term 'service industries'. The name might suggest that they are part of the secondary sector. They are not because they are services and therefore part of the tertiary sector. To add to the confusion, they are commonly found on what are called 'industrial estates'.

How does the economy of a globalised world function in different places?

169

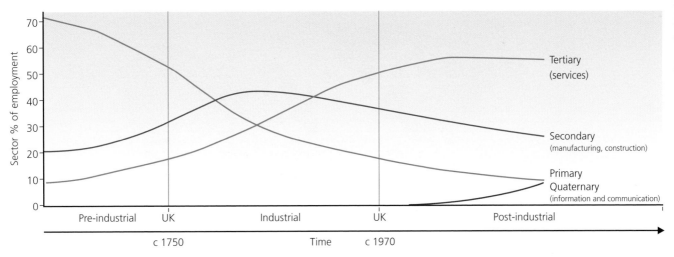

Figure 2: Economic sector shifts over time – the Clark-Fisher model

## The Clark-Fisher model

In 1851 about a quarter of the people in Britain worked in the primary sector – as farmers, foresters, miners or fishermen. The largest employment sector at that time was the secondary one; about half of all employed people were involved in manufacturing. The remaining quarter of the working population were employed in the tertiary (service) sector. Today, however, the employment structure is very different. Only about 2% of the working population are now in the primary sector, and only 27% are in the secondary sector. The tertiary sector is by far the biggest employer – with 68%. The remaining 3% are in the quaternary sector.

The Clark-Fisher model shows how the employment structure of a country changes as it becomes more developed and its economy grows (Figure 2). The sequence of change is divided into three phases:

● **Pre-industrial phase** – the primary sector leads the economy and may employ more than two-thirds of the working population. Agriculture is by far the most important activity.

● **Industrial phase** – the secondary and tertiary sectors increase in importance. As they do so, the primary sector declines. The secondary sector peaks during this phase, but rarely provides jobs for more than half of the workforce.

● **Post-industrial phase** – the tertiary sector becomes the most important sector. The primary and secondary sectors continue to decline. The quaternary sector begins to appear.

The model might be seen as a sort of pathway along which countries pass as they develop. As they do so, not only do their economies grow and the balance of their employment sectors shifts, they become more prosperous and standards of living rise. If we were to take a snapshot of that pathway today, we would see that the world's countries are strung out along it. Some have made little progress along it, while others have gone a long way. Some are moving quickly, others hardly at all.

## Activity 1

In which sector do the following occur: a warehouse; a public park; an internet café; an oil refinery; a plantation; a university?

## Activity 2

1. Approximately when did the UK start the industrial phase?

2. Approximately when did the UK start the post-industrial phase?

3. Describe how the importance of each sector changes over time. Make sure you use numerical data (i.e. percentages) in your answer.

## Watch Out!

The model is based on what happened in developed countries (HICs) like Britain. It may not work in the same way for developing countries (LICs) which may bypass some part of the model. For example, developing countries might encourage tourism and so bypass the industrial phase.

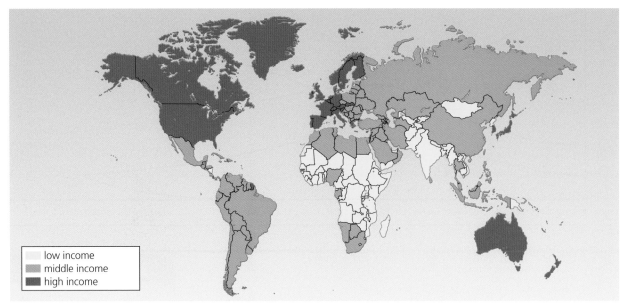

Figure 3: Global distributions of three income groupings

low income
middle income
high income

## Skills Builder 1

Study Figure 3.

Make notes that summarise the distributions of the low-income, middle-income and high-income countries. For example, your notes about the distribution of low-income countries might read: 'Occur largely in Central Africa and in South and Southeast Asia.'

## Top Tip

Remember that some parts of countries might be more or less developed than others. In many developing countries, there are marked contrasts between rural and urban areas.

## Skills Builder 2

In Russia, the primary sector accounts for 11% of all jobs, the secondary sector 29% and the tertiary sector 60%. Draw a pie chart to show these values. How does this pie chart compare with those in Figure 4? Do you think Russia is ahead of or behind China along the development pathway? Give your reasons.

Figure 3 classifies countries according to where they are on the development pathway:

- Low-income countries (LICs)
- Middle-income countries (MICs)
- High-income countries (HICs)

It is still commonplace to divide the countries of the world into **developing countries** and **developed countries**. The problem with this simple scheme is that it ignores the middle-income countries that are making the important transition from the developing to the developed world. Since these countries are clearly developing, they tend to be seen as members of the developing world. We should refer to them as either **emerging** or **newly industrialising countries** (NICs).

### Work and working conditions

In each of these groupings, the balance of the three main employment sectors is different. The primary sector is the most important one in most developing countries. The secondary sector is quite strong in most NICs. The tertiary sector is strongest in all developed countries. To illustrate this point, Figure 4 shows the balance of the three employment sectors in Ethiopia (a developing country), in China (an NIC) and in the UK (a developed country).

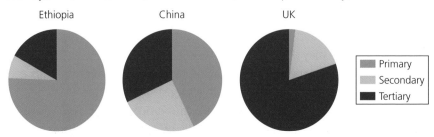

Figure 4: Economic sectors in a typical developing, newly industrialising and developed country

How does the economy of a globalised world function in different places?

171

As each country develops, new activities are added to its economy. For example, as a developing country begins to develop into an NIC, new jobs in manufacturing begin to replace traditional ones in agriculture. As an NIC grows into a developed country, a whole new range of jobs within the tertiary sector will appear, possibly replacing factory jobs. A quaternary sector will gradually develop. Let us take a closer look at those three countries in Figure 4 to see the differences between them in work and working conditions.

**Ethiopia (a developing country)** Three-quarters of the working population work in agriculture. Most of those people are engaged in subsistence farming – they work long hours, often in difficult physical conditions, to produce barely enough food for their survival and that of their families. It is traditional for women to grow the crops and for men to look after the livestock. There is a little commercial agriculture, mainly growing coffee – a major export crop. Both men and women are employed in this activity.

The secondary sector is very small. Foreign investment in textiles and leather is now creating employment. Those jobs are mainly filled by men and are poorly paid. There is a small tertiary sector which provides work for about 10% of the working population; it employs both men and women. Part of this sector is accounted for by tourism. As yet, tourism in Ethiopia is in its infancy, but many developing countries are already attracting large numbers of tourists from developed countries and even NICs.

Such is the general level of poverty in Ethiopia and the shortage of work outside farming that many people are involved in what is known as the **informal sector**. This involves a wide diversity of jobs that are not officially recognised. People (including children) work for themselves, typically on the streets of towns and cities selling goods, providing services (such as fast food, polishing shoes, etc.) and scavenging through waste tips (Figure 5). The people who work in this sector are liable to suffer abuse and exploitation.

**China (an NIC)** China is one of the four main emerging powers, together known as the BRICs (Brazil, Russia, India and China). It is making spectacular economic progress, thanks largely to its secondary sector which is producing huge amounts of manufactured goods that are sold around the world. Most factory jobs involve long hours working in unpleasant, often unsafe conditions. But these workers, both male and female, can earn much more than those farming in rural areas. Much of the farming is now done by older people. Many younger men and women have moved to the rapidly growing towns and cities in search of a better life.

China's primary sector is not just about agriculture. Mining is significant, especially of coal. Frequent mine accidents confirm that mining is a very hazardous occupation due to weak safety regulations.

The rapid growth of towns and cities is being accompanied by the expansion of the service sector. With workers in both this and the secondary sector earning more, there is money to be spent on the goods and services provided by this sector. Working hours are also long; there is little time for leisure. However, the new offices and shops provide a much more pleasant and safe working environment for the men and women who work in them than the factories.

Figure 5: Scavenging on a waste tip

**Quick notes (Ethiopia)**
Harsh physical conditions (including unreliable rains) are partly responsible for Ethiopia's low level of development. The lack of valuable resources also plays a part.

Figure 6: Chinese factory workers

**Quick Notes (China)**
When it comes to development, China's labour force is an important factor. It is very large and hard-working. Many are quick to learn new skills, as well as being ambitious to become part of a consumer society.

*Figure 7: A modern automated factory*

**Quick notes (UK)**

Although the UK is described as a deindustrialised country, it still retains some manufacturing. The clear preference of UK workers is for 'clean' jobs that are secure and well-paid. Technological developments now mean that not all jobs are tied to towns and cities.

## Activity 3

Can you think of some teleworking businesses that could be run from the Hebrides?

**UK (a developed country)** Over the last 50 years, the UK has deindustrialised. It has lost much of its traditional manufacture, such as iron and steel, shipbuilding, car making and textiles. As a result of the so-called **global shift**, these industries have now relocated elsewhere in the world, mainly in NICs. The UK continues to manufacture, but a lot of it is high-tech. So the dirty industrial jobs are gone. They have been replaced by clean, skilled work in state-of-the-art factories. There is much **automation**. Goods are made by machines, and this also reduces the demand for labour.

The primary sector has shrunk as an employer. Mining and fishing have all but ceased, whilst the mechanisation of farming has greatly reduced the need for farm labour. The tertiary sector now dominates the UK's economy; the quaternary sector is now well established.

Working conditions in all employment sectors are good, thanks to strict health and safety regulations. Trade unions also play their part, particularly ensuring that most jobs are paid a fair wage or salary. There is a national minimum wage. Equal opportunities laws mean that there is no discrimination on the basis of gender, age or ethnicity.

Whilst most of the UK jobs are concentrated in towns and cities, new ways of working have emerged. Perhaps the most striking example is what is variously called **teleworking**, telecottaging or telecommuting. Today more than 2 million people are self-employed and work from home. Such home-based work relies on phones, computers and the internet. The activities do not require face-to-face contact with customers and can be done at home. It really does not matter whether you live 10 km or 1000 km from London (or anywhere else) in order to telework. Some of the remote rural areas of the UK, such as the Hebrides in north-western Scotland, have particularly benefited, thanks to the broadband network.

## Globalisation and the global economy

Figure 8 shows the three critical parts in the working of the global economy and the spread of globalisation. They provide the links between the four employment sectors and between different parts of the world:

- **Networks** are the 'spider's webs' linking countries together – for example, transport networks, the telephone, the internet and trade blocs.

- **Flows** are the things that move through those networks – for example, raw materials, manufactured goods, money, migrant workers, information and aid.

- **Players** are the organisations that have a great influence on the workings of the global economy. They include the huge business empires known as **transnational corporations (TNCs)** and a wide range of global organisations.

We will take a closer look at the networks and flows of the global economy later in this chapter (see pages 175–177). But first we will look at the major players.

How does the economy of a globalised world function in different places?

173

## The major players in the global economy

The following are among the most influential international organisations in the spread of globalisation and the growth of the global community (Figure 9):

- The **TNCs** are big businesses promoting still more business (see pages 177–181).

- The **World Trade Organization** (WTO) deals with the global rules of trade between countries. Its main function is to ensure that trade flows as smoothly, predictably and freely as possible.

- The **International Monetary Fund** (IMF) is an organisation of 188 countries, working to reduce poverty around the world by encouraging financial cooperation between countries and by promoting trade and high employment.

- The **World Bank**, despite its name, is not really like a 'high street' bank. It is an important source of financial and technical assistance to developing countries, and its main aim is to reduce poverty.

- The **United Nations** (UN) is perhaps best known as the guardian of international peace and security and the protector of human rights. It also exists to promote the development of poorer countries. This is does through its various agencies such as the FAO, UNCTAD and UNESCO. Both the IMF and the World Bank are really part of the UN.

It is interesting to note that most of the organisations above are concerned with helping the developing world. But they often do so in a way that also benefits the developed world. The reality is that the global economy favours the economically stronger countries, possibly at the expense of the economically weaker ones.

## Some impacts of the global economy

Remember that globalisation is an ongoing process. As a consequence, the global economy is changing all the time. The growing economy is affecting people, particularly workers, throughout the world, but in different ways (see Figure 9 on page 174).

- In the developing world, some things have not changed, such as the importance of the informal sector and the use of child labour. On the positive side, there are opportunities for workers to acquire the skills needed in commercial agriculture, manufacturing and the service sector.

- In the developed world, workers have had to re-skill, as jobs in agriculture, mining and manufacturing have given way to jobs in the tertiary and quaternary sectors. The labour has become more flexible – more part-time working, more self-employment, more teleworking.

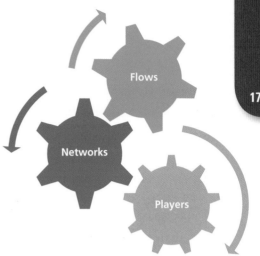

*Figure 8: The main cogs in the working of the global economy*

### Activity 4

Find out what the initials FAO, UNCTAD and UNESCO stand for, and what the agencies do.

- Globally, there have been two broad changes: i) an increase in the number of women in the workforce, and ii) a slow improvement in working conditions (e.g. more attention to health and safety) and terms of employment (e.g. number of working hours, leisure time and pay).

## Decision-making skills

Work in groups and look at the impacts of the growing global economy on working conditions. Decide which you think are the most important and give reasons why.

## Activity 5

In your own words, explain what being 'connected' to the global economy means. Give examples.

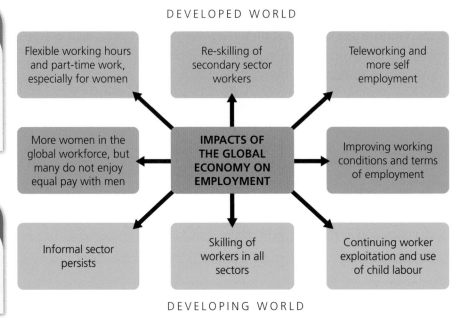

Figure 9: Main impacts of the global economy on employment

### ResultsPlus
### Build Better Answers

#### EXAM-STYLE QUESTION

**Who has benefited most from the growth of the global economy? (6 marks)**

■ **Basic answers** (0–2 marks)
Make sweeping statements to the effect that the whole world has benefited.

● **Good answers** (3–4 marks)
State that the HICs or developed countries have benefited most, but do not give any supporting evidence.

▲ **Excellent answers** (5–6 marks)
Provide evidence to illustrate how the developed countries have benefited, as well indicating that the developing countries are being exploited.

In theory, playing a part of the global economy should benefit every nation and every person in the world. That is true only to a limited extent. Some countries are more 'connected' to the global economy than others – and therefore their people stand to gain more. Because they are among the most powerful players in the global economy, the populations of the rich developed countries are doing particularly well. Sadly, the developing countries are benefiting least, because they are still being exploited by other countries. Within many developing countries, women continue to be treated as second-class citizens, particularly in terms of education, employment opportunities and quality of life.

Workplaces still show many unwanted characteristics. These include:

- Exploitation of workers – women and children, in particular, are paid minimal wages that do not even meet the basic costs of living. There is no trade union protection.

- The working environment is often unsafe. There are many accidents causing injuries or death. Pollution has a bad impact on the workers' health.

- Many jobs are still part of the informal sector and therefore lie beyond the reach of any laws that might be introduced to improve working conditions.

# What changes have taken place in the flows of goods and capital?

## Trade and foreign investment

Let us now take a closer look at flows – one of the cogs in Figure 8 (page 173). Two of the most important flows are trade and foreign investment. We might think of these as the blood circulating around the body of the global economy.

### Trade

The growth in global trade and the global economy go hand in hand. In an era of globalisation, most countries want and need the chance to take part in international trade. Most countries have something which the rest of the world is prepared to buy. Those exports allow a developing country to import what it needs to progress its economic development – machinery, vehicles, fertilisers, and so on.

Unfortunately, world trade does not take place on an even playing-field. There is much talk about free trade and ensuring that goods may be bought and sold across the world without duties, tariffs, quotas or import restrictions. The reality is that goods from developing countries often encounter various forms of trade barrier. The terms of global trade, like the workings of the global economy as a whole, mostly favour the developed countries at the expense of the developing countries.

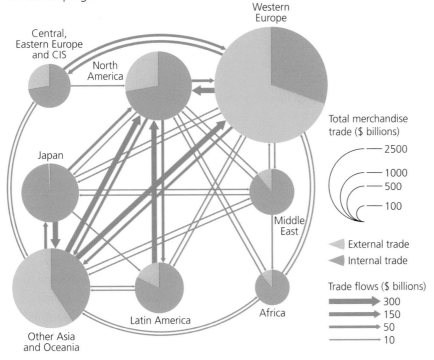

Figure 10: Main flows of global trade

The volume of global trading is expanding, year on year. Figure 10 shows the main flows of trade. Western Europe (the EU), North America (USA), Asia (China and India) and Japan are the main traders. The chief commodities being traded are food, oil and gas, minerals and manufactured goods. Note that within most of the regions, there is a variable amount of internal trade.

## Objectives

- Learn how trade and investment help the growth of the global economy.

- Recognise the factors encouraging the growth of trade and investment.

- Understand the different ways in which TNCs drive the global economy.

175

## Skills Builder 3

Look at Figure 10.

Note the main flows and directions of trade, and suggest in each case what the main commodities might be. Explain why the diagram distinguishes between external and internal trade.

**1500–1840**

**1850–1930**

**1950s**

**1960s**

Best average speed of horse drawn coaches and sailingships was 10 mph

Steam locomotives averaged 65 mph
Steam ships averaged 36 mph

Propeller aircraft 300–400 mph

Jet passenger aircraft 500–700 mph

*Figure 11: A shrinking world*

## Activity 6

Study Figure 11. There are at least two modes of transport missing from this diagram. What are they, and what would be their average speeds?

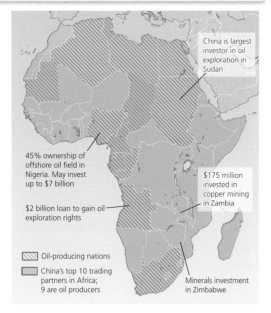

China is largest investor in oil exploration in Sudan

45% ownership of offshore oil field in Nigeria. May invest up to $7 billion

$175 million invested in copper mining in Zambia

$2 billion loan to gain oil exploration rights

Oil-producing nations

China's top 10 trading partners in Africa; 9 are oil producers

Minerals investment in Zimbabwe

(Source: adapted from *Edexcel A2 Geography*, Philip Allan)

*Figure 12: China in Africa*

The growth in global trade has been greatly helped by technological developments in transport. Transport has become much faster. Figure 11 illustrates how the increasing speed of transport has 'shrunk' the world. Furthermore, it is now possible to transport large quantities of cargo at one time. Think of the huge amounts that can be carried by today's supertankers and container ships.

### Foreign direct investment (FDI)

Foreign direct investment is made up of flows of capital which businesses in one country direct towards another country in order to become involved in its business life and markets. In the case of China, the businesses are both private and state companies, but their motives are the same – profit-making. The companies concerned can range in size from small family businesses to giant TNCs. The more specific motives include:

● gaining direct access to foreign markets

● exploiting new sources of energy and minerals

● increasing supplies of food

● taking advantage of cheap labour.

Some of these motives are well illustrated by Chinese investment in Africa (see Figure 12). China's desperate need for oil and mineral resources has focused on Africa. Chinese companies are investing heavily in the search for new sources of raw materials. They are also investing in building the infrastructure (roads, railways and ports) needed to export raw materials back to China. Interestingly, these infrastructure projects are being built by Chinese rather than local workers. There are now close to one million Chinese workers in Africa. Critics argue that all China wants from Africa is resources, and that it is doing little to help African development. Most Chinese investment goes to African governments, TNCs and Chinese companies, not to local companies. Furthermore, China is pushing sales of its manufactured goods in Africa. So Africa is left with its age-old problems of simply exporting cheap raw materials rather than manufactured goods based on those raw materials.

What changes have taken place in the flows of goods and capital?

177

As with trade, the volume of FDI continues to rise. The TNCs are major players, but the general direction of the investment flows is not exclusively from developed to developing countries. In fact, there are big flows of FDI between rich developed countries.

New technology has helped to increase the volume of FDI. There have been major advances in information and communications technologies. This means that investors can keep in close touch with what is happening to their foreign investments and respond quickly to any adverse changes or new opportunities.

It should also be noted that governments are also investing in other countries. More often, the investing is indirect (multi-lateral) and done through an international agency such as the UN. Quite of lot of such investment is in the form of aid.

## Transnational companies

The transnational corporations (TNCs) are thought by many to be the most powerful players in the global economy. They might be seen as the 'builders' of the global economy. In particular, they build the 'bridges' that link together the national economies of the world as they produce goods for a global market.

**The Top 5 TNCs, based on foreign assets (2006)**

| Rank | TNCww | HQ | Industry | Revenue ($bn) |
|---|---|---|---|---|
| 1 | Exxon Mobile | USA | Oil | 377 |
| 2 | Walmart | USA | Retailing | 351 |
| 3 | BP | UK | Oil | 318 |
| 4 | Shell | UK/Netherlands | Oil | 274 |
| 5 | General Motors | USA | Motor vehicles | 207 |

The table makes it clear that these top five TNCs are involved in only three different 'industries' – oil, making motor vehicles and retailing. However, other activities appear in the list of the top 25 TNCs. These include mining (Rio Tinto – UK), publishing and printing (Thomson Corporation – Canada), food, beverages and confectionery (Nestlé – Switzerland), pharmaceuticals (AstraZeneca – UK) and telecommunication (France Telecom – France). But there are also major TNCs that are much less specialised. This is particularly true of Japan-based companies like Sumitomo and Mitsubishi which have wide business interests that stretch across the economic sectors.

The size of these TNCs is emphasised if we compare their revenues with the GDP of some countries. In the same year, Sweden's GDP was $444 bn, Greece's $360 bn, South Africa's $277 bn and Malaysia $180 bn. It is the TNCs' size that gives them so much economic and political influence.

**Quick Notes (China in Africa)**
China is clearly obtaining valuable resources from Africa, but what is Africa gaining in return?

Most TNCs have their HQs in one of the global cities, like London, New York or Tokyo, from where they set up and run their **production chains** (also known as supply chains or commodity chains). The production chains consist of a number of stages involved in the making of a particular product. At each stage, value is added to the emerging product. Figure 13 shows the production chain of a pair of trousers.

```
┌─────────────────────────────────────┐
│        Cotton grown in Egypt         │
└─────────────────────────────────────┘
            ↓ 7,500 km          3,250 km
┌─────────────────────────────────────┐    ┌──────────────────────────┐
│        Cloth woven in Thailand       │ ←  │ Synthetic fibre made in China │
└─────────────────────────────────────┘    └──────────────────────────┘
            ↓ 1,500 km          1,500 km
┌─────────────────────────────────────┐    ┌──────────────────────────┐
│       Trousers made in Bangladesh    │ ←  │ Buttons and zips made in India │
└─────────────────────────────────────┘    └──────────────────────────┘
            ↓ 3,000 km
┌─────────────────────────────────────┐
│ Trousers shipped to Rotterdam (Netherlands) │
└─────────────────────────────────────┘
            ↓ 400 km
┌─────────────────────────────────────┐
│ Trousers distributed to retail outlets in UK │
└─────────────────────────────────────┘
```

*Figure 13: The production chain of a pair of trousers*

**Activity 5**

Research the production chain of either a BMW mini or a personal computer.

Companies set up these 'transnational' production chains for five reasons, which are shown in Figure 15. The two main ones are to reduce costs and increase both sales and profits. For example, the Japanese company Nissan produces cars in Sunderland. This means it is close to the large consumer markets within the EU. Better still, it escapes the tariffs and quotas that make it difficult for foreign companies to export to the EU.

The jeans marketed under well-known labels are made in sweatshops (Figure 14) in countries as distant as Mexico, Bangladesh and China in order to reduce costs. Labour is less regulated in such countries. This means that it is possible for TNCs to pay the workers less, and make them work longer hours – often in sweatshops where there is little regard for health and safety.

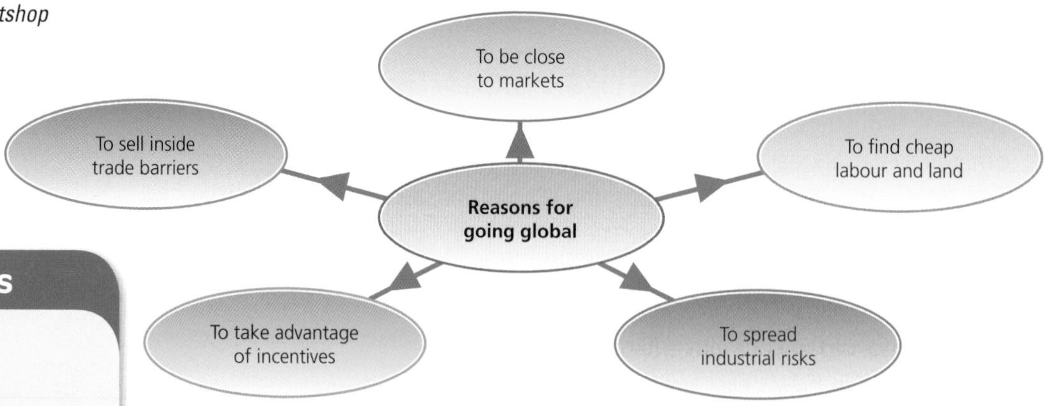

*Figure 14: Inside an Asian sweatshop*

Reasons for going global:
- To be close to markets
- To sell inside trade barriers
- To find cheap labour and land
- To take advantage of incentives
- To spread industrial risks

*Figure 15: Why TNCs operate globally*

**Decision-making skills**

Study Figure 14.

Which do you consider to be the most important reason for TNCs to go global? Give reasons for your choice.

Can you think of any other specific reasons for the existence of TNCs?

Many developing countries are keen to attract the attention of the TNCs. This is done by offering grants and other incentives. For example, countries such as China and Malaysia have set up free trade zones along their coasts to lure the TNCs. Cheap building sites and a good infrastructure are provided. Imports and exports are tax-free. By taking advantage of these 'carrots', TNCs are able to reduce their costs and increase their profits.

What changes have taken place in the flows of goods and capital?

179

One other reason for becoming transnational is that it reduces business risk. If a civil war should break out or a major natural disaster happen in one country, the TNC can simply raise production in another until the situation returns to normal.

The technological developments, in transport and in communication discussed earlier (page 176) have made these long-distance production chains possible. Look at the distances involved in the making of a pair of trousers (Figure 13). The fast transfer of information by the internet and mobile phones means that factories scattered around the globe can keep in close touch with market trends in the UK. Changes in fashions and styles, as well as new orders can be instantly relayed to the factory, whether it is in Dhaka or Shanghai.

Perhaps the greatest impact that the TNCs have had is on the global distribution of manufacturing – the so-called global shift. As a result of their policy of seeking out the cheapest locations for making particular products, factories have been closed down in the UK and other HICs. They have been replaced by new factories (branch plants) set up in these cheap production locations. As a consequence, there has been a global shift in the location of manufacturing. Deindustrialisation in the developed world is matched by industrialisation in the developing world.

Two case studies of TNCs now follow. The first – Nike – is mainly concerned with manufacturing (secondary sector). The second –Tesco – is in retailing (tertiary sector).

## Case study: Nike – a brand and its success

Nike, originally known as Blue Ribbon Sports (BRS), was founded in the American state of Oregon in January 1964. The company initially operated as a distributor for a Japanese shoemaker, selling out of a van at athletic track meetings. Three years later it opened its first retail outlet in Santa Monica, California. The first shoe bearing the now famous Nike brand mark was sold in 1971. In 1978, BRS was officially renamed Nike (the Greek goddess of victory).

Today Nike produces a wide range of sports equipment. Their first products were track running shoes. They currently also make shoes, jerseys, shorts and base layers for a wide range of sports. The company has more than 700 shops around the world and has offices located in 45 countries outside the USA. Most of Nike's products are made in factories located in Asia – in Cambodia, China, India, Indonesia, Malaysia, Pakistan, the Philippines and Taiwan. These factories are the end points of production chains that involve the making of components such as textiles, leather, rubber soles, etc.

Just as Nike's production has gone global, so too has its market and reputation. Its annual turnover continues to rise. The Nike brand mark is now keenly sought by shoppers the world over. Nike also continues to promote its sporting image through the sponsorship of sporting events and sports stars.

## Activity 8

Examine what makes a location 'cheap'.

Is cheapness the only reason why TNCs set up branch plants in developing countries?

## Case study: Tesco – a transnational retailer

The supermarket chain Tesco is one of the few leading TNCs with its head offices in the UK. Currently it is ranked 50th in the global league table. The company started life as a single grocery stall in the East End of London. It did not set up its first self-service supermarket until 1956. It was during the 1970s, 1980s and 1990s that the company really took off to become the largest food retailer in the UK

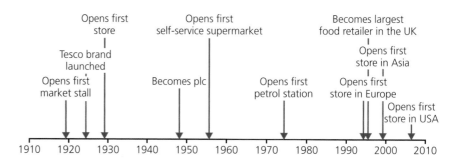

Figure 16: A timeline of Tesco's globalisation

The key to the company's success has been:

- its strategy of diversification into new markets, such as toys, clothing, electrical goods, home products, financial services and telecommunications, in addition to its original business of food;
- **outsourcing** its supplies of foodstuffs, clothing and other goods directly from producers, both in the UK and in LICs such as Kenya, Sir Lanka and Bangladesh;
- globalising its chain of supermarkets (Figure 16). This did not start until the 1990s with the opening of stores in Eastern Europe (Hungary, Poland, the Czech Republic and Slovakia). In 1998, it made its first move outside Europe, opening stores in Taiwan and Thailand, and in South Korea the following year. Tesco's presence in Asia has subsequently spread to China, Japan and Malaysia (Figure 17). Today, 60% of the company's profits come from Asia.

Although Tesco is now an international company, with an annual turnover of £65 billion, its HQ remains in England. It is the third largest retailer in the world and has a clear brand image. You would recognise a Tesco store, whether it was in Shanghai or Prague. When you go inside any of these overseas stores, it is highly likely you will find many of the same products that are on the shelves of your local Tesco. Equally, you will also find products that reflect the particular diets and tastes of the host country (an example of 'glocalisation').

From a single supermarket in 1956, Tesco now has over 6,000 stores and employs over 500,000 people in 14 countries. Truly, Tesco has become a major TNC.

### Skills Builder 4

Go into your local supermarket and look at the labels that tell you where a product has come from. Make a list with two columns – one for the type of product and the other for the country of origin. Analyse your findings, perhaps presenting your findings on an outline world map.

What changes have taken place in the flows of goods and capital?

181

It is interesting to contrast these two case studies. Nike manufactures and sells trendy sports shoes and clothing for a consumer market mainly in the developed countries and the NICs. Outsourcing production in developing countries has created much-needed jobs – and made all the difference to the company's success.

Tesco is also concerned with the consumer market, but at the more basic level of supplying food, clothing and household necessities. Not only are working conditions in its stores good in terms of hours and pay, but the working environment is a clean and bright one – as it should be in any shop selling food! Tesco also outsources many of its products from other countries, so the employment benefits of the company's retailing success are spread quite widely around the world.

Figure 17: A new Tesco store in China

**Quick notes (Tesco)**
Having established itself as one of the leading supermarket chains in the UK, Tesco are now going global.

## Activity 9

Although Nike and Tesco operate mainly in two different employment sectors, make a list of the features that you think they have in common. Suggest some reasons for the similarities.

**ResultsPlus**
**Exam Question Report**

**REAL EXAM QUESTION**

**Choose a TNC that you have studied. Explain the advantages and disadvantages for a country where the TNC has set up a factory. (5 marks)**

**How students answered**
Some students failed to identify any specific TNC, and the impacts were not clearly identified as an advantage or disadvantage.

28% (0–1 marks)

Most students named a TNC and were able to state at least one advantage and disadvantage but without very much explanation.

61% (2–3 marks)

A few students could identify and explain some advantages and disadvantages, and some were able to identify groups who might benefit and groups who might lose. Good case-study evidence was used in these answers.

11% (4–5 marks)

# examzone

# Know Zone
# Globalisation

Globalisation and the growth of the global economy have led to huge changes in the pattern of employment around the world. As countries develop, so the type of work available and the working conditions change. Trade and financial direct investment fuel the growth of the global economy, along with the activities of the key players, the transnational corporations.

## You should know...

☐ What the four employment sectors do

☐ How the balance of the employment sectors changes with a country's development

☐ How working conditions change with development

☐ How international organisations contribute to globalisation

☐ How trade and foreign direct investment help the growth of the global economy

☐ Why trade and foreign direct investment have expanded

☐ How trade and investment favour the developed countries

☐ How transnational corporations drive the processes of globalisation and global shift

☐ How transnational corporations operate

## Key terms

| | |
|---|---|
| Automation | Outsourcing |
| Deindustrialisation | Players |
| Flows | Primary sector |
| Foreign direct investment | Production chain |
| | Quaternary sector |
| Global economy | Secondary sector |
| Global shift | Sweatshop |
| Globalisation | Teleworking |
| Industrialisation | Tertiary sector |
| Informal sector | TTNCs |
| Networks | |

### Which key terms match the following definitions?

**A** The economic activities that provide intellectual services – information gathering and processing, universities, and research and development

**B** A system of linkages between objects, places or individuals

**C** Forms of employment that are not officially recognised, e.g. people working for themselves on the streets of developing cities

**D** Individuals and groups who are interested in and affected by a decision-making process

**E** A process in which a company subcontracts part of its business to another company

**F** The use of machinery, rather than people, in manufacturing and data processing

**G** The sequence of activities needed to turn raw materials into a finished product

**H** The movement of objects, people and ideas between places

To check your answers, look at the glossary on page 321.

**ResultsPlus**
**Build Better Answers**

**Foundation Question:** Explain why transnational companies locate their headquarters in developed countries and many other factories in developing nations. (4 marks)

| Student answer (achieving 2 marks) | Feedback comments | Build a better answer (achieving 4 marks) |
|---|---|---|
| The headquarters are in developed nations because this is where they started up, for example GM in the US.

The factories are in developing countries as this is where products such as palm oil or chocolate grow.

The workforce in these countries works for less.

The main reason is that they often took over existing factories. | • *The headquarters are...* scores 1 mark. This is a clear, correct statement.

• *The factories are...* scores 1 mark. This is one reason for the pattern but there are others.

• *The workforce in...* Although this statement is true, it needs developing to score a mark.

• *The main reason...* is incorrect and does not score any marks. | The headquarters are in developed nations because this is where they started up, for example GM in the US.

The factories are in developing countries as this is where products such as palm oil or chocolate grow.

They also take advantage of the lower costs such as wages in developing countries.

A further advantage is that environmental and safety laws are slack, so there are lower production costs. |

**Overall comment:** The student could have gained an extra mark, but the third point was not sufficiently developed.

**Higher Question:** Explain how transnational companies have led to a global shift in industry and services. (4 marks)

| Student answer (achieving 2 marks) | Feedback comments | Build a better answer (achieving 4 marks) |
|---|---|---|
| The global shift is the movement of economic activity from developed to developing countries.

It has occurred because wages are much lower in developing countries.

There are also very slack environmental laws so again the product is cheaper.

The result is that TNCs have moved all their offices to developing countries. | • *The global shift...* This is a good, clear definition that scores 1 mark.

• *It has occurred...* This statement is correct and scores 1 mark, but there is no mention of TNCs.

• *There are also...* This does not score any marks as it is not well linked to the idea of TNCs.

• *The result is...* This is incorrect as usually it is only branch plants that go transnational. | The global shift is the movement of economic activity from developed to developing countries.

TNCs have been attracted to LDCs because of the low wages/low costs for manufacturing.

TNCs have also outsourced their business functions, such as call-centre operations, which also leads to the shift.

The headquarters remain in developed countries, but the production activities in branch plants are often run by contractors. |

**Overall comment:** Although the reasons given are correct, the answer does not focus enough on TNCs to be awarded more than 2 marks.

# Chapter 12 Development dilemmas

## Objectives

- Recognise that there are different ways of defining and measuring development, and evaluate a range of ways in which development can be measured.

- Explain the development gap and how it has changed over time, and how one Sub-Saharan African country has developed.

- Explain how development can take place, including top-down and bottom-up strategies, and understand that regional differences exist.

## Top Tip

Students who use correct geographical terms wherever possible impress examiners – and score highly.

## How and why do countries develop in different ways?

### Defining development

Development can be difficult to define because it depends on the viewpoint of a person or group of people. For many people, **economic development** is considered most important; this involves ways of making money, such as through industries and trade. In the modern world many national governments still wish for this type of development because wealthier people and businesses can pay taxes, and this money can develop other aspects of a country (e.g. infrastructure projects such as roads, hospitals, schools) and a cycle of growth is created. However, some individuals or small groups (e.g. communities or NGOs) within a country may be more concerned about **socio-political development**, such as education and health care provision, safety, or freedom of speech. Some people believe that economic development creates too many problems, such as pollution and inequality between people (e.g. rich and poor), so they have looked for other ways of defining development.

**Sustainable development** is another way of looking at how a country or region is progressing over time; this considers not only economic and social factors but also the impacts on the natural environment. Development is difficult to measure because accurate data is not available, or because the things people want to consider are difficult to measure (e.g. happiness), or because of the contrasting aims of different cultures (e.g. political or religious beliefs).

### Measuring development

**Gross Domestic Product** (GDP) per capita is one way of measuring development; this takes the money made within a country in a year and divides it by the number of people in the country. Sometimes **Gross National Income** (GNI) per capita is used, which includes wealth made outside the country by its companies and corporations, and things like debt. An advantage of GDP is that data is available for every country with an economic structure. A disadvantage of GDP is that it uses an average amount of money per person and so hides any gap between rich and poor people. (India, for example, has about 60 billionaires – but millions of people below the poverty line.) Another disadvantage is that it does not include important activities that do not involve money (e.g. subsistence farming). According to the World Bank, the GDP per capita in 2011 was $1,521 for Tanzania, $11,719 for Brazil and $35,494 for the UK. (In 2011 Tanzania's GNI per capita was $1,510, Brazil's $11,500, and the UK's $35,940).

The **Human Development Index** (HDI) considers several aspects of development, e.g. income, education and life expectancy, and combines them into one scale. In 2011, out of 187 countries, the HDI ranked the UK as 28th, Brazil 84th and Tanzania 152nd (Figure 1).

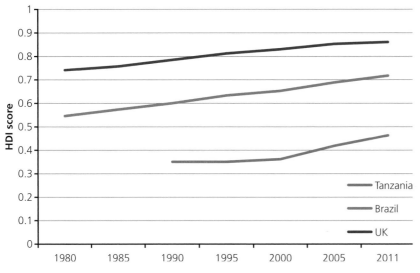

Figure 1: Human Development Index for selected countries from 1980 to 2011

An advantage of HDI is that it covers a wider range of factors – social *and* economic – and so is considered a more accurate measurement of overall development. The disadvantages of HDI are that some data is not available for all countries and it does not consider the natural environment or inequalities within countries. There is now an inequality-adjusted HDI (or IHDI) which reduces the main HDI. For example Tanzania's index is reduced by 28.8%, Brazil's by 27.7%, and the UK's by 8.4%, as shown in this table for 2011:

|  | HDI | IHDI |
|---|---|---|
| Tanzania | 0.466 | 0.332 |
| Brazil | 0.718 | 0.519 |
| UK | 0.863 | 0.791 |

The **Happy Planet Index** (HPI) considers how efficient individual countries are at using their resources to benefit their people, without causing long-term damage to the environment. The advantage of this measure is that it considers sustainability and how well supported a population is by its government – but the economic aspect is absent, and the measure needs to be used with other indicators to give a better view of development. In an assessment of HPI in 2012, Brazil was ranked 21st (score 52.9), UK ranked 41st (47.9), and Tanzania 133rd (30.7) out of 151 countries. The New Economics Foundation, who devised the index, suggests a 2050 target score of 89.0.

The **Gender Inequality Index** considers female participation in the workforce and decision-making processes, the level of education of women, and their degree of control over pregnancy. The lower the score, the lower the inequality. The 2011 index compared with the 1995 index shows that countries are improving: Tanzania has gone from 0.648 to 0.590, Brazil from 0.523 to 0.449, and the UK from 0.243 to 0.209. An advantage of this measure is that it specifically targets a group that are often excluded during the development of a country, but a disadvantage is that the role of women varies considerably between societies, and therefore some of the indicators used may be biased, which will skew the resulting index.

ResultsPlus
**Build Better Answers**

**EXAM-STYLE QUESTION**

**Outline the advantages and disadvantages of using the Human Development Index to measure the development of a country. (6 marks)**

■ **Basic Answers** (1–2 marks)
Only mention and describe one advantage or disadvantage. Poor spelling and grammar, and little or no use of geographical terminology.

● **Good Answers** (3–4 marks)
Identify at least one advantage and one disadvantage. Spelling and grammar are satisfactory.

▲ **Excellent Answers** (5–6 marks)
Offer more than one advantage or disadvantage in addition to the Level 2 answer, such as expansion or explanation, and use a good range of terminology and accurate spelling and grammar.

**Political Freedom** measures political rights and civil liberties, including the freedom of elections, the number of people voting or the number of people with the right to vote, freedom of speech and individual rights (see Figure 2). An advantage is that it uses 7 different measures, combined with 25 key questions judged by experts. In addition, the data has been assembled since 1973, so trends over time can be identified. A disadvantage is a bias towards the 'Western' ideas of freedom.

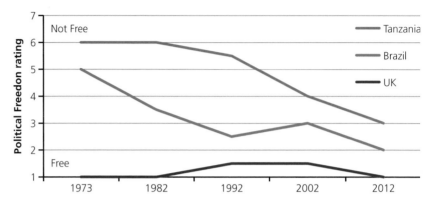

Figure 2: Political freedom rating of selected countries from 1973 to 2012

The **Corruption Perception Index** looks at perceived corruption in governments and their departments. If development is to take place, a government should be working for the people. Perceptions are used because corruption is hidden and difficult to measure but the lack of data is the main weakness of this index. Strengths of this index are that it covers 183 countries and combines different sources of information. In 2011 Tanzania was ranked at 100 with a score of 3.0, Brazil was 73rd (score 3.8), and the UK was 16th (score 7.8), where 0 is very corrupt and 10 is very honest.

The **Environmental Performance Index** considers the health of people and the natural environment using 22 indicators. In 2012 the UK was ranked 9th (68.82), Brazil 30th (60.9), and Tanzania 64th (54.26) out of 132 countries. This measurement has the advantage of showing how countries are looking after their people and natural environment, using a wide range of indicators; however, it does not cover economic factors directly.

## The Millennium Development Goals

The eight **Millennium Development Goals** (MDGs), which were set at a United Nations summit meeting in 2000, provided targets for countries to achieve by 2015. Target 1 for Goal 1 is 'Halve [between 1990 and 2015] the proportion of people whose income is less than $1 per day'. By 2010 Brazil had reduced its level of 'extreme poverty' to 3.6% of the population, but in 2005 Tanzania still had 28% in this category, leaving much to be done by 2015.

### The eight Millennium Development Goals

1 Eradicate extreme poverty and hunger
2 Achieve universal primary education
3 Promote gender equality and empower women
4 Reduce child mortality
5 Improve maternal health
6 Combat HIV/Aids, malaria and other diseases
7 Ensure environmental sustainability
8 Develop a global partnership for development.

One of the development dilemmas facing developing countries is how to improve the well-being of the population while at the same time conserving natural resources and ecosystems. This balance is not easy to achieve, especially over a long period of time, but it is important that the benefits remain for future generations. As a measurement of development, the MDGs have strength because of the wide range of specific factors covered, but a disadvantage is that they are not combined to produce an index, which makes comparing countries difficult. MDGs show considerable progress in south-east Asia, but progress is very slow in sub-Saharan Africa (see the section on Tanzania, page 189).

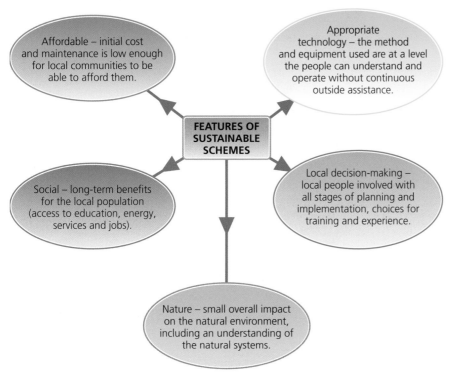

*Figure 3: Sustainability criteria*

## The development gap

Due to geographical factors, such as locational advantages and resources, and historical factors, such as colonialism, the world has always been divided into those that 'have' and those that 'have not' – a development gap. Europe, the cradle of modern industrialisation, has economically dominated since the nineteenth century, joined then by the USA, Japan, and more recently the East Asia region. In the 1980s a North/South divide between developed and developing countries was clear. This has been distorted slightly by the rise of Newly Industrialised Countries (e.g. South Korea, Taiwan), the modifications to communism, and the rise of the BRICS economies (e.g. Brazil, China). These changes are shown by the maps displaying how the proportions of the world's GDP distribution have changed since 1960 (Figure 4).

(a)

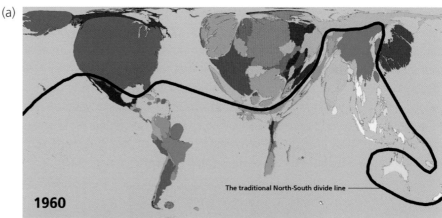

The traditional North-South divide line

**1960**

(b)

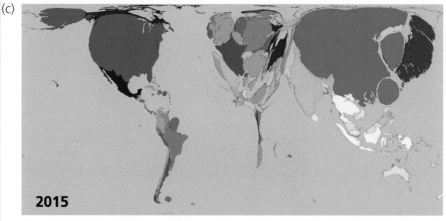

**1990**

(c)

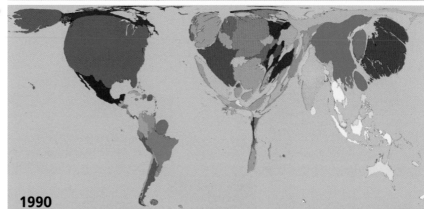

**2015**

Figure 4: World development gap (changes in wealth from 1960 to 1990 to 2015)

## Activity 2

(a) Study the maps in Figure 4 which use GDP information. To what extent was there a development gap in 1960?

(b) Describe how the development gap between world regions has changed since 1960.

(c) Which countries appear to have economically developed the most during this time period?

## Skills Builder 1

Study Figure 5 which uses HDI data from 1975 and 2002. The larger the size of the country, the more development has taken place during the time period.

(a) Describe the changes that have taken place.

(b) Compare Figures 4 (c) and 5. Explain the differences between the two maps.

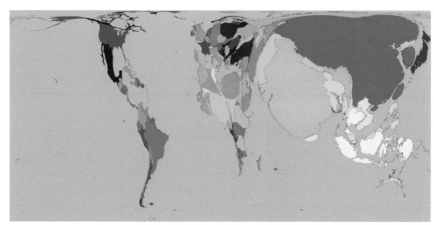

Figure 5: Worldmapper map showing Human Development Index increases from 1975 to 2002

## A Sub-Saharan country: Tanzania

Before 1885 the area now known as Tanzania was mainly a primitive society. Between 1885 and 1961 it was a German and then a British colony and it developed a market economy based on exporting raw materials. After independence in 1961 the country kept this style of economy, based on foreign interests. By 1967, however, it became clear that wealth was not spreading as quickly as had been hoped, and money was leaking to other countries and foreign businesses. Between 1967 and 1985 Tanzania adopted a socialist political and economic approach, with the government taking over industries and reducing external influences. Rural communes were developed and the government subsidised services (health, education, water supply); a long-term advantage was that today's leaders of the country grew up with these benefits. But this development approach did not work either, as the government was earning no money from taxes because they owned the industries, and private business declined during this period. A brief war with Uganda in 1978–79 did not help progress.

In 1981 Tanzania and the World Bank disagreed over a development budget, so a loan was not given to the country and it was forced to change its agricultural policy. From 1986 the International Monetary Fund (IMF) introduced 'structural adjustment policies' for Tanzania to receive steady money flows and boost export growth, mainly farm products; exports did increase but there was also rising inflation and falling GDP, which resulted in more rural poverty, inequality, malnutrition, and deforestation, and made Tanzania even more dependent on foreign aid. Between 1985 and 2005 reforms took place to move Tanzania back towards a market economy with foreign direct investment (FDI), but this time with the decisions made by the Tanzanian government and its people rather than by people outside the country. However, a problem in this phase was that government investment in services declined.

The most recent phase (2005–25), known as Tanzania's 'Development Vision', consists of changing the economic and government structures with the aim of getting the country into the middle-income category, with more investment in infrastructure (e.g. 2,200km of surfaced roads and new bridges). There is a 'National Strategy for Growth and Reduction of Poverty' (2010–15) which concentrates on economic matters, because 34% of the population are under the basic poverty level.

Help is available from organisations such as the United Nations Development Programme (UNDP) – the Millennium Villages Project (2006–11) for example, which had a social emphasis (e.g. primary education for all, women's health, eliminating hunger). An example of a local scheme is the Bumbuli Development Corporation, which promotes local business ideas and farming, as well as encouraging people to report, by mobile phone, incidents of corrupt government officials demanding or accepting bribes.

It is now clear that the people support and contribute to the country's long-term development policy, but Tanzania's experience shows that there is no easy quick way of developing a country (see Figure 6).

### Activity 3

Study Figure 6: Barriers to Tanzania's development today

(a) Which of these barriers are outside the control of Tanzania?

(b) Which of these barriers are being tackled by Tanzania's Development Vision for 2025?

(c) Which of these barriers do you think is the most serious? Why?

(d) Choose 4 other barriers and explain how each may limit or prevent development taking place in a Sub-Saharan country such as Tanzania.

189

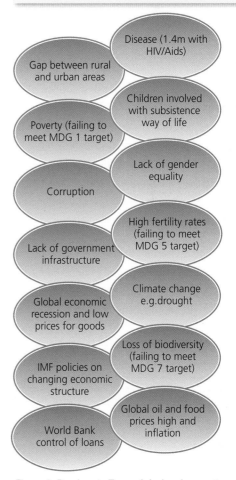

Figure 6: Barriers to Tanzania's development today

Figure 7: Tanzania tax districts

## Skills Builder 2

Look at Figure 7 and the table below.

(a) Choose one of the columns of data 'Tax revenue per person' or '% contribution to total tax revenue', from the table. Consider the range of values and put the data into four categories (e.g. 0–1,500 or 0–0.5). One way to do this accurately is to plot the data on a single-axis scattergraph, or another way is to work out the median and quartiles.

(b) Produce a choropleth map on a copy of Figure 7. Your colour or shading scheme should have a logical sequence (e.g. purple, red, orange, yellow).

(c) Describe the pattern shown by your map.

## Activity 4

Research the United Nations website for information about the Millennium Development Goals (MDGs).

(a) What are the eight MDGs?

(b) How much progress has Tanzania made towards each MDG?

**Tax income from the tax districts of Tanzania, 2005–06**

| Tax district | Tax revenue per person (Tanzanian Shilingi) | % contribution to total tax revenue |
|---|---|---|
| Arusha | 49,816 | 3.16 |
| Dar Es Salaam | 678,928 | 83.19 |
| Dodoma | 2,927 | 0.24 |
| Iringa | 4,993 | 0.37 |
| Kagera | 3,261 | 0.32 |
| Kigoma | 1,417 | 0.12 |
| Kilimanjaro | 36,823 | 2.49 |
| Lindi | 909 | 0.04 |
| Mara | 25,217 | 1.69 |
| Mbeya | 13,553 | 1.37 |
| Morogoro | 17,177 | 1.48 |
| Mtwara | 3,236 | 0.18 |
| Mwanza | 13,842 | 1.99 |
| Pwani | 2,897 | 0.16 |
| Rukwa | 1,276 | 0.07 |
| Ruvuma | 1,520 | 0.08 |
| Tanga | 32,092 | 2.58 |
| Shinyanga | 1,740 | 0.24 |
| Singida | 874 | 0.05 |
| Tabora | 2,093 | 0.18 |

Source: Tanzania Revenue Authority

# How might the development gap be closed?

## Development theories

Development can be considered as continuous in the modern world, apart from **economic recessions** and major catastrophes, but it can be viewed in stages. In the 1960s an economic model was proposed by the American economist Rostow, in which he suggested that all countries would develop in a series of steps or stages, based on what happened in countries such as the UK and USA (Figure 8). Given enough time, all countries would become fully developed. To progress to the next level, a stimulus is needed. Therefore countries invest in strategies, such as large, expensive national **top-down projects** (e.g. the Carajas Project, Brazil) as they believe that development will occur faster. An alternative approach is using small, community based, cheaper **bottom-up schemes**, but the development of the whole country may take longer.

## Objectives

- Recognise that levels of development may vary within a country, between an urban core and rural periphery.

- Explain, using development theories, why societies develop over time.

- Compare the characteristics of bottom-up and top-down strategies and evaluate a top-down project.

191

**High Mass Consumption:** People have more wealth and so buy services and goods (consumer society), welfare systems are fully developed, trade expands.

**Drive to Maturity:** New ideas and technology improve and replace older industries, economic growth spreads throughout the country (i.e. a maturing of industry).

**Take-Off:** Introduction and rapid growth (industrial revolution) of manufacturing industries, better infrastructure, financial investment, and culture change.

**Pre-conditions for Take-Off:** Building infrastructure that is needed before development can take place, e.g. transport network, money from farming, power supplies, communications.

**Traditional Society:** Based on subsistence; farming, fishing, forestry, some mining.

*Figure 8: The Rostow Model*

The **Rostow model** has been criticised for being just based on European countries and overlooking other ways in which a country could develop. It also assumes that all countries start with the same resources and other geographical factors (e.g. population, climate).

## Activity 5

Considering the information presented in this chapter and any of your own research:

(a) In which Rostow stage would you place the UK? Give one piece of evidence for your choice.

(b) In which Rostow stage would you place Brazil? Give one piece of evidence for your choice.

(c) In which Rostow stage would you place Tanzania? Give one piece of evidence for your choice.

**EXAM-STYLE QUESTION**

**Using the Rostow theory, explain why societies develop over time. (6 marks)**

■ **Basic answers** (1–2 marks)
Include only limited detail of steps or stages but no explanation. Poor SPG.

● **Good answers** (3–4 marks)
Use the structure of the model to explain at least one link between steps or stages. SPG is satisfactory.

▲ **Excellent answers** (5–6 marks)
Use the model to explain the links between two or more steps or stages, and use a good range of terminology and accurate SPG.

## Activity 6

(a) How does dependency theory explain why the world is divided into developed and developing countries?

(b) To what extent do you believe that dependency theory is relevant in the twenty-first century?

**Dependency theory** evolved in the late 1960s as a reaction against Westernised economic models of development (e.g. the Rostow model produced in the USA). This theory suggests that the low levels of development in poorer countries (periphery) results from the control of the world economy by rich countries (core). Evidence of this includes the exploitation of developing countries by external forces, e.g. getting cheap resources during a colonial time period, or more recently TNCs employing a cheap workforce. Dependency theory suggests that the unequal pattern of development has been reinforced by:

◉ Rich countries interfering with the internal politics of poor countries

◉ Unbalanced trade, where poor countries sell materials cheaply but buy expensive products

◉ The selling of non-essential products to poor countries (e.g. carbonated soft drink)

◉ Bi-lateral aid being tied to wider agreements so that the rich country gets something in return (e.g. the poor country accepts TNC investment)

◉ Poor countries getting into debt after borrowing too much from the developed world (e.g. about 25% of aid received by African countries each year is used to repay debts rather than build infrastructure).

One dependency theorist, A.G. Frank, suggested that poor countries cannot develop while they are linked to the present world economic (capitalist) system and that each should isolate themselves and make their own products. Other people suggest adopting a socialist system.

However, dependency theory has been criticised because countries that were never colonised (e.g. Ethiopia) remain poor; countries that have tried a socialist system (e.g. Tanzania) found that it did not help; some poor countries have successfully developed (e.g. South Korea); rich country influences today (neo-colonial) may be positive (e.g. aid without ties); and campaigns such as 'Make Poverty History' and 'Free Trade' are examples of positive links between the core and periphery.

### Regional disparity

Development within a country does not take place evenly across all the different regions at the same time. It occurs at different speeds, and so a spatial difference in the economic pattern emerges. This creates a **disparity** – a great difference – between a **core region** and a **periphery**. The rich core is based on the urban areas, which have the majority of the people, services, businesses, industry, and the government headquarters. The periphery is rural – often remote countryside – and involved in the production of raw materials, and so is poor.

### Disparities in Brazil

In Brazil, the south-east of the country is the core (Figure 9), based around the cities of São Paulo, Rio de Janeiro and Belo Horizonte. This region has a hospitable temperate climate, access to ports for trading (e.g. Santos), road and rail networks, and fertile soils for farming (e.g. São Paulo state). Over time these advantages have attracted people (rural to urban migration) and investment in business. So the south-east core has grown.

Figure 9: *Regional disparities in Brazil, according to the Human Development Index*

## Regions of Brazil: GDP per capita in R$ (Brazilian reals), 2005
(Source: IBGE, 2007)

| Region | GDP per capita |
|---|---|
| Piauí (NE periphery) | 3,700 |
| Maranhão (NE periphery) | 4,150 |
| Acre (N periphery) | 6,792 |
| Roraima (N periphery) | 8,123 |
| Rio de Janeiro (SE core) | 16,052 |
| São Paulo (SE core) | 17,977 |
| Brasília Federal District (capital) | 37,510 |

The periphery of Brazil can be found in the north, and in the north-east. Acre and Roraima are regions a long way (3,000 km) from the core, with a wet tropical climate, poor access by land due to the extensive tropical rainforest, sparse population and limited access to the sea for trading. Maranhão and Piauí are sparsely populated, lack resources, are found in a semi-arid climate zone with frequent droughts, and situated a long way (2,200 km) from the core. The table above, which shows the GDP per capita data for different regions, confirms the disparity – the rich/poor divide – that exists in Brazil.

The large disparity within their country encouraged the Brazilian government to build a new capital city – Brasília – inland in 1960 to help spread wealth and economic development. Figure 9 and the table suggest that this had some success. Several large schemes were also implemented in the periphery, such as the Carajás Project (based on the world's largest iron ore deposits), and several **Hydro-electric power** (**HEP**) dams (such as Santo Antonio and Jirau on the Madeira River in Rondônia). Schemes like these partly support the core region (e.g. power supplies), but they can also spread the benefits of the core to the periphery.

### Results Plus
**Build Better Answers**

**EXAM-STYLE QUESTION**

**Look at Figure 9. Describe the regional differences that have resulted from economic development. (4 marks)**

■ **Basic answers** (0–1 marks)
Fail to include geographical terms, such as 'south-east', or the names of states such as 'Roraima' consistently throughout. Statements such as 'the core at bottom of map and periphery at top right' are used. There is little or no evidence from the map and key.

● **Good answers** (2 marks)
Identify the south-east core based around São Paulo and the north-east periphery from Alagoas to Maranhão, name some of the states and make reference to HDI.

▲ **Excellent answers** (3–4 marks)
Also mention the spread of development from the core towards Amazonas via Brasília. They cover range of regional disparity and include places and data.

Governments all over the world usually wish to spread development throughout their countries, because:

◉ People in the periphery could be left with a lower quality of life than those in the core.

◉ Conflicts could arise between the people of different regions (e.g. between rich and poor, or between local people and development companies).

◉ Lots of poorer people from the periphery could move into the core region.

◉ Overcrowding and lack of jobs could arise in the core.

◉ Resources in the periphery could be left undeveloped, so slowing the development of the whole country.

◉ If the periphery remains undeveloped, the government receives less tax money to help improve the whole country.

## Types of development

A country can try different ways to prompt development. Some use large-scale technology, often advanced, and these schemes are usually expensive. In the past, countries wishing to develop often borrowed money from the World Bank, or involved companies from other more developed countries to help them start up a big scheme. The decisions about these big schemes were made by the national government and, where necessary, new regulations or laws were passed to enable these schemes to go ahead. Local people who lived near to where the scheme was to take place were often not involved in the process and had great difficulty influencing the decisions being made. Such an approach is known as '**top-down**' (Figure 10).

National government

External groups (e.g. World bank, TNC's)

Local people

▢ Decision made here
➡ Major influence
➤ Minor influence

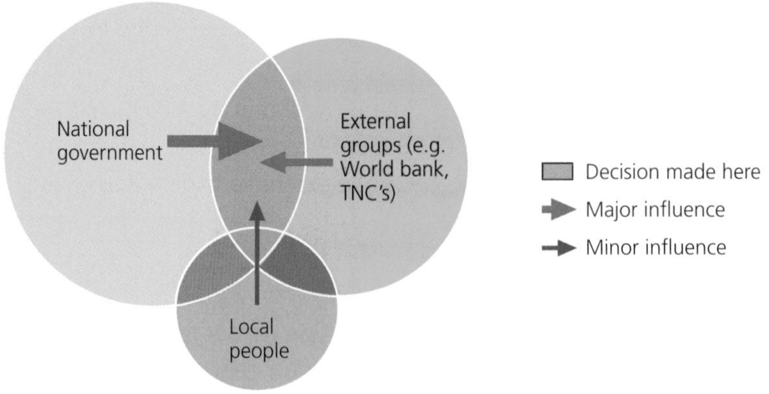

*Figure 10: Top-down decision-making*

This approach was adopted by Brazil, which has made considerable progress since the 1960s. It is one of the 'BRICS' countries (Brazil, Russia, India, China and South Africa) which have been recognised as having made significant economic improvements.

---

### Activity 7

(a) Why would a developing country wish to get rid of disparities within the country?

(b) Name three ways in which a country may reduce the disparity between the urban core and rural periphery.

(c) Why may both top-down and bottom-up schemes be needed to bring about long-term economic development in a developing country?

### Activity 8

Study Figure 10.

(a) Describe the roles of different groups of people in the top-down decision-making process.

(b) For a country, how could debt be a result of this top-down decision-making process?

### Activity 9

Compare Figure 10 and Figure 14 (on page 200). What are the main differences between the decision-making processes for a top-down and a bottom-up approach?

### Common problems with large schemes

- Getting the country into debt, because of the large amounts of money borrowed from the World Bank or International Monetary Fund to pay for the expensive schemes.

- Conditions may be attached to the loans, leading to some external control over the economy or other aspects of development (such as spending on health and education).

- Using machinery and technology, rather than providing unskilled jobs for local people.

- Being energy-intensive and expensive to operate after construction.

- Relying on external links and technology rather than internal links and **appropriate technology**.

- Becoming 'growth poles' which take resources away from peripheral areas.

Many people and organisations believe that development schemes that use intermediate technology are more appropriate and more beneficial to peripheral areas of a country. This is sometimes called a '**bottom-up**' approach to development, because local people are fully involved in the process and decision-making. Many contemporary aid programmes use such an approach, raising the level of technology for local people so that their lives get better.

Dams are common large schemes in developing countries, e.g. the High Aswan Dam on the Nile in Egypt, Akosombo in Ghana on the Volta river, or the Three Gorges Dam on the Yangtze in China. This is because as a country begins to develop economically, energy is needed for businesses, industry and housing. A developing country cannot afford to import expensive fossil fuels, and will use cheaper alternatives. Those countries with major rivers have the opportunity to build hydro-electric power (HEP) schemes, in which the force of the water turns turbines in a dam built across the river. As the turbines turn they generate electricity. This is a cheap source of energy because the water is free, so, although the initial cost of building a dam can be high, the electricity costs can be kept low for businesses and industries (so helping them to make greater profits and expand) and low for people (so helping them improve the quality of their lives).

Brazil has several major rivers, including the largest in the world, the Amazon. Several of the Amazon's tributaries, as well as other rivers, provide HEP opportunities. The Santo Antonio dam is the first top-down HEP scheme for fourteen years – the last one was the Xingo Dam on the São Francisco river, in 1994.

### Activity 10

In which ways is water an important resource for a low-income country?

### Decision-making skills

Working in groups, choose one large scheme – an airport or an industrial project, perhaps – and use Google to find out a bit more information on it. Then, draw a diagram to show the environmental impacts of your chosen scheme, e.g. impact on land, air and water.

## Activity 11

Study the tables on pages 197 and 198.

Explain how the Santo Antonio Dam, a top-down project, has met (a) the needs of Brazil, and (b) the needs of local people.

## Skills Builder 3

Using Figure 11 and an atlas, describe the location and characteristics of the Madeira scheme.

## Decision-making skills

Write a report/PowerPoint in which you investigate the environmental, social and economic impacts of the Madeira scheme. Do you think that the Brazilian government should have said yes? Why/why not?

## Top-down development and bottom-up development

The priority of top-down schemes is to help the whole country, especially urban cores. But without careful management, the impacts on small areas of the rural periphery can be severe. An alternative approach is bottom-up schemes which use a smaller, more appropriate technology, and maximise benefits for the rural periphery. But such schemes do not consider the urban cores directly. This is the 'development dilemma' that faces developing countries – which type of development is better for them?

## The Santo Antonio dam

The Santo Antonio dam is part of the Madeira River Project – the largest project in the Amazon region's history (Figure 11). It is one of the 'Integrated Regional Infrastructure for South America' projects and includes four dams, a navigation channel, three highways (e.g. BR364) and electricity transmission lines. The Madeira River basin covers over 1.5 million km² in three countries – Peru, Bolivia and Brazil. The main river is formed by the confluence of the Guaporé, Mamoré and Beni tributaries, creating an average discharge of 23,000 m³ per second. The Santo Antonio dam is 5 km upstream from Pôrto Velho, the capital of Rondônia. The dam is designed to produce 3,150 MW of electricity and is costing $5.3 billion to build. It is being constructed by Consorcio Madeira Energetica, which includes Brazilian, Spanish and Portuguese banks, the Brazilian construction company Odebrecht, and Furnas (Brazil's electricity company).

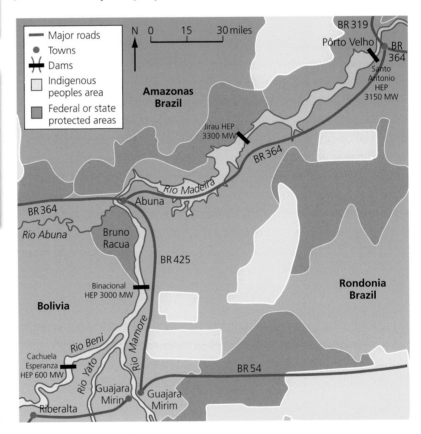

Figure 11: The Madeira River scheme

## Impacts of the Santo Antonio Dam on different groups of people (positive +, or negative −)

### Residents of Porto Velho

+ $30 million given to improve the sewerage system in Porto Velho.
+ Residents closely monitored for mercury poisoning.
+ Expansion of agriculture and agribusiness in the area provides new jobs.
− Too many immigrants (up to 100,000) may enter the area, causing a strain on services and living space.
− An increase in malaria is likely, because of the increase in water area.

### Local farmers and fishermen

+ Social support programmes put in place to help rural communities.
+ Health, education, leisure, safety and sanitation infrastructure improved.
+ Fish channels in the dam created to allow fish to migrate.
− An estimated 3,000 people forced to leave their homes, and some towns have disappeared (e.g. Mutumparana).
− The riverbank traditional way of life may be lost (e.g. subsistence fishing) affecting 5,000 fishermen.
− Commercial fishing (worth $1 billion) is at risk from the dam, because the preferred fish catch will be disturbed. 2,400 fishermen may lose their jobs.
− Irrigation water will be lost and fisheries affected downstream of the dam.

### Indigenous tribes

+ Two Indian reservations paid for by the consortium.
− Indigenous peoples' lands are at risk from flooding and erosion.
− Migrant workers may upset the delicate relationship with indigenous peoples (e.g. the Oro Bom), including contact with remote tribes in voluntary isolation (e.g. Katawixi).

### Poor people living in South-East Brazil

+ 20,000 jobs created and 100,000 people attracted to the region.
+ An education and training centre for immigrants and job seekers established.

### Brazilian government

+ A waterway for barges will be created, by-passing rapids on the river, making it easier to transport soy, timber and minerals from the region.
+ Better infrastructure (roads and waterways) in the centre of South America (Brazil–Peru–Bolivia hub) provided, so products can be taken to the Pacific or the Atlantic.
+ Essential energy supplies for the development of Brazil and Rondonia created (e.g. the Madeira scheme will supply 8% of Brazil's electricity).
− Political conflict between Brazil and Bolivia has arisen because part of Bolivia will be flooded by dams in Brazil. For example, a Bolivian wildlife reserve (Bruno Racua) will be affected.
− The Madeira scheme is very costly ($22 billion).

Figure 12: Protests against building the Santo Antonio dam

**Build Better Answers**

### EXAM-STYLE QUESTION

**For a named example of a top-down development strategy that you have studied, explain how successful it has been in meeting the needs of the developing country. (6 marks)**

■ **Basic answers** (1–2 marks)
Do not include the name of the scheme and only make general points, such as 'created wealth and jobs in Brazil'.

● **Good answers** (3–4 marks)
Name an appropriate scheme and country, and make detailed points, such as specifying the number of jobs created, the amount of electricity created by a dam and how this was used by industries in a specified place.

▲ **Excellent answers** (5–6 marks)
Are based on a wider range of information with linkage between the points and explanation.

## Results Plus
**Build Better Answers**

### EXAM-STYLE QUESTION

**For a named example of a top-down development strategy that you have studied, explain how successful it has been in meeting the needs of the local people. (6 marks)**

■ **Basic answers** (1–2 marks)
Do not include the name of the scheme and only make general points, such as 'created wealth and jobs in Brazil'.

● **Good answers** (3–4 marks)
Name an appropriate scheme and country, and make some detailed points, such as specifying the creation of jobs in businesses and farming, and improvements to services and infrastructure.

▲ **Excellent answers** (5–6 marks)
Are based on a wider range of information with linkage between the points and explanation.

## Activity 12

(a) What are the benefits of involving local people and communities in the decision-making process?

(b) Are there any ways in which this scheme is helping the development of Peru as a whole?

### Environmental and conservation groups

+ HEP is a renewable form of energy and avoids the need for oil or nuclear power.
+ The area flooded is minimised by using a 'run of the river' technology (using the natural flow rate of the river rather than the release of water from a reservoir).
+ Two forest reserves will be paid for by the consortium, with $6 million made available for environmental police.
− Investor banks may be in breach of the 'Equator Principles' (which say that only socially responsible and sound environmental management schemes should be supported).
− Soya expansion in the twenty-first century is a major cause of deforestation in Brazil, and the dam is opening up even more areas. 80,000 km² could end up being converted to farmland, in an area rich in biodiversity.
− **The Environment Impact Assessment** was based on insufficient data and so was superficial and lacked thoroughness. There were no measures to help solve a third of the impacts listed.
− River food webs will be affected. Some fish will gain from the changes to habitats (e.g. piranha-caju) but others will not (e.g. the dourada). River-bottom feeders (e.g. bagres) and scaly fish (e.g. pescadas) are affected, as well as freshwater shrimp, which form the base of many food chains.

### Businesses in South-East Brazil

+ The scheme will produce the cheapest electricity in Brazil.
− There is too much reliance on HEP in Brazil already (76% of electricity).

## How might countries develop more sustainably in the future?

### Bottom-up schemes and their effects at a local scale

In contrast to large HEP schemes such as Santo Antonio, some low-income countries have decided that their remote rural peripheries would benefit from 'micro-hydro' schemes – a 'bottom-up' approach. **Micro-hydro schemes** are defined as those with a generating capacity under 100 KW. Most use a 'run of the river' method like the Santo Antonio dam, but on a much smaller scale. Water is diverted from a stream to a high point on the valley side and then dropped through a pipe to turn a turbine (Figure 13). Much of the equipment can be made in developing countries.

Micro-hydro projects are considered to be appropriate to the local skills level, and they help spread technology to the rural periphery. They are low-cost and involve local people in all stages of the scheme, from planning to daily operation.

### Micro-hydro schemes in Peru

On the eastern side of the Andes are some of the poorest areas of Peru – 44% of the population live on less than $2 a day. This rural periphery has steep slopes, poor road networks and a scattered agrarian population. The low population density makes it uneconomic to provide grid electricity. However, rainfall is high and there are plenty of streams and rivers.

A charity called 'Practical Action' has ten full-time staff working on micro-hydro in Peru and it has helped install nearly fifty schemes, providing electricity to 30,000 people. Part of the costs are paid for by the local community, who are also responsible for management and maintenance. There is also a government 10-year plan to provide 58 more micro-hydro schemes, with possible funding from Japan.

One community to benefit is Chambamontera, with 60 families. The village is in the far north of Peru, near the border with Ecuador. It is more than two hours' drive from the nearest town and 1,700 metres above sea level. The farming in the area is well organised and some services are available, but electricity is needed to improve the social and economic environment. This micro-hydro scheme started in September 2008 and when it is finished, it will generate 15 KW from a water flow of 0.035m³/s. Meetings were held with villagers to share information and an agreement was drawn up, which included the village commitment of labour, materials and some money. From this a legal community group was formed to make decisions and, if necessary, to obtain loans. A socio-economic survey of the community was completed and this helped to establish the needs of the local people and confirm their ability to maintain the electricity service. The total cost of the scheme is £34,000 (55% coming from the Matthiesen/OrbisPictus Foundation, 22% from other donations, 17% from Practical Action and 6% from the village community).

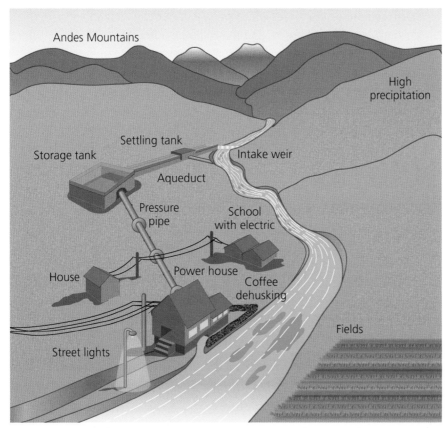

*Figure 13: Peruvian valley with features of micro-hydro*

## ResultsPlus
### Build Better Answers

**EXAM-STYLE QUESTION**

**Using named examples compare the main features of a top-down project and a bottom-up project. (6 marks)**

■ **Basic answers** (1–2 marks)
Do not name examples and only offer one or two comparison points, or they make a series of statements with no comparison.

● **Good answers** (3–4 marks)
Names examples and structures the answer with linked comparison points and some factual detail.

▲ **Excellent answers** (5–6 marks)
Are detailed throughout, with several comparison points and facts, and with appropriate use of geographical terminology.

## Skills Builder 4

Study Figure 13. Use evidence from the sketch to describe the impacts that this micro-hydro scheme has had on the river and the valley.

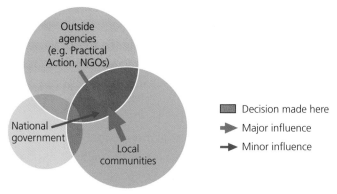

*Figure 14: Influences on decision making in a bottom-up approach*

## Impacts of the Chambamontera micro-hydro project on different groups of people (positive +, or negative –)

### Villagers

+ Reliable electricity supply provides refrigeration (e.g. milk coolers), lighting, entertainment, communication, computer access for families, and street lighting.
+ Health care is improved as electricity allows the storage of medicines (e.g. vaccines) in refrigerated conditions.
+ Electricity for schools (Figure 15) means that they can use equipment (e.g photocopiers, audio-visual).
+ Electricity means that kerosene lamps can be replaced. The lamps produced unhealthy fumes which people (especially women and children) were breathing in, and they created a fire risk.
+ Electricity powers machinery for small businesses (e.g. carpentry, battery charging).
+ Piped water can drive machinery directly e.g. for coffee de-husking and processing (Figure 13).
– Poor people still have to pay for the electricity, which is metered.
– Some villages have doubled in size, causing population pressures.
– Initial capital cost is high for a poor village, e.g. £500 per household, so loans may be needed and this can lead to debt.

### Peruvian government

+ Young adults and teachers are encouraged to stay or return to the village from large urban areas because now there are opportunities, so relieving the problems of rapid urban growth.
+ Cheaper electricity than a large HEP scheme (only $0.14 per KW) and with low running costs.
– Some specialised equipment has to be imported (e.g. load controllers from Canada) which creates dependency and expense.

### Environmental and conservation groups

+ Avoids flooding a large area of land which would destroy ecosystems and habitats.
+ Avoids the need for villagers to burn wood from local trees, so reducing deforestation and soil erosion.
– A small storage dam is required, along with construction in the valley, which alters the flow of the river and spoils the scenery (Figure 13).

## Activity 13

Study Figure 14.

(a) What is meant by a bottom-up approach to development?

(b) Describe the roles of different groups of people in the bottom-up decision-making process.

(c) Why must outside agencies be careful to ensure that local people do not become dependent on their help?

## Activity 14

1. Compare the characteristics of the Santo Antonio dam (a top-down project) with the Chambamontera micro-hydro project (a bottom-up scheme).

2. Using the sustainability criteria shown in Figure 3, on page 187, assess the sustainability of the Santo Antonio dam project, and the Chambamontera scheme.

*'We are so happy to have light, we feel very thankful and proud. Our neighbouring communities envy us. At first they thought our work was in vain but now they are asking us what they can do to obtain light.'*

**Alfredo Sarango, Pampa Verde, Peru**

(a) Make a copy of Figure 16 and add the following four labels in the correct boxes: 'Little investment in industry or services', 'Low level of consumer demand for services', 'Low economic output and no community amenities', 'No savings or investments by people'.

(b) Make a list of the ways in which the cycle can be broken so that the social and economic environment can be improved.

(c) How would a top-down scheme and a bottom-up scheme help to break this cycle?

*Figure 15: Electricity for a Peruvian school from micro-hydro*

## Bottom-up schemes and sustainability

Many rural peripheries of developing countries are stuck in a **poverty cycle**, which is extremely difficult for them to break out of. Achieving sustainable development (see Figure 3, on page 187) therefore becomes a problem for these areas. Large schemes that focus on national needs (often the economic needs) may overlook the rural poor and do little to improve the lives of people living a long way from the core. Small-scale schemes, on the other hand, may be more appropriate – and more sustainable.

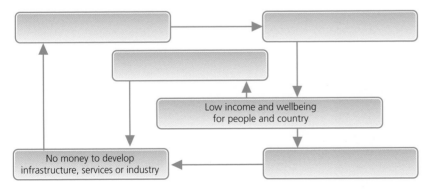

*Figure 16: The Poverty Cycle*

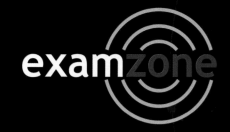
# examzone
## Know Zone
## Development dilemmas

Areas develop at different rates, leading to disparities within and between countries. There is a major dilemma as to how development is defined and whether bottom-up methods are better than top-down practices.

## You should know...

- [ ] How to define development in contrasting ways
- [ ] How to explain why societies develop over time
- [ ] That the rate that areas develop varies spatially both in countries and in general
- [ ] How this uneven development leads to disparities with a developed core region and a more rural backward periphery in many countries and regions
- [ ] How, at a world scale, disparities can lead to different rates of development
- [ ] How there are different ways development can take place
- [ ] What the features of bottom-up developments are and how they compare to top down strategies
- [ ] What the advantages of sustainable development are
- [ ] How appropriate technology can play a vital role in development
- [ ] What the barriers are to the development of Tanzania
- [ ] What the impacts of the Santo Antonio dam are on different groups of people

## Key terms

Appropriate technology
Bottom-up approach
BRICS
Core region
Dependency theory
Disparity
Environment Impact Assessment
Gross Domestic Product (GDP)
Gross National Income (GNI)

Happy Planet Index (HPI)
Hydro-electric power
Human Development Index
Micro-hydro schemes
Millennium Development Goals (MDGs)
Periphery
Poverty
Poverty cycle
Rostow model
Top-down approach

### Which key terms match the following definitions?

**A** The use of fast flowing water to turn turbines which produce electricity

**B** The outer limits or edge of an area, often remote or isolated from its core

**C** A great difference – e.g. between parts of a country in terms of wealth

**D** Equipment that the local community is able to use relatively easily and without much cost

**E** Development projects that originate in local communities rather than in central government or external agencies

**F** A method of evaluating the effect of plans and policies on the environment

**G** A set of processes that maintain a group or society in poverty

To check your answers, look at the glossary on page 321.

**Foundation Question**: For a Sub-Saharan African country that you have studied, describe the barriers that are holding back its development. (4 marks)

| Student answer (achieving 2 marks) | Feedback comments | Build a better answer (achieving 4 marks) |
|---|---|---|
| Tanzana is a developing cointrie in east Africa | • *Tanzana is...* A correct example of a located Sub-Saharan country (1 mark). | Tanzania is a developing country in East Africa. |
| This countrie is very poor | • *is very poor...* is correct but the point is not developed enough. A fact was needed. | The country is very poor; in 2005 28% of the population lived on less than $1 a day. |
| Many children are involved with working on farms and feching and carrieing water. | • *Many children....* is correct as they do not attend school regularly to learn how to improve their lives (1 mark). | Many children are involved with a subsistence way of life rather than attending school. |
| Tanzana is a colonie of the UK. | • *Tanzana is...* is not correct. Tanzania became independent in 1961. | A recent barrier to Tanzania's development has been the World Bank refusing to give a loan. |

**Overall comment:** In this question explanation is not required and the candidate does well to stick to description. A correct country is chosen and a couple of good statements are made. However, further detail was required, and a big improvement in spelling and grammar was needed.

- - - - - - - - - - - - - - - - - - - - - - - - - - - - - - - - - - - - - - - - - - - - - - - - - - -

**Higher Question:** Using a named example, examine the impacts of a large top-down development project on different groups of people living near the site. (6 marks)

| Student answer (achieving 3 marks) | Feedback comments | Build a better answer (achieving 6 marks) |
|---|---|---|
| The Santo Antonio dam in Brazil has affected people in Porto Velho and local farmers and fishermen. | • *The Santo Antonio dam...* names and locates an example of a top-down project. It also names affected local groups of people. (1 mark) | The Santo Antonio Dam, in Brazil, has affected people in Porto Velho, and local farmers and fishermen. |
| The dam has attracted many migrants, which has strained local services, and put pressures on housing. | • *The dam has...* A negative impact on local people as there are more people trying to use the same services or living space. (1 mark) | The dam has attracted up to 100,000 migrants, so more people are trying to use services or find somewhere to live. |
| The creation of a reservoir means that there is a larger breeding area for mosquitoes and so malaria may spread. | • *The creation of...* Another correct negative impact, as the health of local people may suffer. (1 mark) | An increase in malaria amongst local people is likely because the reservoir provides a breeding area for mosquitoes. |
| 24,000 fishermen will not be able to fish anymore. | • *24,000...* An incorrect fact with a vague statement (perhaps rushed). | Social programmes were set up for local people (e.g. health and education), and $30m was spent on improving Porto Velho's sewerage system. |

**Overall comment:** Explanation was required and is given in all but the last statement, and it is a well-structured answer, but it is unbalanced as there are no positive impacts. The SPG is very good.

# Chapter 13 The changing economy of the UK

204

## Objectives

- Identify changes in the primary and secondary sectors.

- Identify which industries have declined and why and which have grown and why.

- Examine changes in the tertiary and quaternary sectors.

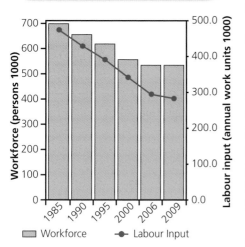

Figure 1: UK Agricultural Labour 1985–2009

**Employment in forestry and primary wood processing (thousands)**

| Year | Forestry | Sawmilling | Panels | Pulp & Paper | Total |
|------|----------|------------|--------|--------------|-------|
| 2006 | 11 | 13 | 5 | 14 | 43 |
| 2007 | 13 | 12 | 5 | 12 | 42 |
| 2008 | 12 | 11 | 5 | 13 | 41 |
| 2009 | 14 | 9 | 4 | 13 | 40 |
| 2010 | 14 | 10 | 6 | 13 | 43 |
| 2011 | 9 | 6 | 3 | 11 | 29 |

## How and why is the economy changing?

### Changes in the UK industrial structure: The primary sector

#### Agriculture

UK agriculture has changed dramatically in the past 50 years, and this has meant that farms and farmers have changed even more. The changes in UK agriculture can be summarised as:

- Increased mechanisation of farming with the introduction of newer and bigger machines such as combine harvesters and tractors

- The new machines needed bigger fields in which to operate successfully, so field boundaries were removed to make bigger fields

- Increased use of chemicals such as fertilisers to increase yields

- Increase in farm sizes to make them more efficient.

The mechanisation of agriculture, the increased use of chemicals, and the development of larger and more specialist farms have all contributed to the decline in employment in agriculture. Figure 1 shows the decline in UK farm labour between 1985 and 2009. The decline in labour input has been higher over the same period showing that there has been a move from full-time to part-time workers. The percentage of part-time workers increased from 21% in 1984 to 43% in 2004. In 2005 the number of part-time workers exceeded the number of full-time workers for the first time.

So within the primary sector, agriculture continues to decline in relative importance because of the decline in employment, from 690,000 in 1990, to 499,000 in 2009.

#### Forestry

The woodland area in the United Kingdom in 2012 was 3.1 million hectares, compared to 2.8 million hectares in 1999. Thirteen thousand hectares of new woodland were created in the UK in 2011/12 and 10.3 million green tonnes of UK roundwood (softwood and hardwood) were delivered to primary wood processors and others in 2011, representing a 3% increase from the previous year. Clearly forestry is not typical of **primary industry** in the UK because it is growing in production. But it is declining in employment – from 43,000 in 2006 to 29,000 in 2011. However these statistics conceal other changes:

- Increasing mechanisation of tree felling and sawmilling

- Increased use of larger, more efficient machinery, needing fewer workers

- Growth in importance of recreation in forest areas, so environmental issues are now as important as production and output.

### Mining and quarrying

The UK quarrying industry extracts a range of minerals from rocks, including sandstone, limestone, chalk, and coal as well as sand and gravel. Seven large companies account for 85% of the products from quarrying. The quarrying industry is in decline in terms of employment and numbers of quarries. In 2000 there were over 1700 quarries, employing 15,300 people. By 2011 these figures were 1,000 quarries employing only 11,200 people. The reasons for the decline are:

● Exhaustion of the mineral deposits in some areas

● Growing unpopularity of mining and quarrying with people who live near developments. Local protests can lead to long delays whilst environmental surveys and other impact surveys are undertaken, so developers may abandon a project.

● Bureaucracy involved in getting permission to mine or quarry, especially in some rural areas of the UK

● Increasing costs for preventing environmental damage and contributions to habitat restoration

● Increased mechanisation, with use of newer and bigger machinery such as trucks, so fewer people are needed

● It is often cheaper – and relatively easy – to import the products of mining and quarrying from other countries.

As a result, the decline of mining and quarrying in terms of employment and output also helps to explain the relative decline in importance of the UK primary sector.

### Fishing

As with other parts of the UK primary sector, fishing has declined – from 15,400 employees in 2001 to 12,400 in 2011. Similarly, the fishing fleet contracted from 7,721 vessels in 2001 to 6,444 in 2011. This decline is due to:

● Fishing quotas introduced by the European Union which control the number of fish that can be caught. The purpose of the quotas is to conserve fish stocks.

● Selling of fishing quotas to foreign boats/owners

● The decline in the numbers of fish as a result of **overfishing** in the past, so some vessels have been scrapped and jobs lost

● The replacement of older fishing vessels with newer, bigger and more modern vessels which often employ fewer people than the older vessels.

So, in summary, all areas of the primary sector have seen a decline in relative importance, and with it a decline in employment and often a decline in production. The main reasons for these changes are:

● Mechanisation (Figure 2) and changes in production

● Competition from other countries as a result of globalisation

● Failure to modernise or to invest sufficiently

*Figure 2: Modern trawlers are highly mechanised*

- Exhaustion of seams of minerals which it is no longer economic to exploit

- Over-use of a resource, such as overfishing.

## Changes in the UK industrial structure: The secondary sector

The UK was the first country to experience the Industrial Revolution, from 1750 onwards. The invention of the steam engine, which used coal to drive new machinery, led to the development of new factories in which the steam power and machinery could be housed. At its peak in 1914, over half the UK workforce was employed in manufacturing industry – the secondary sector. At this time the main industries were steelmaking, engineering, shipbuilding, metal smelting, and cotton and woollen textiles. The focus of these industries was the coalfield regions of Clydeside, South Wales, north-east England, Lancashire, Yorkshire and the West Midlands.

The UK was also, however, the first country to experience **deindustrialisation** – a decline in the relative importance of manufacturing industry, which led to a sharp fall in industry's share of the workforce, from 46.9% in 1964 to 15.6% in 2011. This fall began in the 1950s but increased in the 1980s and 1990s (see table below). The main reasons for the decline of UK heavy industries were:

- The impact of **globalisation**, particularly the exposure of UK manufacturing industries to competition from abroad, where many countries had much cheaper labour costs. So factories in India, Pakistan, Brazil and China were able to produce textiles, ships and even steel cheaper than the UK.

- Poor investment throughout the period from 1960 to 2000. Manufacturers in the UK did not invest enough in new machinery, new technologies and new products, so they were not able to develop new products quickly or cheaply enough. They then lost out to competition from abroad.

- The poor international image of some UK industries. These firms often had poor labour relations and so were in the news for strikes, or else they failed to deliver goods on time and on budget.

- Failure to see the need to start producing higher quality, more expensive textile goods rather than mass producing cheap goods.

- Too many small firms which were not able to afford the modernisation necessary in industries like textiles.

### Case study: The decline of the UK shipbuilding industry

In the 1950s the UK was the world's most important shipbuilding nation. British yards produced one-third of all the world's shipping. Since 1956 however, Japan and South Korea – and many other countries – have overtaken the UK, which now ranks a lowly 25th as a shipbuilding nation. The decline of shipbuilding in the UK is shown very clearly in the tables. The reasons for this dramatic decline are:

- Competition from other countries such as South Korea

- Changes in government support policies

- Lack of modernisation and investment in new machinery and techniques, such as covered yards in UK shipyards

**Changing industrial structure in the UK, 1964–2011 (% of total employment)**

|  | 1964 | 1983 | 1990 | 2005 | 2011 |
|---|---|---|---|---|---|
| Primary | 5.1 | 3.0 | 2.1 | 1.1 | 1.0 |
| Secondary | 46.9 | 35.4 | 26.6 | 16.6 | 15.6 |
| Tertiary | 47.8 | 61.4 | 71.3 | 82.2 | 83.3 |

Source: ONS (2012) Labour Market Statistics, September UK National Accounts

- World demand for ships has fallen, with fewer but bigger and more specialised ships being demanded, and more competition from airlines and railways

- Failure to deliver on time

- Ships have become bigger, so shipyards have to be very big to accommodate them – and many UK yards were too small

- Lack of forward planning.

### Impacts of the decline

As a result of declining employment in the secondary sector, by the 1970s there were massive job losses in places like central Scotland, South Wales and north-east England where there were few alternative jobs. So there were severe economic problems in these areas with rising unemployment, together with poverty and deprivation. This could result in a landscape of closed factories (Figure 3), abandoned mines and houses, and widespread pollution of rivers, canals and lakes.

### Growth in the secondary sector – the case of the UK vehicle industry

Not all UK secondary industries suffered a decline in employment or production. The UK vehicle industry, for example, is still successful, although it has changed considerably. In 2010 about 175,000 people were employed directly in the UK automotive manufacturing industry. The firms (most of which are foreign-owned) produced 1.5 million passenger vehicles and 203,000 commercial vehicles in 2010. The industry had a turnover of £54 million and generated £26 million of exports.

In the 1970s the UK vehicle manufacturing industry was one of the most profitable sectors of the economy. However the period after 1970 saw major changes in the industry:

- Imported cars from Europe and Japan began to take an increasing share of the market.

- UK firms were not able to compete on price and quality, and imports increased from 24% of all new cars sold in the UK in 1972 to 87% of all new cars sold in 1980.

- Many firms were taken over by foreign companies, such as Renault and Peugeot.

- Governments gave financial support to firms such as the Rover Group at Longbridge (Birmingham) to help them to modernise and stay in production, but in the end the losses were too great and the company went into liquidation.

- UK factories were overmanned and slow to introduce new technology.

- Factories closed (e.g. Linwood in Scotland and Speke near Liverpool).

**Percentage of world ships built in UK**

| | |
|---|---|
| 1913 | 59 |
| 1950 | 38 |
| 1960 | 16 |
| 1970 | 6 |
| 1980 | 3 |
| 2011 | 0.3 |

**Numbers employed in UK shipyards**

| | |
|---|---|
| 1980 | 78,000 |
| 1985 | 45,700 |
| 1990 | 12,760 |
| 2000 | 3,548 |
| 2011 | 896 |

**UK shipbuilding output**

| | Gross weight (tonnes) |
|---|---|
| 1970 | 1,300,000 |
| 1975 | 987,000 |
| 1980 | 850,000 |
| 1990 | 210,000 |
| 2000 | 98,000 |
| 2010 | 16,000 |

## Activity 1

(a) Using the data in the tables, draw line graphs of the declining employment and output of UK shipbuilding.

(b) Draw a bar graph of the UK's declining share of world shipbuilding.

*Figure 3: Derelict factories and houses in Stoke on Trent*

By 2000, vehicle manufacture in the UK was dominated by foreign-owned firms such as Ford, Nissan, Toyota, and General Motors. The UK was still an attractive place in which to manufacture vehicles, as can be seen when Nissan opened their new factory in Sunderland in 1986 and when companies like BMW chose to build the new Mini in Oxford and Tata Motors expanded Jaguar Land Rover in 2010.

The expansion in vehicle production is linked with the attractions of the UK to foreign manufacturers. These include: financial incentives, such as tax breaks to locate in the UK; a cheap, skilled labour force often in areas of high unemployment; and large flat sites available close to motorways and ports for the movement of parts and raw materials. In addition, because the UK is a member of the European Union (EU), cars manufactured here can be sold throughout the EU without import duties – and this is a big attraction to foreign manufacturers. The output of vehicles, however, varies with the changes in demand, often linked to economic factors such as recession.

*Figure 4: Robots operating in the Land Rover factory*

The vehicle manufacturers have been more successful than many secondary sector industries because:

◉ They invested in new technology and new models

◉ They produced high-value products which required a lot of skill, and which could compete with products of firms in other countries

◉ They established reputations for reliability and quality.

As with many other industries, the automotive manufacturing industry employs many thousands more people in firms which supply components to the main assembly line companies. The success of the manufacturers in introducing new technologies and new models has therefore had a big impact on the survival and growth of the component manufacturers, thereby creating a circle of self-supporting growth. The main danger with this situation is that if manufacturers fail to remain competitive in a global market the numbers of people who may lose their jobs is much more than simply those employed in the assembly factory.

## Changes in the tertiary and quaternary sectors

◉ Tertiary industries grew rapidly from the 1970s because people needed more services (such as education, health, retailing, business services).

◉ At the same time quaternary industries also began to grow, but they have grown particularly in the last 10 years. This is because firms are investing in larger and larger research departments to develop new products more quickly, such as mobile phones.

◉ There has been a big growth in retailing, in the financial services sector and in business services generally (see table on page 209). One of the fastest growing areas of quaternary industry is research related to information technology, particularly with the rapid spread of computers, mobile phones and other IT-related gadgets. This growth is related to changes in technology and fashions, and it shows how industries adapt to meet changing demands.

The growth of tertiary and quaternary services in the UK is due to:

- The impact of globalisation on the economy which has meant that UK companies have focused on those things they can do best, such as the provision of financial services. For example, London is still one of the world's three main financial centres.

- The development of new technologies which have allowed firms a much wider choice of where to locate the different parts of their firm.

- Rising living standards which have created a demand for more and better services.

- Increased competition within sectors such as retailing, where large supermarkets such as ASDA (owned by Wal-mart in the USA) Tesco, Waitrose and Sainsbury's now dominate trade in the UK

## Case study: Financial services in the UK

The UK financial services has been the fastest growing area of tertiary services since the 'Big Bang' of 1986 when the UK government deregulated the financial markets. This meant that banks and building societies began to offer a wider range of services to their customers, from basic accounts, services such as bank cards, credit cards, loans, insurance, investment services, accountancy, and financial advice. Many people have established themselves as financial advisers, whilst the larger banks have bought a wide range of other companies to broaden the services they offer to the public. This would include the purchase of estate agents, as well as solicitors. All of this activity led to a financial boom in the 1990s and early 2000s and with it massive new construction projects of headquarters for banks and other insurance companies.

Figure 5: Share dealing in London is big business

Employment similarly expanded up to 2008. The financial crisis which began in 2008 began a period of job losses for the financial services sector as some banks had to be supported financially on a large scale by the government to prevent them collapsing. In the period 2008–11, several thousand jobs were lost in the UK financial services sector.

**Changes in employment in some UK tertiary sectors, 1981–2011**

|  | 1981 Employment (thousands) | 2011 Employment (thousands) | Change 1981–2011 (thousands) |
|---|---|---|---|
| Hotels and catering | 4,172 | 6,558 | +2,386 |
| Transport | 987 | 1,429 | +442 |
| Communications | 438 | 1,109 | +671 |
| Banking, insurance and finance | 1,738 | 6,241 | +4,503 |
| Education and health | 4,508 | 8,382 | +3,847 |

## Activity 2

Study the table on the left. Between 1981 and 2011:

(a) Which sector increased its labour force most?

(b) Which sector increased its labour force least?

(c) What were the three main sectors of growth?

## Changes in the structure of the UK workforce in the last 50 years

We can classify employment in several different ways:

- Full-time/part-time
- Average wage
- Temporary/permanent
- Male/female
- Executive/skilled/semi-skilled/unskilled.

### How and why the balance of types of employment has changed

There are now many more part-time workers than in the past because firms have found it cheaper to employ people part-time. In 2011 there were 21.2 million people in full-time employment in the UK (13.6 million men and 7.6 million women) and 8 million in part-time employment (2 million men and 6 million women). More people now want to have the flexibility to work part-time. This often applies to women who have to balance work with looking after a family.

But for many people, part-time work is the only work they can find. The Trade Union Congress (TUC) said almost 600,000 men were working part-time in May 2012 while looking for full-time positions, compared to 293,000 at the end of 2007. The official figures for May 2012 show that 1.4 million workers and self-employed people were working part-time because they could not find full-time employment. This was the highest figure since records began in 1992.

The numbers of women in different parts of the workforce has varied in the period 1970–2011, but the main increase has been in the professional category, as the table on the left shows.

From 1971 to 2011 the economic activity rate among women rose from 59% to 74% whilst in the same period the male economic activity rate fell from 95% to 75%. This was a time of the decline of male-dominated heavy industries, such as steelmaking and shipbuilding, and the rise of service industries, such as health, social services and education.

- In 2000–10 there were many women in UK industries, but the recession which started in 2008 led to a decline in the numbers of women employed. This is because firms have to save money at a time of economic recession, so they greatly reduce (or lose completely) the part-time labour force. The majority of the part-time workers are usually women – often those who are balancing a career with looking after a family.

- Average wages in the UK vary from job to job and from place to place. Following the recession of 2008 wages generally grew slowly. Since the 1970s wage rates in the UK have tended to stay flat relative to the cost of living. This is a result of the increasing importance of globalisation in UK industry and the desire for firms to retain their competitive position by keeping wage costs down.

---

**Watch Out!**

Questions about tertiary and quaternary industries will ask why they grew as well as which ones grew. So make sure you include the reasons for the changes as well as naming the industries which have changed.

---

### Trends in employment, by gender (%)

|  | 1970 | | 2011 | |
| --- | --- | --- | --- | --- |
|  | Men | Women | Men | Women |
| Professional | 24.9 | 18.9 | 31.5 | 35.9 |
| Clerical | 7.6 | 34.5 | 5.5 | 23.4 |
| Sales | 6.8 | 7.4 | 11.3 | 13.0 |
| Manufacturing | 48.1 | 17.9 | 37.9 | 9.2 |
| Service | 8.2 | 20.5 | 9.9 | 17.4 |
| Agricultural | 4.5 | 0.8 | 3.8 | 1.1 |
| Total | 100.0 | 100.0 | 100.0 | 100.0 |

---

**Top Tip**

Most descriptions of industrial decline in the UK talk about which industries declined but often fail to mention that some industries have survived and even grown – such as the manufacture of cars, electrical and optical equipment. This is a good point to include in an exam answer.

• There are now many more temporary jobs than permanent ones in the UK. This is a result of the rationalisation by companies seeking to maintain their competitive edge in a global market by moving to more temporary contracts, which gives them the flexibility to respond rapidly to changing economic circumstances.

## Two contrasting UK regions: North East England and South East England

### North East England

The industrial structure in this region used to be dominated by coalmining, iron and steel production, shipbuilding and chemicals. However, these industries have all declined rapidly in the twentieth and twenty-first centuries (see table below) due to foreign competition and high land and labour costs and the exhaustion of coal seams in North East England. For example in 1947 there were 108,000 miners in the region in 127 pits, but by 1994 the last mine was closed and employment fell to just 55. In 1971 manufacturing accounted for 40% of employment in the region, but by 1996 this had fallen to 24%, with a loss of 95,000 jobs.

Despite these massive changes in employment structure the North East is still an important manufacturing area. Part of the reason for the growth of industry in the region was the establishment of the **Enterprise Zone** in the 1980s which gave relief from local business rates, reductions in corporation tax, and 'fast tracking' of the planning process, as well as subsidies for capital spending. These all helped to attract new firms to the area. However there is some disagreement as to how far these were completely new jobs and how far they were just jobs moved from other local areas. One report found that by 1987 some 4,300 companies had been established in the eleven original Enterprise Zones in the UK. These companies employed 63,000 people, but only 13,000 of them were in genuine new jobs. More recently, in 2011, the government announced another new Enterprise Zone in the North East, covering parts of Durham, Sunderland, South Tyneside and Northumberland, so clearly there is still a need for government support for industry in the region.

In terms of current industrial structure the chemicals industry, especially pharmaceuticals, is still important in the region but it now employs fewer people, due to automation and improved technology. The Nissan car factory, opened in 1986 is now an important employer (about 4,000 people) and a key part of the recovery of North East England after the de-industrialisation of the 19070s.The growth of public sector employment has also been very important. By 2012 the public sector employed 263,000 people in the region, accounting for 23% of all employment. Other newer industries include the manufacture of oil and gas platforms for the North Sea fields. Tyneside is becoming a centre for scientific research into stem cells as well as an entertainment focus while Sunderland is becoming a centre for quaternary industry, science and high technology. However unemployment is high (11.6% in 2011) compared to South East England (6.3%).

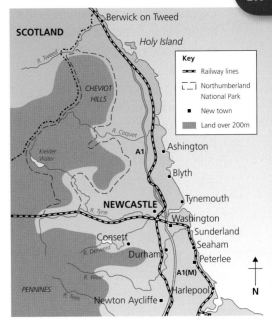

Figure 6: North East England

## Employment change in North East England, 1971–2010

| | 1971 | 1981 | 1996 | 2010 |
|---|---|---|---|---|
| Energy and water | 45,500 | 36,900 | 7,700 | 4,500 |
| Manufacturing | 201,300 | 151,300 | 116,300 | 87,300 |

212

### EXAM-STYLE QUESTION

**Explain the main differences in the industrial structure and workforce of two contrasting UK regions. (8 marks)**

■ **Basic answers** (1–3 marks)
Talk a little about the industrial structure of each region but fail to discuss employment. Do not offer reasons for the points described and there is no comparison between the two selected regions.

● **Good answers** (4–6 marks)
Talk about the industrial structure of each region in some detail and include aspects of the workforce, but this is not strong. Do not compare the two selected regions or offer many reasons for the differences. Clear structure and relevant use of geographical terminology.

▲ **Excellent answers** (7–8 marks)
Give a detailed analysis of both industrial structures and workforce with a clear structure to the answer and explain the similarities and differences between the regions. Well communicated with excellent use of geographical terms.

### South East England

South East England is a centre for service industries such as health, education and transport. In addition there are important oil refineries at Southampton. This is an area of light industry, which includes new industries in the M4 corridor such as electronics and light engineering. Oxford is a centre for car production. There are also many financial and business service industries located in offices in the region. Unemployment is low (6.3% 2011) and prosperity is high in comparison to the North East.

The region is attractive to industry for several reasons:

◉ It is the focus of the UK's motorway and rail networks. 70% of UK freight is carried by roads in South East England, so firms located there can easily assemble raw materials and distribute their finished products.

◉ Four major airports (Gatwick, Heathrow, Stansted and Luton) make for excellent communications with other countries.

◉ Important ports, such as Southampton, allow the movement of heavy, bulky goods, and Tilbury and Southampton are major centres of UK container traffic.

◉ Very large local market of 19 million people, many with high incomes.

◉ Large skilled local labour force.

◉ Close to the power and decision making based in London, where over 60% of UK's thousand biggest companies have their headquarters – and where Parliament and other national bodies are located.

◉ Proximity to the rest of the European Union, with easy access via the Channel Tunnel.

◉ Government action which has encouraged (by grants to move and restrictions on expansion on site) some firms and some offices (in the 1970s) to move away from central London into other towns in the South East, such as Basildon, Newbury and Swindon.

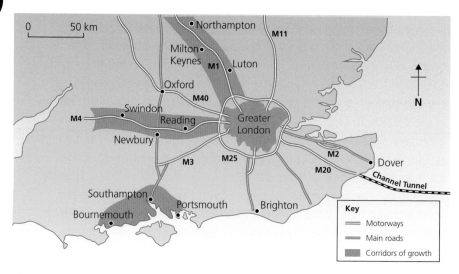

*Figure 7: South East England*

# What is the impact of changing work on people and places?

## Environmental impacts – a case study of the Glasgow area

In 1918 Glasgow was 'the second city of the Empire' (after London) because of the importance of its manufacturing industries. The city had developed coalmining, iron and steel production, and heavy engineering industries (especially shipbuilding) and textile industries. During the 1950s these manufacturing industries declined dramatically because new technologies were being developed and there was now competition from abroad. So factories and shipyards were forced to close, especially during the 1980s. In those that remained, mechanisation and automation meant that fewer workers were needed. As a result, employment in manufacturing industries declined from 307,000 in 1971 to 121,000 in 1991. This left a landscape of derelict factories and shipyards and decaying areas of tenements and housing. Although employment grew in the service industries, any new manufacturing plants were sited away from the older locations near the city centre and in the East End and Riverside districts. Manufacturing now took place in new industrial estates on greenfield sites on the edge of the urban area.

### Positive impacts of deindustrialisation

- There are areas of land available for development.

- There was a decline in the consumption of water in the industrial processes.

- There was a reduction in the demand for energy to power machines.

- There was a reduction in noise and air pollution with the factory closures.

### Negative impacts of deindustrialisation

- There are large areas of derelict land. In 2011, 30% of the vacant or derelict land in Glasgow had been derelict or vacant since 1985. So the problem remains and is still a major issue. However the total amount of vacant and derelict land has declined by 21% between 1996 and 2011.

- There are many empty factory buildings.

- The infrastructure has decayed, especially the roads.

- The areas of polluted land are hard to clean up.

### Regeneration in Glasgow

Since the 1980s Glasgow has experienced widespread regeneration in a number of areas. As part of a policy to promote the city as a tourist destination, there has been considerable government investment in the arts and culture, with new museums and art galleries being developed, and Glasgow was named the European City of Culture in 1990. This new tourism has led to increased employment in the area and boosted the income of the city. It has also led to slum clearance schemes and the redevelopment of areas with new flats and houses.

New shopping centres were built at St Enoch, Princes Square and the Italian Centre, together with new office blocks. Old warehouses were converted to luxury flats. In 1985 the Scottish Exhibition and Conference Centre was inaugurated and in 1988 Glasgow housed the National Garden Festival. In 1990 the New International Concert Hall was opened. All these events and cultural resources, including the Scottish Opera, the Royal Scottish National Orchestra and the Citizens Theatre became central to the city's success.

The old docks on the River Clyde have been redeveloped, partly by private investment, to create shops, offices and riverside apartments. These developments too have created jobs in service, retail and construction, as well as leading to the creation of many more green spaces within the urban area. Elsewhere in the city, government support has enabled construction of facilities including an athletes village for the Commonwealth Games in 2014.

### Positive impacts of redevelopment

● There is now less derelict and vacant land.

● New jobs have been created and new homes built.

### Negative impacts of redevelopment

● Traffic flows have increased, bringing more noise and air pollution.

● The new homes tend to favour the richer people who can afford them.

● There is increased demand for energy from the new service industries.

## Greenfield and brownfield development

There are differing views about proposals for new economic developments. Some people think that new developments should be on **greenfield sites** – areas of land that have not been developed before. Others want to protect these 'green' areas and they argue that new developments should be on **brownfield sites** – sites that have been developed before but where buildings such as factories have now been closed and perhaps demolished. The two examples below show how both brownfield and greenfield sites are being developed in parts of the West Midlands.

### Costs and benefits of a brownfield site development in Longbridge
The proposed plan is a £1 billion redevelopment programme for the site of the former Rover car factory in Longbridge in Birmingham. Part of this is the construction of a £70 million town centre for Longbridge. A new Sainsbury's store will be part of a major development of 165,000 square feet of retail space. This will bring 25 new shops and restaurants, together with 40 apartments overlooking the new two acre park. There will also be a new hotel. Currently 200,000 people live within a 10-minute drive of this new development. Bournville College, with its 15,000 students, has already relocated to its new £66 million campus on the site.

*Figure 8: View of smart new flats in Riverside area of Glasgow*

*Figure 9: Redeveloped Glasgow docks*

There are various costs and benefits associated with the development:

- Three new green parks are being created to improve the quality of the local environment.

- 10,000 new jobs will eventually be created.

- Shop owners in nearby Northfield will lose a lot of trade and some have already closed.

- The development eases pressure on the nearby green belt.

- Local house prices are already rising as the development grows. Local people may no longer be able to afford them.

- 98% of house building in Birmingham in 2011 was on brownfield sites.

- Other local shops in surrounding towns such as Bromsgrove fear that the new development will mean a loss of trade for them.

- Unemployment in the area is still high following the closure of the car factory in 2005, so many young people are looking forward to getting a job for the first time.

- There are concerns over increased traffic problems in the area as a result of the new shops.

### Costs and benefits of greenfield site development in Dudley, West Midlands

There is a plan to build 334 homes in Pensnett, an area on the edge of Dudley. There will be 27 two-bedroom flats, and 126 three-bedroom, 128 four-bedroom and 8 five-bedroom houses. There is green belt countryside around the site, so this is a pleasant environment. There are various costs and benefits associated with the development:

- The development has been criticised because it will add to traffic congestion in an area where existing roads are already congested.

- The loss of green space has been criticised locally, because this is a heavily built-up area.

- The site will require new infrastructure such as drains, electricity, access roads.

- It is easier to build here because there are no old buildings to demolish and no pollution to clean up.

- Using greenfield sites is not sustainable as there is huge pressure on the local green belt.

- There is a big demand for new homes in the area, and the construction will create new jobs.

215

*Figure 10: Longbridge's new Technology Park*

**ResultsPlus**
**Build Better Answers**

**EXAM-STYLE QUESTION**

**Using named examples, examine the costs and benefits of developing both brownfield and greenfield sites. (8 marks)**

■ **Basic answers** (1–3 marks)
There are descriptions of both greenfield and brownfield sites with accuracy. Some costs and benefits are outlined but they are unbalanced and lacking detail. Some structure to the answer and some relevant use of geographical terms.

● **Good answers** (4–6 marks)
There are descriptions of both greenfield and brownfield sites with examples although they may be more general locations. Some range of costs and benefits are explained in some detail, but unbalanced. Clear structure to the answer and relevant use of geographical terms.

▲ **Excellent answers** (7–8 marks)
There are detailed descriptions of both greenfield and brownfield sites, with examples. A range of costs and benefits are explained in detail and the answer moves towards which is better. Clear structure to the answer and excellent use of geographical terms.

**Top Tip**

When writing about new jobs created by green technology, be careful to mention a range of jobs from those in water and waste management to green transport, as well as those in renewable energy.

*Figure 11: An offshore wind farm*

## Activity 4

1. How much energy do European countries expect to come from renewable sources by 2020?

2. What is the 'third industrial revolution'?

3. Which sector is likely to generate most new green jobs by 2020?

4. Why do you think there may not be as many new jobs in water management and waste recycling as predicted?

5. Why might the growth of green jobs create problems for workers in other sectors?

# The diversification of employment in the UK

The employment and the economy of the UK is currently being changed radically by a series of factors:

- The growth of the green employment sector
- The development of the digital economy
- Education and research
- Foreign workforce.

## Employment in the 'green' sector

One of the areas of employment that experts predict is likely to grow in the future is the **'green' sector** – the jobs connected with making a sustainable environmental future. But just how many jobs will be created by the new green technologies?

## The theory

Europe has pledged to cut carbon dioxide emissions by 20% by 2020, and to obtain 20% of its energy from renewable sources, such as wind or water. There are also important developments to reduce the carbon footprint of all transport, construction and electricity firms. This should create new jobs and has been called the 'third industrial revolution'. So where will the new jobs be created?

- Most of the new jobs will come from the generation of renewable energy. For example, huge new offshore wind farms are being built or are planned for the coasts around the UK (Figure 11). Firms making new wind turbines and solar power generators as well as hydro-electric power stations could create 2.5 million new jobs by 2020, according to an EU-funded report in 2003. One problem with this is that no one really knows if these new jobs will materialise. Another problem is that in some places such as the coalfields of Germany and Belgium even more coal mines will have to close – and this means that there will be job losses there.

- There may be new jobs in water management. As people become aware of the need to use water carefully – and to reuse it where possible – more workers may be needed.

- There may be new jobs in waste management. For example, as recycling gains momentum more firms will need extra workers to deal with recycling. But there may be only a few jobs because the work will rely heavily on machines.

- Green transport may also create new jobs. As cars and lorries become more environmentally friendly there may be a demand for more skilled workers to design better types of transport.

## The development of the digital economy

Digital technology, in the form of the personal computer and the internet, has already made huge changes in work, education, government, leisure and entertainment, as well as generating new market opportunities and having a major economic impact. In the UK, ownership of personal computers reached 76% in 2009, compared with 29% in 2000. The use of mobile phones reached 78% of all UK households in 2011, and 49% of all 7–18 year-olds had access to the internet. The growth of new digital wireless networks, mobile devices and positioning technologies is the start of the next shift in digital technology, as it becomes embedded into the public spaces, architecture and all our daily lives (see table).

**Computer ownership, 2009**

| Country | Computers per 100 people |
|---------|--------------------------|
| Israel | 122.0 |
| Canada | 87.6 |
| Switzerland | 86.5 |
| Netherlands | 85.4 |
| USA | 76.2 |
| UK | 75.8 |
| Australia | 75.7 |

The use of handheld computers, mobile phones, digital cameras, satellite navigation, embedded sensors and a range of increasingly interconnected devices is the beginning of a shift towards a world of computing everywhere that will ultimately see people served by many thousands of computers. This will be a key part of innovation for our future digital economy, and it already underpins views of the future of transportation and healthcare and has the potential to transform other sectors, including the creative industries and financial services. This growth in the importance of computing will have a major impact on the ways in which people work, travel, learn, entertain themselves and socialise.

There are costs and benefits associated with the digital economy:

◉ Many transactions can be done very swiftly with fewer mistakes.

◉ Speed and connectivity (to all parts of the world) are key advantages.

◉ Greater reliability of some systems, such as healthcare.

◉ Potential dangers to young and vulnerable people from access to the internet and social networking sites, where they may meet undesirable people.

◉ Growing areas of criminal activity – hacking into accounts etc.

## Education and research

As education and research become more and more important in the development of all aspects of life, but particularly economic development, so the growth of schools, colleges and universities is linked, having an increasing impact on the UK economy. Research in UK universities is contributing to the development of the economy by developing new products such as new treatments for diseases and new drugs to combat disease.

There has been an increase in funding for research which focuses on new technologies, some of it biomedical, some of it IT-based. These developments will create new jobs and even new industries. For example, in North East England the number of companies with biotech interests has grown from 28 in 2003 to 54 in 2011. These companies employ over 5,000 people and contribute £700 million to the UK economy. The area has interests in biofuels, and stem-cell and regenerative medicine research, and is closely linked to the five universities in the region.

**Foreign migrants in UK industries, 2011 (workforce share, recent migrants)**

| | % |
|---|---|
| Elementary process plant (e.g. industrial cleaning, packers, bottlers, fillers) | 23.6 |
| Process operatives (e.g. food processing) | 13.0 |
| Elementary cleaning | 9.8 |
| Elementary goods storage | 9.3 |
| Elementary personal services | 9.1 |
| Assemblers & routine operatives | 8.2 |
| Information & communications technology | 8.1 |
| Elementary construction | 7.7 |
| Food preparation trades | 6.5 |
| Research professionals | 6.3 |

*Source:* Labour Force Survey 2011

*Figure 12: Workers picking vegetables in the Vale of Evesham*

## Activity 5

1. How might the digital economy start to change the financial services and creative industries?

2. Why have so many foreign workers come to the UK to find a job?

3. Why might foreign workers face problems from workers in the UK?

### Foreign workforce

In the last few years there has been a big increase in the number of workers from other countries who are engaged in the UK economy. One example of this is the large number of workers from countries like Portugal, Romania and Poland who have come to work in UK agriculture (Figure 12). Many of these people work harvesting fruit and vegetables, and sometimes moving from place to place as the harvest time of different crops occurs in different parts of the UK. Other foreign workers are in industrial occupations, often associated with processing or cleaning products. In 2011 there were 2.6 million foreign workers in the UK, and the Confederation of British Industry was saying that these foreign workers had essential skills which the UK economy needed. The top ten industries involved are shown in the table.

## The impact of changing working practices

One of the main changes to take place in working in the UK in the last 20 years (see page 210) has been the growth of flexible working. But flexible working means different things to different people. It can mean:

- Flexi time – choosing when to work (around a core period)

- Compressed hours – working agreed hours over fewer days

- Homeworking – working from home

- Zero-hours contracts – no specified hours, but always available to meet the needs of the employer.

### Homeworking

Many employees can now work from home (or away from the office) using ICT to stay in touch with the office. This is ideal for workers whose job requires them to spend a long time on the computer and who do not need regular face-to-face contact with people. A survey in 2011 established that, in the UK, 7% of people work from home. The main reasons why so many people are changing to homeworking are:

- About 2 hours commuting time or more is saved each day.

- People can work as and when they want at home.

- There are no other workers to consider.

- Workers can care for family members at home whilst still working.

- Firms can save money on office space and costs such as heating etc.

- ICT has made it possible to do much work at home rather than in an office.

### Teleworking

This is people who work using ICT, wherever they are. Some may be at home, but others, sometimes called 'nomad workers' or 'web commuters', use mobile telecommunications technology to work from coffee shops or other locations. The reasons for **teleworking** are the same as those for homeworking – for firms it saves expensive office space, and for workers it gives flexibility and choice.

## Self-employment

This is people who work for themselves. In recent years more and more people have chosen to set up their own business and become self-employed. Some of these people work from home, some have shops, factories and offices. These people have done this because:

● They prefer to work for themselves rather than for a firm.

● Some are people who have been made redundant and who choose to start their own firm.

● Some are people who have a good idea for a product or service but cannot get a big company to be interested, so they start their own company.

● Some are young people with a good new idea and they want to try it out – such as the creator of Facebook, or Bill Gates, the man who started Microsoft, or Anita Roddick who started the Body Shop.

The impacts of these changes in working are:

● There should be a reduction in rush-hour traffic as more people either work from home or vary the times they travel to work.

● Hot-desking – the practice whereby people use any desk in an office that is available and do not have a permanent desk – has reduced the size of some offices, and this saves money.

● Firms can employ more people on flexible working so that they are better able to meet targets and deadlines.

● Many people work flexibly who would not be able to work at all if they had to stick to strict hours, so there are more people in the workplace.

● People do not waste time commuting to and from work.

● Firms can continue to employ people whose circumstances have changed and who would otherwise have had to leave.

● In rural areas like northern Scotland there has been a decline in jobs in farming, fishing and forestry, so this is a good opportunity to work either from home or a café.

● The popularity of flexible working has meant an influx of incomers – people new to the area in places like northern Scotland and north Yorkshire. Some are retired people but many are younger people seeking a 'better' lifestyle.

● Third party contracted labour – the company does not employ labour directly but employs people through an 'agency'. The company has reduced overheads (costs associated with admin, employment taxes, sick pay, holiday pay, redundancy, etc.) and just pays the agency. The agency employs the staff.

### ResultsPlus
### Build Better Answers

**EXAM-STYLE QUESTION**

**Explain the main changes in industrial employment in the UK in the last 20 years. (4 marks)**

■ **Basic answers** (0–1 marks)
Only talk about some of the changes, especially decline, and do not mention more than two industries.

● **Good answers** (2 marks)
Describe the main changes, but do not offer much explanation for them.

▲ **Excellent answers** (3–4 marks)
Describe the changes but also offer explanations for them.

# Know Zone
# The changing economy of the UK

The UK economy has changed greatly over the past 50 years, as result of globalisation and government policies. These changes have been linked to changes in the structure of the workforce and this is evident in places like North East England and South East England. These changes in employment have both positive and negative impacts on the environment. The one constant factor is that employment is changing and will continue to change.

## You should know...

- ☐ The nature of the changes in the industrial structure of the UK economy
- ☐ The reasons for these changes, especially the role of governments and globalisation
- ☐ That employment has changed in the last 50 years
- ☐ That changes in employment vary from place to place within the UK
- ☐ That changes in employment are growing
- ☐ The reasons for the differences in industrial structure and employment between South East and North East England
- ☐ The environmental impacts of changes in employment
- ☐ That some environmental impacts are positive and some are negative
- ☐ The impact of de-industrialisation and economic diversification in Sandwell in the West Midlands
- ☐ The factors affecting the growth of the UK economy, such as the digital economy and green employment
- ☐ That there are costs and benefits to developing both greenfield and brownfield sites
- ☐ That changing working practices, such as homeworking and teleworking, are affecting employment and the economy

## Key terms

Big Bang
Brownfield site
Component manufacturers
De-industrialisation
Derelict land
Digital economy
Enterprise Zone
Flexible working
Globalisation
Greenfield site
Green sector
Homeworking
Industrial structure
Overfishing
Primary industry
Quaternary industry
Quotas
Secondary industry
Teleworking
Tertiary industry

### Which key terms match the following definitions?

**A** A piece of land which has been used and abandoned and is awaiting development

**B** The decline in industrial activity in a region or an economy

**C** Land on which factories or houses have been demolished

**D** The proportion of people who work in primary, secondary, tertiary or quaternary jobs

**E** The process, led by transnational companies, whereby the world's countries are all becoming part of one vast global economy

**F** The part of economic activity that pays attention to environmental issues

**G** A piece of land that has not been built on before, but now is waiting for development

**H** Taking too many fish (or other organisms) from the water before they have had time to reproduce and replenish stocks for the next generation

To check your answers, look at the glossary on page 321.

**Foundation Question:** Using a named example, explain the advantages and disadvantages of developing greenfield sites. (6 marks, plus 3 for spelling, grammar and punctuation)

| Student answer (achieving 3 marks) | Feedback comments | Build a better answer |
|---|---|---|
| Greenfield sites have lots of space for building | This is an introductory sentence that does not contain enough detail to score marks. | Greenfield sites may have planning restrictions. |
| It is expensive to build new roads in the green site | This is a valid point and scores 1 mark. | Infrastructure may be expensive and it may contribute to urban sprawl. |
| Green field sites have lots of space for new developments | This scores 1 marks because it is correct. | Greenfield sites allow for a more spacious layout. |
| Builders prefer to build on green sites | This is very general and lacks detail. | Land is cheap and is often close to motorways. |

Total available for spelling, punctuation and grammar = 3 marks. Marks achieved: 1.

**Higher Question:** With the use of examples, explain the growth of tertiary and quaternary industries in the UK. (8 marks, plus 3 for spelling, grammar and punctuation)

| Student answer (achieving 2 marks) | Feedback comments | Build a better answer (achieving 3 marks) |
|---|---|---|
| People need more services like schools and hospitals. | This is an introductory sentence that contains enough detail to score 1 mark. | Tertiary industries grew to provide services such as health, education, retailing and business services for the growing UK population. |
| Banks and building societies began to offer more financial services after the big bang in 1986. | This is a valid point on inputs with a good range of examples and scores 2 marks. | The financial services sector grew rapidly after the Big Bang of 1986 when financial services in the UK were deregulated. This sparked massive growth in financial services such as banks, building societies and insurance companies to meet the growing need for a wide range of services such as investment and loans. |
| Quaternary services have also grown especially research. | This scores 0 marks because although it is correct it does not explain the growth. | Quaternary services have grown rapidly because firms have to invest in a great deal of research in order to keep developing new products such as mobile phones. Research related to IT is one of the fastest growing parts of quaternary industry because more people now use IT as part of their everyday lives. |
| Now more people have computers and IT gadgets so there is growth in these areas. | This scores 1 mark because it is correct. | |

Total available for spelling, punctuation and grammar = 3 marks. Marks achieved: 2.

# Chapter 14 Changing settlements in the UK

222

## Objectives

- Recognise the economic, social, political and demographic processes operating in urban areas.

- Investigate the impacts of these processes on UK urban settlements.

- Explain how the processes have led to variations in the quality of life for people in cities.

*Figure 1: Many older shipyards have closed*

*Figure 2: Industrial estates are often sited on the edge of urban areas*

## How and why are settlements changing?

### Changes in UK urban areas in the last 50 years

Towns and cities in the UK have experienced a series of major changes in the last 50 years as a result of different processes.

### *Economic processes*

One of the main changes has been the decline of older, traditional industries such as iron and steel making, engineering, textiles, shipbuilding and motor manufacture. This decline is a result of growing international competition, especially from Japan and Korea, government policies, and a failure to introduce newer higher-tech machinery and equipment (See Chapter 13). The UK shipbuilding industry had produced 60% of the world's shipping in 1913 but it then declined dramatically. In 1976 the UK still produced 134 vessels but by 2011 the total was down to just four. The decline was due to competition from countries like Japan, South Korea and (most recently) China – countries who are able to use the very latest technology in ship construction. The UK shipbuilding industry was slow to introduce mechanisation and to give up old working practices. So some towns and cities, such as Glasgow, Sunderland and Wallsend, experienced more yard closures (Figure 1) and rising unemployment.

The decline in traditional industries left many inner city areas by the 1980s with empty, derelict works and factories – and decaying housing. Some cities, like Birmingham, still had car making but even this collapsed in 2005 when the industry moved out of the city. This meant that even more local factories which had made the components for the cars, vans and lorries also closed. So unemployment rose and people moved away in search of work.

By contrast, London has experienced rapid economic growth in the last 50 years, partly as a result of the development of the financial services industries in places such as London Docklands. In addition, newer, light industries such as printing, publishing and light engineering were attracted to London by the size of the market, its excellent communications network, its large, skilled labour force and, after 1994, its easy access to the markets of the rest of Europe via the Channel Tunnel. This economic growth attracted more and more people to live in London which had lots of jobs. So population grew in and around London and the demand for houses and flats grew with it. The new industries preferred locations just outside the urban areas, close to the green belt where there was more space and cheaper land. So industrial estates grew up in these locations on the edge of urban areas (Figure 2).

## Political processes

Towns and cities were affected by political processes too. After the 1980s, governments began to try to regenerate decaying inner-city areas by using **Enterprise Zones** (see page 211, North East England). These zones set about redeveloping the areas by demolishing older houses and empty factories, building new offices and creating more open space. The London Docklands development (see Figure 3) is one example of this **redevelopment** process which created many new jobs in the growing financial services. It also included 'luxury apartments', which began a process or re-urbanisation, as younger, richer people chose to move back into areas close to the city centre. The attraction was easy access to the jobs, entertainments and shops of central London. A major part of the growth of London in the period after 1986 was the government **deregulation** of the financial markets. The government introduced these measures in order to make financial institutions in London more competitive with those in other world cities, especially New York. The deregulation was introduced on 27 October 1986 and was known as the 'Big Bang' because of the 'explosive' growth in financial activity as a result of the changes. The effect of the changes was to strengthen London as a major world financial centre and to create an economic boom which led to massive new office, apartment and retail developments in the Isle of Dogs and Canary Wharf areas of the old London docklands.

The 1980s also saw the privatisation of many industries such as steel, airlines, telecommunications, water, gas and electricity. These changes had a major impact on key industries like steel, with factories closing or being modernised and re-organised. Also part of the political processes affecting towns and cities were the planning policies in the period. Throughout the period from 1970 to the present, planning has had a major impact on the shape, form, function and economic makeup of towns. Planning legislation throughout this time has focused on:

- redeveloping and **rebranding** inner city areas

- establishing **green belts** between towns to prevent them merging into one large urban area

- Improving **infrastructure** such as roads and more recently railways

- Protecting historic buildings, often in city centres

- Reviving shopping areas in city centres – but also encouraging out-of-town retail developments.

## Social processes

Social processes affected urban settlements through the processes of redevelopment. Cities like Birmingham decided to redevelop its older inner city areas. In order to do this, the local residents had to be moved away from their old decaying houses and flats into new estates in the suburbs. Birmingham created five new Comprehensive Development Areas (CDAs) with new flats and more open space. In some cities, however, many people from the inner city areas have had to move further away – in some cases to one of the many New Towns, such as Telford in Shropshire.

Figure 3: Office blocks in Canary Wharf, London

Throughout the period since 1980 a major trend has been the increasing number of single-person households. Some of these are students seeking places to live – and many larger houses have been divided up into student flats. There are also growing numbers of divorced people who are seeking accommodation in flats or small houses. More women have gone out to work in the last 20 years. All these changes in society have affected the demand for homes – their number, their type and their location. Some aspects, such as access to good roads and parking have become more important and others, such as garden size, less important. Settlements have changed as the number and type of homes being built has responded to all these changes in social processes.

### Demographic processes

Demographic processes have also affected urban areas – especially the movement of large numbers of people away from city centres into the suburbs or even further into villages in the surrounding countryside. In addition, many cities have seen an influx of people from other countries attracted by the prospect of better-paid jobs. Many new immigrants are prepared to live in cramped, sometimes sub-standard houses and flats, in order to get established in the workforce. Areas such as Handsworth in Birmingham have become famous for the variety of shops selling exotic fruit, vegetables and food introduced by the newcomers (Figure 4). More recently the influx of immigrants has been from Eastern Europe (see table below). The newcomers have also tended to be young people who have subsequently started families. As a result, population growth in many cities has been rapid.

*Figure 4: Fruit and fashion shops in Handsworth, Birmingham*

**Net immigration to the UK 1996–2005 (thousands)**

|      | From the Commonwealth | From Eastern Europe |
|------|-----------------------|---------------------|
| 1996 | 28                    | 47                  |
| 1997 | 18                    | 50                  |
| 1998 | 33                    | 72                  |
| 1999 | 8                     | 8                   |
| 2000 | 6                     | 101                 |
| 2001 | 11                    | 101                 |
| 2002 | 11                    | 101                 |
| 2003 | 14                    | 164                 |
| 2004 | 26                    | 166                 |
| 2005 | 22                    | 125                 |

The result of all these processes is the complex ethnic mix that is found in many cities. In Birmingham for example, 34% of the population are non-white, compared to the 30% non-white in London and 7.7% in Liverpool. The figure for Liverpool is low because at the time of the peak immigration the job opportunities in the city were poor. Minority communities tend to have their own momentum of growth – so in cities like Birmingham they grow around similar communities, once a core area has been established. The table on page 225 shows the ethnic mix in Birmingham in 2011.

**Birmingham's population in 2011**

| Ethnicity | Population |
|---|---|
| **White**<br>British<br>Irish<br>Other White | 627,100<br>24,100<br>22,200 |
| **Mixed**<br>White and Black Caribbean<br>White and Black African<br>White and Asian<br>Other mixed | 16,600<br>2,100<br>8,400<br>5,600 |
| **Asian or Asian British**<br>Indian<br>Pakistani<br>Bangladeshi<br>Other Asian | 61,600<br>114,000<br>23,700<br>13,800 |
| **Black or Black British**<br>Black Caribbean<br>Black African<br>Other Black | 44,900<br>17,100<br>4,200 |
| **Chinese or other**<br>Chinese<br>Gypsy and Traveller<br>Other Ethnicity | 12,600<br>54,750<br>12,100 |
| **Total** | 1,064,850 |

**Activity 1**

Study a large town near you.

(a) Try to map the areas of housing, shopping, industry and recreation for the town.

(b) Now try to explain the pattern you have mapped in terms of economic, political, social and demographic processes.

## Variations in the quality of urban residential areas

The economic, social, political and demographic processes have led to a pattern of different quality residential areas in UK cities.

### Access to recreational areas

Better-off residential areas in cities tend to have good access to open spaces like parks, country parks, and open green space for recreation. This is because the people planning these areas realised the importance of giving people access to recreational areas. In poorer areas of cities, such as Aston in Birmingham, houses and flats are often crammed close together with little open space between them. This is because many of these areas were built in the nineteenth and early twentieth centuries when towns were growing very quickly as large numbers of people left the countryside to find work in towns. The pressure was to build houses as quickly as possible, not to create open space for recreation (Figure 5). In the recent past, many inner city areas such as Islington and Hackney in London have undergone rapid changes, including 'gentrification'. This is the process by which wealthier (mostly middle-income) people move into, renovate, and restore housing and sometimes businesses in inner cities. This process has led to improved access to recreation areas in these inner city areas. Ease of access to recreational areas is also linked to income, in that people with good incomes can access recreational areas easily and quickly, often by car.

*Figure 5: Terraced houses in Aston, Birmingham*

### Access to amenities

**Amenities** are things like parks, playgrounds and golf courses, as well as health clubs, restaurants, shops, community centres, cinemas and theatres. Some residential parts of cities have good access to these amenities. These are usually places near city centres or in the parts of cities where people have more money. Amenities like restaurants and cinemas are located in city centres because routes and public transport services converge there, so people can reach them easily. In the same way, richer areas of cities have more people who will pay to go to health clubs, restaurants and luxury good shops.

There are also 'disamenities' – things people do not want to be near, like rubbish tips and air-polluting incinerators. So one result is that there is competition and conflict to either live close to or away from amenities and disamenities. For example, more affluent areas try to 'capture' as many services and amenities as possible in order to have easy access to them.

### Access to services

There are many different types of services in residential areas, such as transport, electricity, waste disposal, water supply, and medical services. One example of this is the access to medical services, such as doctors. This affects the quality of the urban area. In the UK, research shows that 0.75 km is the upper distance limit of walking for mothers with school-age children. Access to medical services is linked to: distance from the surgery; access to a car; and the quality of public transport. So single parents on low incomes, for example, living in high-rise flats with no car have very poor access to medical facilities.

Most General Practitioner (GP) surgeries are located in older, more middle-class areas or higher socio-economic areas. This is reinforced by two factors. First, doctors tend to live and work in well-established, high quality areas, and secondly many council estates were built without suitable places for surgeries. The result is that there is an uneven distribution of medical resources. This is called the inverse care law, in which the availability of medical services is inversely proportional to the needs of the population. Figure 6 shows the accessibility to medical services in the different parts of Birmingham, based on an 'accessibility index'.

| | |
|---|---|
| Sutton Coldfield | 135 |
| Erdington | 101 |
| Perry Barr | 99 |
| Hodge Hill | 88 |
| Yardley | 90 |
| Sparkbrook | 76 |
| Hall Green | 100 |
| Selly Oak | 132 |
| Northfield | 106 |
| Edgbaston | 98 |
| Ladywood | 75 |

**Accessibility index**

| | |
|---|---|
| Over 130 | Excellent access to medical services |
| 100–129 | Good access to medical services |
| 70–99 | Below average access to medical services |
| Below 70 | Very poor access to medical services |

*Figure 6: Accessibility to medical services in Birmingham*

### Access to different types of housing

Areas of expensive housing in cities tend to be either on the outskirts of the city, where houses are bigger and there is better access to more open space and recreational areas, or in the city centre where land is expensive but where high rise flats give access to all the amenities of the city centre. The city centre flats are often the result of urban redevelopment schemes which have seen older, derelict areas demolished or rebuilt with high rise flats. In places like London, old warehouses have been redeveloped as luxury flats (Figure 7). In some parts of inner city areas there are 'gated communities', which are only accessible via a gate which serves to filter who can enter the area. This is often a reaction to high crime rates, and in many areas the numerous wall-mounted cameras testify to a community which is afraid. Cheaper housing tends to be in older areas or some council estates with poorer access to amenities, services and open space.

## Changes in UK rural areas in the last 50 years

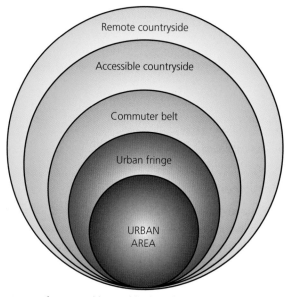

Figure 8: The four types of countryside outside the urban area

Change is everywhere and that includes the UK's countryside. 'Change' is a big umbrella under which many things happen. The nature of countryside change varies from place to place. The most important factor is distance from a major city. According to this, it is possible to recognise four different types of countryside, as shown in Figure 8.

● The urban fringe – This type of countryside is being quickly lost to urban growth, particularly where there are no planning controls such as green belts.

● The commuter belt – This is countryside, but the settlements within it are used as dormitories by urban-based workers and their families.

● The accessible countryside – This is beyond the commuter belt, but within day-trip reach. Still very much a rural area.

● The remote countryside – This takes the best part of a day to reach from a city. Almost totally rural.

Figure 7: An old London warehouse, converted into luxury flats

**Build Better Answers**

#### EXAM-STYLE QUESTION

**Explain the different quality of housing and services in residential areas in UK cities. (4 marks)**

■ **Basic answers** (0–1 marks)
Describe the different types of housing but do not discuss services.

● **Good answers** (2 marks)
Describe the different types of housing and services in UK cities but do not explain them.

▲ **Excellent answers** (3–4 marks)
Both describe and explain the different quality of both housing and services.

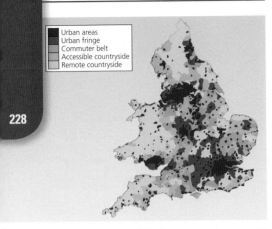

*Figure 9: The distribution of the four types of countryside outside the urban area in England and Wales*

## Skills Builder 1

Study Figure 9 and refer to an atlas.

(a) Where are the main areas of remote countryside?

(b) What are the physical features of these remote areas?

Figure 9 shows the distribution of these four types of countryside in England and Wales.

These four types of countryside are being affected by a number of processes of change. As shown in Figure 10, some of the processes span from the city to the remote countryside, whilst others involve just two or three of the countryside zones. Remember that these processes are working in two directions – a few towards the city, the majority outwards from the city.

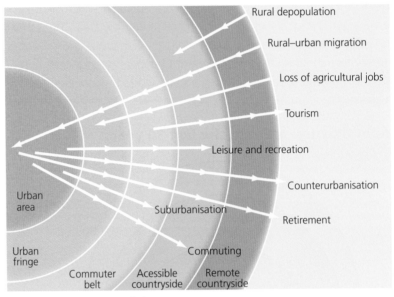

*Figure 10: The processes of change*

### Types of rural settlement

Commuter villages have grown up in the commuter belt for people who are able to afford to live in rural villages and commute to jobs in towns. Belbroughton in the West Midlands is a typical example (Figure 11). Belbroughton has developed from a small village with a population in 1960 of 603, to a large commuter village of 2,400 people in 2011. It has developed as a result of:

● being close to Birmingham (17 km away) and the Black Country

● having good road and rail links to Birmingham and the Black Country

● having space in the fields around the old village to build new houses

● people wanting to move out of Birmingham and the Black Country into a rural area where they hope to find cleaner air and less crime

● farmers willing to make money by selling their farms.

In the past, Belbroughton was a centre of scythe making and sharpening for local farms. Up until the 1960s it had a school, post office, pub, numerous shops and some farms within the village. Now there are lots of new houses and lots of new residents. The new residents often have two or more cars and they tend not to shop so much in the village but go to supermarkets nearby. So some shops have closed, but restaurants have opened catering for the new villagers.

*Figure 11: Belbroughton High Street*

**Retirement communities** have also developed in parts of rural areas which have attractive scenery, such as coasts or hills. These settlements are usually in the accessible or remote rural areas. Currently about 10% of people who retire from work choose to go and live in villages and small towns. These villages and towns (Figure 12) have developed as a result of:

● the hope by people who retire that living in the country will give access to cleaner air

● the hope that there will be a strong sense of community to which the newcomers can belong

● the hope that there will be less crime and danger in the countryside

● the growing affluence of some retired people

● the fact that most people are living longer and are seeking a better quality of life.

The retired people do not always mix with local people, and they can create a strain on local health services because more people need treatment. Retired people can afford to pay more for houses and flats than local people so house prices rise and young local people may be unable to afford to buy a home in their own village.

Remote rural villages and small hamlets grew up in the past when more people worked in farming. They have declined because:

● people have left because local jobs have disappeared as farming has become more mechanised

● some farmers have left farming, especially in hilly areas where soils are thin and acid, so farming is challenging and restricted to hill sheep or cattle farming

● they do not attract commuters or retired people because they do not have access to good, fast roads and they are a long way from major towns and cities.

### East Anglia – A prosperous rural region?

East Anglia consists of Norfolk, Suffolk, Cambridgeshire and part of Essex (Figure 14 on page 230) and is home to nearly 3 million people. East Anglia is a region with a generally high quality of life and low levels of deprivation. The rate of unemployment is lower than the national average, and East Anglia is, in general, one of the UK's more prosperous rural areas. For example, in terms of the Index of Multiple Deprivation (see table on page 230) it is clear that over half of the very highest areas of multiple deprivation (the '1% areas') are in North West England and that only 2% are in the East of England (which contains East Anglia).

Figure 12: Many retired people move to the seaside

Figure 13: Cockington in Devon

### Activity 2

Figure 13 shows an example of the type of English village landscape that has been much admired for well over one hundred years.

(a) State three reasons why some people would see this as an ideal place to live.

(b) State three reasons why some people would not want to live here.

(c) Identify three features in the photograph that make it an unusual view of an English village in the twenty-first century.

The reasons for East Anglia's relative prosperity are:

1. A prosperous agriculture. This is a region with large flat areas, much of which used to be marsh (locally called 'fenland') which has been reclaimed. This has created an important agricultural area based on the fertile soils of the reclaimed land. East Anglia produces much of the English crop of cereals, such as wheat and barley, as well as crops such as peas and potatoes. There are some very large farms in the area which have become even bigger as farms have adopted newer and bigger machines and used more intensive farming methods.

2. The growth of traditional industries such as those based on food and drink, and the development of new industries, particularly biotechnology and Information and Communications Technology firms, especially in and around Cambridge. These firms have been attracted here by the scientific expertise of the people of Cambridge and the good access by rail and road to London and the South East. Other new industries include the manufacture of wind generation equipment. Part of East Anglia is also in the London–Stansted–Cambridge corridor, one of four growth areas in the UK which in the 1990s attracted new industrial growth, much of which is based on financial and business services offices. In turn, this has led to a strong demand for more new houses and the services they need.

3. Growing employment in ports such as Felixstowe (the largest container port in the UK) and Ipswich. New docks have been built at these ports to take even bigger vessels, and as the ports grow so does the number of available jobs, either in the ports themselves or in industries attracted by the ports.

4. Good transport links by road, rail and air with London, the rest of South East England and mainland Europe. This has attracted firms to locate in the area as it is close to London but the land and houses are less expensive.

5. The growth of tourism, based on the coast and the Norfolk Broads (an area of flooded peat workings now popular for boating holidays (see Figure 15)) has also helped to make this a relatively prosperous area.

However, not all of East Anglia is prosperous. There are significant pockets of deprivation, particularly in some of the more remote rural areas of Suffolk.

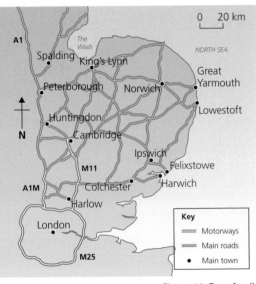

Figure 14: East Anglia

Figure 15: Boating holidays have made the Norfolk Broads a popular place for tourists

### Index of Multiple Deprivation scores (2010) for the English regions

| Region | Percentage of the most deprived areas (the '1% areas') |
| --- | --- |
| North East | 12% |
| North West | 52% |
| Yorkshire & Humberside | 17% |
| East Midlands | 5% |
| West Midlands | 9% |
| East of England | 2% |
| London | 0% |
| South East | 3% |
| South West | 2% |

### The Highlands of Scotland

The Highlands of Scotland consist of the area north and west of the Highland Boundary Fault (Figure 16). This is a huge area covering half of the landmass of Scotland but it has only 4% of the total population (about 210,000). The highlands are a region with a quality of life which varies from place to place. There is lower than the UK average unemployment but prosperity is generally low. The reasons for this are:

● This is a region where agriculture is limited by steep slopes, high mountains, including Ben Nevis (1344 m) the highest mountain in the UK, and thin acid soils. It is also an area which gets a lot of rainfall (over 1500 mm per year) so farming is restricted to hill sheep and cattle rearing, with small patches of better agricultural land for oats, barley and wheat. In the past, more people lived here on small farms (or 'crofts' as they were called) but in the eighteenth and nineteenth centuries the large landowners drove the crofters away from the land, because they wanted to create large sheep farms. This was called the 'Highland Clearances'. Since then the outward migration of people has slowed, but this is still a difficult area in which to earn a living, and young people are still leaving for better paid jobs in the big cities. This means that there are far fewer young people living in the Highlands than in East Anglia, for example.

● Fishing used to be an important industry here but has also declined as the fish stocks shrank. The growth of fish farming has done a small amount to slow the exodus of people but there are relatively few jobs.

● Tourism has developed in places like Aviemore (skiing) and some coastal areas, and this has helped to provide jobs and to slow the loss of population but many of these jobs are seasonal.

● People in the Highlands sometimes complain about the long distances they have to travel for services such as hospitals, schools, shops and libraries and this explains why some people leave the area.

● There is little industry in the Highlands. There are a few distilleries producing whisky, plus some clothing factories and industries linked to timber or fish processing, but these are automated industries requiring relatively few workers and, in the face of foreign competition, many are now contracting.

● Although there is a growth in people working from home (Figure 17) using teleworking, the numbers are still relatively small.

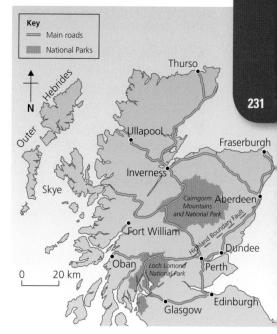

Figure 16: The Highlands of Scotland

Figure 17: Working from home in the Highlands of Scotland

## Decision-making skills

(a) Working in pairs, write down five features of your local area that you think are positive and five things that you think are negative.

(b) Suggest two ways in which your list might be different if you were aged 40.

## Objectives

- Know why there is a rising demand for urban living space in some cities.

- Understand some of the economic, social and environmental impacts of the demand for more urban living spaces.

- Evaluate the successes and failures of strategies to improve urban and rural areas.

Figure 18: Offices in central Leeds

Figure 19: New flats overlooking the River Aire in Leeds

# How easy is it to manage the demand for high quality places to live?

## Leeds and the impacts of rising demand for good living spaces

### Economic impacts

Leeds in West Yorkshire has undergone rapid economic growth in the last 20 years. This economic growth has attracted lots of new firms to the city, many of them located in the new office developments in and around the city centre. Leeds has had an image of itself as the 'Barcelona of the north' and between 1988 and 1995 the Leeds Development Corporation spent £72 million building new offices and shops. After this, in the period 1995 to 2010, £6.7 billion of major construction schemes were built. At the same time the population of Leeds has increased from 696,000 in 1981, to 715,000 in 2001, and 798,000 in 2011. The most recent growth in population parallels the economic growth which is changing the face of the city.

The attractions of Leeds are:

- Intensive redevelopment schemes, especially in the city centre creating new space for offices and shops

- Excellent communications with the rest of the UK and with mainland Europe via the motorways (M62, M1, M18), rail and the Leeds Bradford airport

- A large well-trained and educated labour force

- Three universities in the city

- A large affluent local market in the city itself and in the rest of West Yorkshire

- Its status as a regional centre for northern England which has attracted the regional headquarters of large firms.

The result has been the rapid growth of employment in the city. Specific 'quarters' have now emerged within the city centre, such as the financial quarter, the shopping quarter and the entertainment quarter. Initial economic growth has led to the city becoming very attractive to new firms which in turn has led to even more economic growth. Much of the new growth is housed in new skyscraper buildings (Figure 18).

### Environmental impacts

The rapid economic growth of Leeds in the last 20 years has meant a huge demand for new houses and flats. Much of this new development has taken place on **brownfield sites** which were the location of older industries such as brewing. Large blocks of new flats (Figure 19) were built, either in converted warehouses, often close to the River Aire, or on redevelopment sites close to the river and the canal. In these ways the environment of many central parts of Leeds has been improved by the construction of new homes. In other areas, such as Headingley and Hyde Park, nineteenth-century houses have been updated and redeveloped for students attending the universities in the city.

In other cases, the rapid growth in the demand for high quality places to live has put pressure on the green spaces in and around the city. Leeds is fortunate to have some major parks (such as Roundhay Park) and green spaces, as well as a green belt around the city. There has been a lot of pressure for developments close to these scenic areas and in the areas close to water (rivers and canals). So far the city planners have been able to resist building too much new development on either the green belt areas or the remaining green areas in the city. However a 2012 report said that there was a need for 70,000 new homes in Leeds between 2012 and 2028. Only 16% of this building could be accommodated on brownfield sites, so the rest would have to be in the green belt.

### Social impacts

As more and more people have moved to Leeds in search of new and better jobs, so the social pattern of the city has changed. Leeds is now a place with many young people. This is because many of the people moving to the city for work are often young people at the start of their careers and many of the entertainments such as nightclubs in the city reflect this. In the same way, many of the new homes which have been built on redevelopment sites near the city centre are one- or two-bedroomed flats aimed at young people with no children who want to live close to the attractions of the city centre. Other people who have moved into Leeds are students attending courses at one of the three universities in the city. Distinct student areas, (such as Headingley) have emerged, where nineteenth-century houses have been modernised and converted into flats for students. Many of these stay on in Leeds after they have qualified, having found jobs in the city.

## Strategies to improve urban areas

A range of strategies have been adopted to try to improve urban areas.

### Comprehensive redevelopment

Comprehensive redevelopment generally means 'knock it all down and start again'. It was the approach adopted in cities like Liverpool and Birmingham in the 1950s and 1960s. Some central areas in these cities had large areas of sub-standard housing, so slum clearance was a main aim. In the process of knocking down the slums people had to be moved out to estates on the edges of the city, and many small businesses closed for good. The slums were replaced by big tower blocks of flats which had amenities that most of the older houses lacked, such as inside toilets. However the tall blocks of flats in Birmingham soon began to have problems of damp and condensation. Often the construction was poor quality and so buildings deteriorated quickly. Flats lacked any soundproofing and they were poorly insulated. Residents on the upper floors had to rely on lifts for access, especially the elderly and parents with prams and pushchairs. Often the lifts were vandalised or broke down. So many people were afraid of being mugged in the corridors and they felt trapped in the flats. There were few play areas and no community facilities for young people. In recent years some of the tower blocks of flats have been demolished to be replaced by high-density low-rise housing. In other places close to the city centre tower blocks have been redeveloped as luxury flats and sold to young professionals (Figure 20).

**ResultsPlus**
**Build Better Answers**

**EXAM-STYLE QUESTION**

Explain the pressures that can result from the demand for access to high quality urban residential areas in one UK city. (4 marks)

■ **Basic answers** (0–1 marks)
Use words such as 'difficult' and 'hard' with reference to finding available land but do not explain why. Some students talk about the pressures, not their impact.

● **Good answers** (2 marks)
Identify at least one pressure in a city and the impact this has, but not more pressures.

▲ **Excellent answers** (3–4 marks)
Explain at least two pressures such as the need to build on brownfield sites and the impact this has on city centres and the social mix of cities, and the pressure to build on greenfield sites and the opposition to this from local people whose living space would be adversely affected.

Figure 20: New and refurbished flats in central Birmingham

234

*Figure 21: The Merry Hill Centre, Dudley*

### Enterprise Zones

In the 1980s the government began to use Enterprise Zones as a way of improving many inner city areas. The Enterprise Zones were given powers within their areas which largely bypassed local planning controls. They were designed to promote industrial growth, and firms locating there had a range of financial benefits including grants and loans. One of the main Enterprise Zones was at Dudley in the West Midlands. Here between 1984 and 1990 the site of a derelict steelworks was redeveloped to create a large shopping centre (the Merry Hill Centre in Figure 21) with 260 shops, offices and restaurants, 5,500 employees, and a turnover in 2009 of over £500 million. This development still attracts over 50,000 people a day and more than 70,000 at weekends. However the results locally have not all been positive. Nearby town centres such as Dudley and Brierly Hill were seriously affected by Merry Hill, with shops closing and unemployment rising. Traffic in the area is congested with the many visitors, and the new jobs are often only part-time.

### Urban Development Corporations

**Urban Development Corporations** (UDCs) were property-led regeneration agencies for inner city areas. They operated in areas where there was serious economic decline and a lot of derelict land. The UDCs were able to buy land and attract private investment to rebuild the infrastructure and develop attractive office and housing developments. One of the largest and arguably the most successful UDC was London Docklands. This rebranding (changing names to start a new identity) used flagship projects such as Canary Wharf to attract investment. New roads, new offices, houses, flats and shops have been built and the area is seen as a big success in terms of redevelopment.

However, some experts argue that although it is claimed that Docklands 'created' over 40,000 jobs, 20,000 of them were not new at all, they were just jobs moved from other parts of London. Other issues are that the new jobs did not match the skills of the local people, many of whom remained unemployed. Population decline has been replaced by population growth, but the new, mostly young people have very different lifestyles to local people. Local people cannot afford the high cost of the new houses and flats. In response the Development Corporation agreed to build 1500 cheaper homes and to make 25% of the jobs available for local people. Critics still argue that 'two docklands' exist side by side, the new enclave for the richer newcomers and the poorer council estates for the original tenants. The issue here is has this 'rebranding' been successful, and was there a balance between meeting the needs of the locals and those of incomers? Could more have been done with the money?

### Action for Cities, City Challenge Schemes

Governments continued to try to regenerate inner city areas through 'Action for Cities' and later 'City Challenges Schemes'. These regeneration projects invited local authorities to bid for government money to spend on inner city projects in their area. One such project in the Hulme area of Manchester spent £24 million over ten years from 1992 to 2002, which included building and improving housing, developing more job-creation and training projects, and reversing the loss of population. The population grew 3% over the period (compared to 0.2% for Manchester as a whole) and deprivation improved relative to other areas of the city. But the planned new small shops never materialised, and Hume remained an area of poverty in 2002. Deprivation was still an issue in terms of employment, education and child poverty.

### Recent developments

A 'New Deal for Communities' was established in 2001 to try again to regenerate inner city areas. In Aston in Birmingham, for example, £54 million was spent on a scheme was called 'Aston Pride' between 2001 and 2012. New health centres were built and money was spent on education for unemployed people and schemes which found jobs for 1328 people. Other projects helped improve flats and houses and started training schemes for unemployed men and women. One of the results was a big improvement in crime figures, with burglary down 54% and robbery down 28%.

## Strategies to improve settlements in rural regions of the UK

Different strategies can be used to improve the quality of settlements in rural regions of the UK to make them sustainable. People living in rural areas in the UK are faced with many challenges:

- Out-migration. Many young people leave rural areas like northern Scotland to find work and to find housing they can afford.

- There is a lack of jobs. Agriculture, fishing and forestry – the traditional jobs in more rural areas – all need fewer people as they become more mechanised or as they contract (and they are relatively poorly paid).

- Some rural areas, such as parts of Mid Wales, have poor communications, with little or no bus services and difficult roads.

- There is often a lack of shops, pubs and post offices (Figure 22), due to changes in shopping habits as more people seek to shop in supermarkets and larger stores which are able to offer cheaper goods.

- The lack of easily accessible entertainment. For people living in Kington in Herefordshire, for example, the nearest cinema is 36 miles away, in Shrewsbury.

- Some small schools in rural areas, such as parts of Shropshire and Herefordshire, may have only three or four staff to cover everything. As a result they may struggle to deliver the full range of subjects in the National Curriculum.

- The high price of housing. In Cornwall the average house price is eight times the average annual income, which means that many local people cannot buy a house.

- Some areas, such as parts of Herefordshire, do not have a broadband connection to the internet, something that is essential for people working from home.

235

### Activity 3

(a) What do you think a person living in Aston in 2001 and still unemployed 10 years later would think about the Aston Pride project?

(b) What might a school leaver getting help to find a job in 2010 think about the Aston Pride project?

(c) How would you try to measure the success or failure of projects like this?

(d) Why do you think the challenges of inner city areas are so difficult to resolve?

Figure 22: Many rural post offices have closed

**EXAM-STYLE QUESTION**

**Using examples, explain the challenges faced by rural areas in the UK. (6 marks)**

■ **Basic answers** (Level 1)
Mention the problems created by outmigration but include few other challenges.

● **Good answers** (Level 2)
Include locations and discuss challenges but do not give many reasons for these challenges.

▲ **Excellent answers** (Level 3)
Clearly identify challenges, locate them and explain and discuss causes.

### Rural development schemes: the Eden Project

One approach to improving rural living spaces is the development of large-scale projects such as the Eden Project in Cornwall (Figure 23). This project has converted a series of former china clay quarries, using two huge enclosures consisting of adjoining domes that house thousands of plants. The domes consist of hundreds of hexagonal and pentagonal, inflated, plastic cells, supported by steel frames. Each enclosure illustrates a natural biome, the first dome contains the plants found in a tropical environment, and the second the plants in a Mediterranean environment.

The advantages of this type of development are:

● Jobs for local people in an area with high unemployment and few alternative forms of employment. The project employs 600 people directly.

● The project attracts over 1 million visitors each year. This brings big benefits to local hotels, restaurants and shops who have far more customers. A 2011 survey showed that the biggest increase in customers was in travel and accommodation but there were also increases in jobs in catering. St Austell, the closest town, had the highest positive effects, followed by Penzance and Plymouth.

● The project is generating over £150 million indirectly to the region annually.

● It prioritises local suppliers for its annual £7 million spend, and this too brings big benefits to the area.

● The project calculates that it has secured 200 local jobs in addition to the people it employs directly.

However, as with all such projects, there are some disadvantages:

● Traffic in the area has become worse as a result of the large number of visitors.

● Some local firms argue that recruiting staff has become more difficult because of the competition for staff from the Eden Project.

● The St Austell area benefits, but areas not close to the project lose out.

● Some other local firms and attractions claim a drop in their visitor numbers and turnover as result of the attraction of the Eden Project.

● There are still many people in poverty and there is still significant deprivation in the area.

How easy is it to manage the demand for high quality places to live?

237

*Figure 23: The Eden Project attracts more than 1 million visitors each year*

### *Planning policies for rural areas*

Rural areas are under pressure from:

● The growth and expansion of towns and cities

● The desire to build new shops, offices and factories on greenfield sites

● The need to build new roads and new rail lines

● The demands of more people with more leisure time who want to enjoy the countryside in lots of different ways. Some people want peace and quiet, while others want to drive 4-wheel-drive vehicles across the countryside.

### *Green belts*

So the government tries to ensure that the planning process puts a limit on urban growth. It has done this through establishing green belts around towns and cities. Green belts are made up of mostly open space, agricultural land and woodland that is protected by planning controls. Green belts have five main functions:

● To check the unrestricted sprawl of large built-up areas

● To prevent neighbouring towns from merging into one another

● To assist in safeguarding the countryside from encroachment

● To preserve the setting and special character of historic towns

● To assist in urban regeneration, by encouraging the recycling of derelict and other urban land.

*Figure 24: A green belt is an ideal location for a riding school*

## How successful have green belts been?

- Green belts have had some success in containing urban sprawl.

- Much of the land in green belts is degraded and actually brown-belt or brownfield land. As much as 20% of London's green belt consists of this type of land, with sewage works, power stations and landfill sites.

- Some people argue that jobs have been lost because people do not try to build new developments where planning controls are so strict.

- There is huge pressure to release land for more housing.

- There is extra pressure on rural land just beyond the edge of the greenbelt for new shops, offices and homes.

### National Parks

National Parks are areas of countryside where scenery and wildlife are protected. The first parks were created in the 1950s when there was a fear that some of the UK's best scenery would be damaged, so conservation – the protection the environment – is very important. Now there are 15 parks (Figure 25).

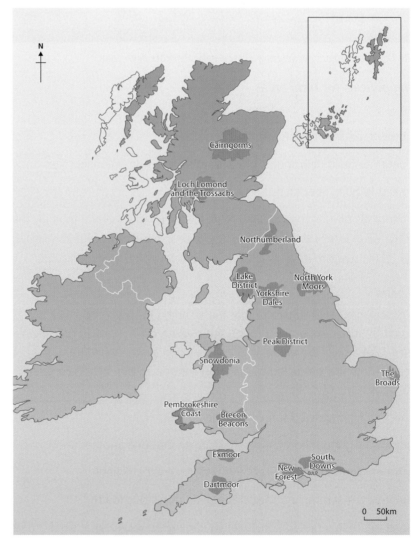

*Figure 25: Britain's National Parks*

How easy is it to manage the demand for high quality places to live?

239

The main aims of National Parks are:

- To preserve and enhance the natural beauty of the countryside

- To provide places for recreation

- To protect the social and economic well-being of the people living or working in the parks.

How successful have the National Parks been?

- Some national parks, such as the Yorkshire Dales, have large quarries in them, extracting the limestone for roadstone and cement. The quarries create noise and dust (from blasting) and can create problems with large lorries on narrow roads. Spoil heaps from quarries can be eyesores, and farmland and wildlife may be lost.

- In other parks, such as the Lake District or Snowdonia, so many visitors are attracted to the main viewpoints in the park that footpaths become eroded by the sheer number of feet using them (Figure 26), and roads can be filled with the noise and fumes of traffic. Other issues are farmers finding their animals harmed by dogs, and too many boats on some lakes (like Windermere) disturbing the peace and quiet.

National Parks have put in place plans to overcome these challenges, such as restricting quarrying within the parks, and managing different parts of a park for different activities, such as one area for noisy activities, another zone for peace and quiet. They have also undertaken a lot of work to repair footpaths damaged by erosion. In addition, park managers argue that they have been successful in protecting wildlife and habitats within the parks. In the Cairngorms National Park, for example, the numbers of deer are managed to maintain healthy sizeable herds. On Dartmoor, the upland heath, blanket bog, valley mire and grass moor have all been conserved and protected.

### Activity 4

1. Draw up a matrix to show 'Successes of National Parks' and 'Challenges in National Parks'.

2. Explain why you think National Park planners have or have not been successful.

*Figure 26: It can get very crowded at the top of Snowdon*

There have been many changes in both rural and urban settlements over the last 50 years as a result of government policies, as well as economic, social and demographic changes. Settlements in some places have grown rapidly, whilst in other they have declined as people have left. The result is a set of wide variations in the quality of residential areas. In areas where there is a great demand for high quality places to live there is pressure on the environment. In other places the challenge is to improve the quality of the area, both urban and rural.

## You should know...

- [ ] The economic, social and demographic processes operating in urban areas
- [ ] The impact of these processes on urban settlements
- [ ] How the different processes have led to variations in the quality of life for people in cities
- [ ] The factors that affect access to recreational areas, amenities, services and different types of housing in urban areas
- [ ] The challenges facing different types of rural settlements from the urban fringe to the remote countryside
- [ ] The different types of rural settlement
- [ ] Why there are variations in the quality of life and deprivation in East Anglia and the Highlands of Scotland
- [ ] Why there is a rising demand for urban living space in Leeds
- [ ] What are the main economic, social and environmental impacts of the demand for more urban living space in Leeds
- [ ] What are the successes and failures of strategies to improve urban areas
- [ ] What are the challenges facing people in rural areas
- [ ] The advantages and disadvantages of the Eden Project
- [ ] The successes and challenges of National Parks and green belts

## Key terms

Accessible rural areas
Amenities
Brownfield sites
Commuter villages
Deregulation
Enterprise zones
Green belt
Infrastructure
New Towns

National Parks
Rebranding
Redevelopment
Remote rural areas
Retirement communities
Urban development corporations

### Which key words match the following definitions?

A Countryside within easy reach of urban areas

B Adopting a new name in order to create a better impression of an area

C Rural areas that are distant from and thus little affected by urban areas and their populations

D Places where most of the people are retired

E An area designated by the government to promote economic growth and where firms get grants and relaxed planning regulations

F Local places people want, such as theatres, restaurants, hotels

G Places of mostly farmland where development is strictly controlled

H Reducing or abolishing state regulations on institutions such as banks and insurance companies

To check your answers, look at the glossary on page 321.

**Foundation Question:** State three ways in which economic processes have affected urban areas in the UK in the last 50 years. (6 marks)

| Student answer (achieving 2 marks) | Feedback comments | Build a better answer (achieving 6 marks) |
|---|---|---|
| Factories closed and people moved away so some parts of towns became derelict. | This is a really good start and scores 2 marks. | Older industries in some cities have declined and so people lost their jobs and factories have closed. |
| Some new industries grew in London. | Needs to explain why the new industries grew there, so this is only part of the answer, so no mark. | Some cities like London have had rapid growth as a result of the development of the financial services. |
| New industry have made towns grow quickly. | This is partly correct but misses the point about the location of new industries, so no mark. | New industries preferred to locate on the edge of towns so these areas have grown rapidly. |

Total available for spelling, punctuation and grammar = 3 marks. Marks achieved: 2.

- - - - - - - - - - - - - - - - - - - - - - - - - - - - - - - - - - - - - - - - - - - - -

**Higher question:** Explain the environmental impacts of the rising demand for high quality residential areas in one UK city. (8 marks)

| Student answer (achieving 4 marks) | Feedback comments | Build a better answer (achieving 8 marks) |
|---|---|---|
| Leeds has had some very fast economic growth so more houses are being built. | This scores no mark as it does not look at the environmental impact of growth. | More houses and flats are built on brownfield sites in Leeds. |
| Flats have been built on brownfield sites. | Good part of the answer, scores 2 marks. | Nineteenth-century houses have been upgraded and improved. |
| More houses are being built in the green belt | Just gets to the point but could have linked it to the pressure for more homes. Scores 2 marks. | There is increased pressure to build on green spaces including the green belt. |

Total available for spelling, punctuation and grammar = 3 marks. Marks achieved: 2.

# Chapter 15 The challenges of an urban world

242

## Objectives

- Examine the global patterns of urbanisation.

- Contrast world cities in the developed and developing worlds.

- Identify the social and environmental challenges faced by cities.

## How have cities grown and what challenges do they face?

### World urbanisation

Urbanisation is the movement of people from rural areas to towns and cities – and it is happening all over the world. Between 2011 and 2030 the world's population living in cities is expected to increase from 3.6 billion to 5.1 billion. Almost all of this expected growth will be in the towns and cities of the developing world. In 2011, the number of people living in cities in developing countries was 2.7 billion, and by 2030 there will be 4.1 billion. By contrast, the urban population in developed countries will only grow from 1 billion to 1.05 billion. Figure 1 shows how the percentage of the population living in towns and cities for different parts of the world has increased from 1950 to 2007, with a projection to 2030.

The urbanisation now taking place in developing countries is very different from that which took place in developed countries.

## Activity 1

Study Figure 1.

(a) In Europe, what was the percentage of people living in towns and cities in 1950?

(b) How will this change by 2030 ?

(c) In Asia, what was the percentage of people living in towns and cities in 1950?

(d) How will this change by 2030?

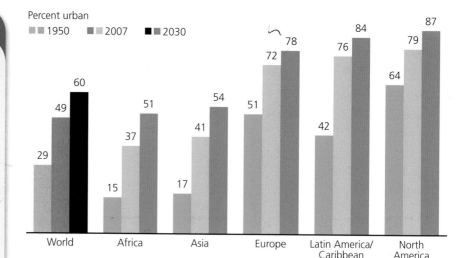

Figure 1: Percentage of world population living in towns and cities in 1950, 2007 and 2030

### Urbanisation in developed countries

- In the developed world, towns and cities grew in the nineteenth century during the Industrial Revolution. There was an Agricultural Revolution at the same time, so new machinery meant fewer farm workers were needed. So people moved to towns where there were jobs in new factories.

- Towns grew at rates of 10% per year.

- Cities continued to grow into the twentieth century as a result of rural depopulation.

● 90% of the UK population lives in towns or cities (Figure 2).

● Some people are now leaving cities to live in villages where they think the quality of life is better. This is called **counterurbanisation**.

● Some large cities like Birmingham are growing again (this is called re-urbanisation) as younger people choose to live close to city centres.

### Urbanisation in developing countries

● On a world scale, urban growth is now concentrated in developing countries.

● By 2010 there were 25 megacities (urban areas with over 10 million inhabitants) – and over half were in developing countries.

● The rate of urbanisation in developing countries is much faster than in nineteenth-century Europe.

The main reasons for the growth of cities in developing countries are:

● **Natural increase** (birth rates higher than death rates) due to high fertility rates (large numbers of children per woman). So cities are dominated by young children (Figure 3).

● Migration from the countryside as a result of **push factors** (lack of jobs in the countryside, increased mechanisation of farming, crop failures, harsh rural life) and **pull factors** (job available in towns, better health care and services such as piped water and electricity in cities, higher wages, better entertainment).

Figure 2: Central London

## ResultsPlus
**Build Better Answers**

### EXAM-STYLE QUESTION

**Explain the rapid growth of cities in developing countries in recent years. (8 marks)**

■ **Basic answers** (1–3 marks)
Only talk about the rapid speed of growth in cities in developing countries, and do not offer an explanation.

● **Good answers** (4–6 marks)
Offer rural to urban migration as the main reason for the increase and do not include population change.

▲ **Excellent answers** (7–8 marks)
Identify two reasons for the rapid growth of developing cities and include examples of push and pull factors.

Figure 3: Cities in the developing world have lots of young children

244

*Figure 4: Central Paris*

## World cities and megacities

'**World cities**' are places such as London, Paris (Figure 4) and New York (in developed countries) or Beijing, New Delhi and Singapore (in developing countries).

Their main characteristics are:

● They have the world's main stock exchanges and major stock markets.

● They are centres of political power.

● They have the headquarters of the transnational companies (TNCs) and the firms that provide financial services.

● They are the main centres for the world's media organisations, e.g. BBC, Thomson, Reuters, Al Jazeera.

● They have world-class cultural institutions (theatre, opera, ballet) and major sporting teams.

● They are usually centres of tourism.

● They have mass transit transport systems – light rail, metro.

● They have one or more major airports.

● They are where the rich and powerful usually live.

## Megacities

Megacities are cities with over 10 million people. They may have more than one main centre but they are recognisable as one continuous urban area. Examples in the developed world include:

● New York

● Los Angeles (Figure 5)

● Osaka.

In the developing world, examples include:

● Guangzhou

● Jakarta

● Seoul

● Shanghai

● Mexico City.

*Figure 5: Freeways in Los Angeles are often very busy*

Megacities are major centres of economic activity, including manufacturing industries as well as financial and service industries. But they do not have the same global importance as world cities, because they are not the centres of government and they do not have headquarters for all the major national and international companies. They will also generally have fewer cultural outlets than world cities.

### *Spatial growth of megacities*

The spatial growth of megacities in the developed world differs from that of megacities in the developing world.

In the developed world the megacities usually have distinct urban zones, based on their spatial growth over a long time:

- The Central Business District (CBD) is at the centre, with its shops and offices.

- Then there is an 'inner city' zone with mixtures of old decaying housing and factories and brand new developments.

- Then there are the suburbs which tend to be mostly residential (houses) with a few supermarkets.

In developing countries, apart from a central CBD, surrounded by older housing, the spatial growth pattern of megacities tends to be different:

- The layout is much less regular than in megacities in developed countries.

- Land uses are not separated so well, because growth has been so fast and planning controls are often weak.

- People set up homes on any patch of land they can find because there is such a shortage of housing. These spontaneous or squatter settlements, with mostly slum housing, quickly grow into shanty towns.

- There may be distinct sectors of land use, such as expensive houses stretching out along a particular road.

### *The population of megacities*

- In megacities in the developing world – like Mexico City – there are high fertility rates (large numbers of children per woman) and high rates of natural increase (birth rates higher than death rates). The result is a dominance of young people under 25 in these cities.

- By contrast, in megacities in the developed world – like Osaka – the population structure is dominated by older people, because fertility rates are lower, as are rates of natural increase.

- Many people in megacities in the developing world live in squatter settlements because they cannot afford better housing. In 2012 over 1 billion people, nearly a seventh of the world population were living in squatter settlements.

- In the developed world, the megacities still have some areas of poor, slum housing.

**Key**

| | | | |
|---|---|---|---|
| CBD | Inner city | Suburbs |

*Figure 6: A simplified diagram of the layout of developed world megacities*

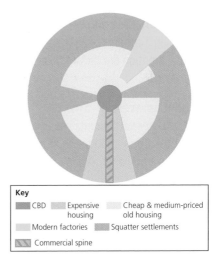

**Key**

| | |
|---|---|
| CBD | Expensive housing | Cheap & medium-priced old housing |
| Modern factories | Squatter settlements |
| Commercial spine | |

*Figure 7: A simplified diagram of the layout of developing world megacities*

### Activity 1

**Use the table below to compare megacities in developed and developing countries**

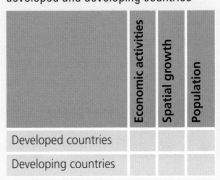

| | Economic activities | Spatial growth | Population |
|---|---|---|---|
| Developed countries | | | |
| Developing countries | | | |

## ResultsPlus
### Build Better Answers

**EXAM-STYLE QUESTION**

**Explain what squatter settlements are and why they are found in developing cities. (6 marks)**

■ **Basic answers** (0–2 marks)
Only talk about the nature of squatter settlements in cities in developing countries, and do not offer an explanation for their growth.

● **Good answers (3–4 marks)**
Talk about building materials used in squatter settlements and mention the lack of homes or the lack of land rural.

▲ **Excellent answers** (5–6 marks)
Explain what squatter settlements are and the reasons for their growth in terms of rapid population increase and migration plus a lack of land and homes.

# Challenges facing cities – like London and Paris – in the developed world

## Food

The provision of food to a large city like London or New York is a major challenge. Most food is not grown in London itself but has to be produced elsewhere in the UK and the world. Then it has to be transported to the city by road, rail or air. For example, London gets fresh vegetables like beans from Kenya. This all adds to the cost of the food and also the **carbon footprint** of the city because of the fuel burned in the transportation. In order to try to reduce this dependence on imported food, people are being encouraged – in New York, for example – to buy more locally produced food.

## Water

A second major challenge for cities in the developed world is the provision of safe, clean water. Demand for water in many cities is far greater than supply. For example, each person in London consumes 161 litres of water each day compared to the average for the UK of 149 litres. It is made worse in London by the increasing number of single-occupancy homes which, on average, consume 78 litres per person more than a household of four people. Add to this the fact that the rainfall in London is lower than the national average and this creates a sizeable challenge. So water has to be brought from areas much further away, such as the Lea Valley. In addition, a desalinisation plant has been built to draw water from the River Thames and treat it to make it available to the population. This plant provides water for nearly one million people.

## Energy

Providing enough gas and electricity is a major challenge. At the moment electricity is generated in power stations around the UK which burn gas, oil and coal or use nuclear power. All these sources of fuel have major problems. For example coal pollutes the air and adds to global warming and the carbon footprint, as to a lesser extent do oil and gas. Nuclear power is potentially dangerous as it may release radioactivity. Yet demand for electricity in cities like London and New York is huge because there are so many homes and businesses there. So vast amounts of resources have to be used to generate enough electricity for the city.

## Transport

London is the focus of the UK's road and rail network, but with so many people and so many businesses, transport is a major challenge. Some 8.6 million people live in London's 1706 square kilometres. There are almost 3 million private cars in London as well as 8,500 buses and 18,000 black cabs, plus coaches, lorries and motor cycles. Car ownership in London is 34 per 1000 people so despite the 408 km of underground rail network there are a lot of vehicles on the roads. So traffic congestion particularly at peak times is a major issue (Figure 8). In addition, all the vehicles generate exhaust fumes which create high levels of nitrogen dioxide and other pollutants in the air.

*Figure 8: Rush hour traffic in central London*

Technology is improving vehicle exhausts which is helping to keep pollution levels down but still 41% of the nitrogen dioxide pollution in London comes from cars and motorcycles. Road traffic contributes 22% of London's carbon dioxide output. The introduction of the congestion charge in 2003 which makes vehicles entering central London pay a charge, has led to a reduction of 13% in the levels of nitrogen dioxide in the city centre. Some 80% of the congestion charge is spent on public transport (buses) and 20% on roads.

### Waste

The 8.6 million people and the thousands of businesses in London all generate a lot of waste, which has to be removed from the city or disposed of safely. London generates about 20 million tonnes of waste each year, much of which is still buried in **landfill** sites. This is an expensive and wasteful way to dispose of waste. The annual cost to London of sending waste to landfill is £260 million (2011) and by 2014 this will be £280 million. In addition, the green waste and food which goes to landfill generates methane as it decays in the ground and this adds 460,000 tonnes each year to the greenhouse gases in the atmosphere above London. So London has embarked on a major scheme to reduce the amount of waste generated by the city by 20% by 2031 and to increase recycling and composting by 45% by 2015 and 50% by 2020.

## Challenges facing cities in the developing world

### Housing

Simply trying to provide enough housing for everyone is a major challenge in developing world cities. The cities are growing so fast that they just cannot cope with the numbers of people looking for a home. So people choose to build their own homes on any spare land they see, using scrap materials such as cardboard, corrugated iron and plastic. The houses may be a fire hazard, and the land may be very steep, or a swamp, or prone to flooding. Most of the homes are unplanned and have little or no electricity, water supply or sewerage facilities. Up to 50% of a city's population may live in these squatter settlements. Sewage often pollutes the water supply, leading to diseases such as diarrhoea, typhoid and cholera. There is often no organised refuse collection, so rubbish builds up and is a breeding ground for disease. Many people suffer from malnutrition and food may be in short supply.

### Transport

Traffic congestion is very bad in cities like Mexico City and Beijing. The roads in many of these cities were never built to take so much traffic, so there are frequent holdups and congestion. Car ownership has grown very rapidly in many cities. In Beijing, for example, there were 3 million cars in 2006. By 2011 this figure was 5 million. The growing number of cars has created gridlock in some parts of the city and air pollution has become a major issue. In 2008, the government had to ban some cars before the Olympic Games in order to improve the air quality. In 2011 the government restricted the number of new car registrations to 240,000 (a third of the previous year's total) in an attempt to improve conditions. The large number of vehicles generate serious air pollution, creating health issues such as asthma and bronchitis. In some cases dense smog settles over the whole of Beijing (Figure 9)

Figure 9: Smog is a big problem in Beijing

### Water supply and pollution

The provision of adequate water supplies and preventing the pollution of those supplies are vital challenges for cities in the developing world. The United Nations estimates that 1 billion people do not have access to adequate supplies of water and 2 billion do not have access to adequate sanitation facilities. Dhaka is a major city of 9 million people in Bangladesh. The city is not able to supply enough safe water, so people have to drink from pools of water on the ground. This water is often polluted, and drinking this type of water is responsible for 2 million deaths worldwide each year. Open water attracts mosquitoes which may transmit malaria. It is also a breeding ground for snails, which can carry diseases such as schistosomiasis, a disease which affects the intestines. In addition, water can easily be polluted by animal and human waste, by fertilisers, by industrial chemicals and by the run-off from towns. The challenge for the authorities is to increase access to water treatment facilities. In India, for example, a 2010 survey by the Central Pollution Control Board found that towns and cities treat less than 30% of their sewage, and discharge 26 million m$^3$ of untreated wastewater into rivers and coastal waters. Similarly, the water in the canals of Bangkok in Thailand is so badly polluted that its biochemical oxygen demand (BOD), a measure of water quality, is equivalent to the BOD of the water in sewage-treatment works.

### The informal economy

In the developing world, a major problem in cities is unemployment or underemployment. Underemployment is when people do not have full-time, continuous work, but instead work seasonally or temporarily. Some people can get jobs in, for example, the car factories of São Paulo in Brazil. This is employment in the formal sector – with permanent jobs and regular pay. But most people cannot get these jobs, so they work in the informal sector. They shine shoes, give haircuts, sell water, carry luggage, take photographs to sell or they make food to sell. They have no shop or office, but they often set up on a street corner (Figure 10).

*Figure 10: A street seller*

# How far can these challenges be managed?

## Managing challenges in the developed world

### Reducing cities' environmental impact (eco-footprint)

People all over the world have recently become concerned about the impact that cities are having on the environment. They are particularly worried about the pollution of air, land and water by the things we do in our homes, factories and offices. One way of measuring the impact we have on the environment is an 'ecological footprint' (**eco-footprint**). The eco-footprint of a city looks at how much land is needed to provide it with all the energy, water and materials it uses. The footprint calculates how much pollution is created by burning oil, coal and gas, and it works out how much land is needed to absorb the waste created by the people of the city. The purpose is to work out how sustainable any city is and especially what changes we need to make to improve the quality of life for people now and in the future.

Eco-footprints are expressed in terms of how much land is needed to support the lifestyle of the people and to deal with the pollution and waste they create. The UK average is 5.3 ha (hectares) per person. But this average varies from place to place. For example, towns in the countryside generally have a lower eco-footprint than towns close to other cities. This is because they may produce more of their own food (and possibly power) and they may generate less waste.

*Figure 11: Landfill sites are used to get rid of many cities' rubbish*

The contrasts are even greater between different countries (Figure 12). As you can see, the USA has a very large eco-footprint because it uses so much power and food and generates so much waste. The UK has a larger eco-footprint than Germany, because Germany has a very sophisticated recycling programme. Developing countries like Kenya and India have much smaller eco-footprints. This is because they grow much of their own food and they may use local wood for cooking, heating and lighting. In addition, the people get around by walking or on bicycles rather than by using cars, so they create a lot less pollution of the environment.

## Objectives

- Examine how the environmental impact of cities in the developed world can be measured by their eco-footprint, and how this can be reduced.

- Examine how self-help schemes, NGOs and urban planning can help to manage social and environmental challenges in cities in the developing world.

- Discuss the successes and failures of efforts to develop cities that are less polluted.

## Activity 2

Study Figure 11.

(a) Do you think this employs many people?

(b) Why are sites like this used to get rid of rubbish?

## Top Tip

Remember that eco-footprints measure the amount of land that would be needed to support people's lifestyles and dispose of the pollution and waste created. They are a useful way of calculating the impact of people's lifestyles on the environment.

## Skills Builder 1

Look at Figure 12.

(a) Which country has the largest eco-footprint?

(b) Which country has the smallest eco-footprint?

(c) What is the size of Germany's eco-footprint?

(d) What is the size of the world's eco-footprint?

## Activity 3

(a) Why is using a shower rather than a bath a good idea?

(b) Why is it a good idea to wring or spin dry wet clothes before putting them in the tumble dryer?

(c) Why is it a good idea to wait for a full load before using the dishwasher?

(d) Why should we not leave TVs on standby?

(e) Why do we need to turn off lights when leaving a room?

(f) Carry out a survey of the class. How many people do half of the things on the list to save water and electricity? How many people do one-third of the things on the list? How many people do none of the things on the list?

(g) Some homes do not have a dishwasher, washing machine or central heating. Some homes do not have a shower. How could people in these homes reduce their eco-footprint?

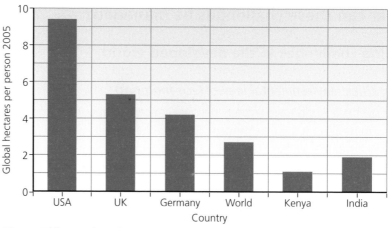

Figure 12: The eco-footprints of different countries and the world in 2009

### Reducing York's eco-footprint

York is a typical British city which is keen to reduce its eco-footprint. York's eco-footprint is 5.4 ha per person – slightly above the UK national average. The city's attempts to reduce its eco-footprint are focused on (1) reducing energy use and (2) reducing the amount of waste it creates.

Figure 13 shows that energy makes up 24% of York's eco-footprint. Domestic gas consumption makes up most of this, with electricity second. The local council have produced a series of tips for people to save energy in their homes, because home heating accounts for 60% of the carbon dioxide emissions that come from household use. The tips include:

1. Turn down the thermostat. A cut of 1 °C can save 10% off the bill.
2. Don't leave the fridge door open.
3. Use the shower rather than the bath – it uses 40% less water.
4. Don't put wet clothes in the tumble dryer, wring them or spin them first.
5. Use the 30 °C cycle on the washing machine not the 40 °C.
6. Don't fill the kettle if you don't need to heat all the water.
7. Wait for a full load in the washer or dishwasher.
8. Fix dripping taps.
9. Don't leave appliances like TVs on standby.
10. Turn off lights when you leave a room.
11. Close curtains at dusk to conserve heat.

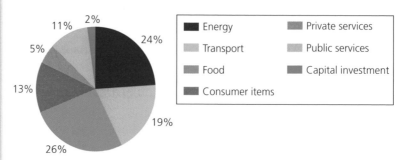

Figure 13: How York's eco-footprint is made up (Data provided by SEI, using REAP baseline 2004)

York is also trying to reduce its eco-footprint by reducing the amount of waste it creates. In 2011, York generated 85,000 tonnes of waste. The manufacture of products and the disposal of waste needs energy, which in turn produces greenhouse gases. Worse still, as the waste decays it produces methane, a greenhouse gas. However, York now recovers 70% of the methane from its landfill sites and it is used to generate electricity.

How is waste reduction happening?

- Recycling is very important. People have different coloured bins and they sort their waste. This has led to a big reduction of 30% since 2006 of the amount of waste going to landfill (Figure 14).

- Shops and businesses now sort their own waste.

- Only 16% of commercial waste is recycled and this is mostly office paper. So offices are now asked to use electronic communication like email and text.

- People are encouraged to buy fewer pre-packaged goods, as this creates a lot of waste.

Figure 14: Yorwaste deals with waste in York

### The potential for more sustainable transport in Amsterdam

Amsterdam is already one of the most cycle-friendly cities in the world, but this is just part of its serious attempt to develop an even more sustainable transport system, based on its electric trams and metro (public transport) and electric vehicles (private transport).

Cycling is the main form of transport in the city centre, accounting for 40% of all commutes (Figure 15). Cyclists in Amsterdam benefit from an infrastructure that allows them to be the dominant form of transport. There are cycle lanes everywhere and thousands of bicycle parking places. The canal streets in Amsterdam are traffic calmed, with cycle lanes wide enough for two cyclists abreast. The main street junctions have a priority position for cyclists to help them keep moving safely ahead in traffic. The local traffic laws mean that in almost any accident involving a car and a bicycle, the car driver is considered to be at fault.

*Figure 15: There are thousands of cyclists in Amsterdam*

In addition, Amsterdam already has an extensive electric tram system which serves both the city centre and the suburbs. There is also a four-line metro system which is underground in the city centre but then above ground in the suburbs. There are plans to extend the network in the future.

Amsterdam is looking to develop sustainable transport still further by encouraging the use of electric private vehicles. A new subsidy scheme will repay local businesses up to 50% of the cost of purchasing an electric vehicle. Three billion euros have been set aside to cover these costs and to stimulate the project. New recharging stations are being erected all over the city (and throughout the Netherlands), many using 'green electricity' (electricity generated from green sources such as wind or solar power). At one 'park and ride' hub on Amsterdam's outskirts, the parking fee includes free recharging for electric cars.

## Managing social and environmental challenges in the developing world

There are a range of approaches to improving the quality of urban life in developing countries. These involve national governments, city governments, non-governmental organisations (NGOs like Oxfam) and local self-help schemes. In Curitiba in Brazil, for example, the government is producing a model for a greener urban future.

### Activity 5

(a) List as many ways as you can think of that Amsterdam could develop still further its potential for sustainable transport (think about car sharing, electric transport for buses and taxis, bicycle loans, greater use of water transport).

(b) What are likely to be the main problems with trying to develop this potential?

### Curitiba – a role model for a greener urban future

An urban plan guided the growth of Curitiba in southern Brazil, which now has a population of 2.2 million. The plan was based around several axes crossing the city which operate as high-speed, one-way, three-lane roads, with the central lane reserved for express buses. The success of Curitiba is based around its public transport policy, centred on buses. There is an 'Integrated Transportation Network ' in which bus lanes are interconnected through twenty terminals. The system is fast, efficient and cheap. The buses run very frequently, some as often as every 90 seconds. The buses stop at cylindrical clear-walled stations with turnstiles, steps and wheelchair lifts. Instead of steps the buses have extra wide doors and ramps that extend out to the station platform when the doors open (Figure 16). This system of same-level bus boarding results in a typical average time at the bus stop for the bus of only 15 to 19 seconds.

The system transports 2.6 million people every day, and is used by 70% of the city's people. One fare allows passengers to travel anywhere on the system. Buses now use **alternative fuels**, especially natural gas, which creates less pollution. Curitiba uses 30% less fuel per person than the eight other Brazilian cities of the same size. So now this pattern of public transport and of land use is a model for other countries, because it has greatly reduced greenhouse gas production and prevented traffic jams (and their pollution) within the city.

Figure 16: A bus stop in Curitiba

### Activity 6

(a) What is the main cause of Curitiba's success?

(b) Why do buses stop for such a short time at stops?

(c) What are the benefits of using alternative bus fuels?

(d) Why is Curitiba a model for other countries?

### The work of NGOs

Non-governmental organisations are concerned with a range of issues, such as human rights as well as trying to reduce poverty or prevent disease. NGOs raise money by campaigning or applying to governments or other organisations for grants, and in developing countries they work to improve the quality of life for people in cities. For example:

- In Sri Lanka, the Urban Green Partnership Programme helped to create 300 home gardens in three cities – Matale, Moratuwa and Badulla. This has helped local people feed themselves, and now the local council requires all new buildings to incorporate green space in their design.

- In Kachhpura (Figure 17) – a small town close to Agra in India – one of the problems is that sewage and waste water drain into the main channel that runs through the town, on its way down to the river Yamuna. A proper sewage plant is planned but so far this has not been built. So an Indian NGO, the Centre for Urban and Regional Excellence (CURE) has now provided a low-tech temporary solution which can cope with the disposal of the sewage and waste water. This simple plant requires little technical maintenance and does not depend on electrical or other forms of power.

*Figure 17: Kachhpura, India*

### Self-help schemes

Rocinhna is a neighbourhood in Rio de Janeiro in Brazil. Here, 70,000 people live packed close together in houses built on a steep slope. It used to be no more than a favela, with unsafe homes built from plywood and cardboard. But the local authority is now helping the people to help themselves. It provides the building materials, such as breeze blocks and cement, and the local people provide the labour. This means that more and more houses in Rocinha are built using solid, stable materials, meaning that they can withstand floods and mud slides much better than before.

## Developing a less polluted city: Mexico City

Mexico City – with its 21 million people – is one of the world's largest cities. It is also a city with three serious environmental challenges, which it is vigorously addressing: air pollution, waste disposal and water pollution.

### *Air pollution: the problem*

In 1998 the United Nations described Mexico City as 'the most dangerous city in the world for children'. This was because air pollution was so serious that young children in the city were vulnerable to chest and breathing infections – and they were dying much earlier than in other cities. Over 1000 deaths and 35,000 hospital admissions were attributed to air pollution in 1998. The main sources of air pollution were vehicle exhausts, industrial emissions and emissions from power stations. The danger arises from:

- Nitrous oxides, which in bright sunshine undergo a chemical change and are converted into nitrogen dioxide. This lethal substance accumulates especially in cities like Mexico City which are surrounded by mountains, to form a photo-chemical smog.

- Ozone, which is created by toxins from vehicle exhausts which react with air in bright sunlight. Ozone sensitises the body to irritants such as pollen and house dust, causing symptoms similar to hay fever.

- Carbon monoxide, which comes from the incomplete combustion of fuels. It causes headaches, fatigue and even death.

- Sulphur dioxide, which is produced from car exhausts. It irritates eyes and can make breathing difficult.

### *Trying to solve the air pollution problem*

Mexico City has introduced various measures to tackle air pollution, including:

- Providing funds for the spare parts needed by buses in the city. Many buses were unreliable and badly maintained which adds to air pollution.

- Changing the legal formula for petrol and diesel so that they now contain fewer pollutants.

- Prohibiting drivers from using their cars on one day per week.

- Building a new $2 billion underground train line which will reduce the number of cars and reduce average commuting times from 150 minutes to 78 minutes.

- Introducing bigger, more efficient articulated buses along four routes to speed commuting and reduce pollution (Figure 18).

- Starting up a bike share scheme – Ecobici – with 275 stations and 4,000 bicycles.

- Moving – or closing down – the worst polluting factories.

*Figure 18: One of the new articulated buses in Mexico City*

**The main advantages and disadvantages of Mexico City's air pollution measures**

| Measure | Advantages | Disadvantages |
|---|---|---|
| Funds for spare parts | Will reduce air pollution and improve reliability of the bus services. | Cost of parts is high because they have to be imported. Buses are old so they will wear out eventually. |
| Changing formula for petrol and diesel | Fewer pollutants emitted into the air from all vehicles. | This is expensive and some critics argue that money would be better spent on less polluting engines. |
| Prohibiting drivers from using their cars one day per week | Reduced traffic and therefore reduced air pollution. | Drivers get round it by buying two cars. |
| Underground line | Reduced air pollution as more people use it rather than using their cars. | Very expensive. |
| Articulated buses | Fewer buses on the road, so less air pollution. | Expensive, but not as expensive as an underground line. |
| Ecobici | Reduces congestion and air pollution. | No good for people with a long commute. |

Figure 19: A rubbish mountain in Mexico City

### The waste disposal problem

Mexico City produces 13,000 tonnes of rubbish every day – but the waste collection system can only remove 9,000 tonnes each day. The rest is dumped on open ground, and in waterways, streets and drains, where it clogs the system. In 2012 the biggest waste dump in the city, Bordo Poniente was closed, but no alternative was suggested. This resulted in a rubbish mountain (Figure 19) and a refusal by surrounding towns to take Mexico City's waste. The closed site had been used since 1985 and was receiving 12,000 tonnes per day, of which 7,000 tonnes came from other towns close to the city. Worse still, 70 million tonnes of waste had been buried underground at Bordo and this was causing serious water and air pollution. In 2012 Mexico City was sending 3,600 tonnes of rubbish per day to landfill, 3,000 tonnes were going to make compost and 800 tonnes was used to make plastic bottles. The rest had no designated destination.

### Trying to solve the waste disposal problem

Mexico City has introduced various measures to tackle the waste disposal problem, including:

- Encouraging more recycling

- Building a new plant to burn some waste to generate electricity

- Encouraging more composting

- Burying it in new landfill sites.

## The main advantages and disadvantages of Mexico City's waste disposal measures

| Measures | Advantages | Disadvantages |
|---|---|---|
| More recycling | A major part of the long-term solution – reducing waste that goes to landfill. | Many people are slow to adopt recycling and it is only part of the solution. |
| Building a new plant to burn waste | Reduces waste going to landfill and reduces water and air pollution. | May add to air pollution, and this is not the best use of the waste material. |
| More composting | A big part of the long-term solution. Reduces waste going to landfill so reduces air and water pollution. | There is a limit to how much waste can be composted. |
| Burying it in new landfill | Will ease the problem in the short term. | May add to air and water pollution and not a long-term solution. |

### The water pollution problem

Water supply and pollution are major challenges for Mexico City. As the city has grown, the problem has arisen as a result of population growth and the over-exploitation of underground water supplies. This has been made worse by a failure to recycle and a failure to build sewage-treatment plants. For many years the only solution to the lack of water was to pump water up from the 514 underground aquifers. This has now gone so far that the land surface of the city is sinking at the rate of 9 cm per year. This is causing water and gas pipes to fracture and roads to crack, as the aquifers start to dry up. Worse still, each person in Mexico City uses 320 litres of water each day and this excessive consumption is causing problems for sewage plants which cannot cope with the volume.

### Trying to solve the water pollution problem

Mexico City has introduced various measures to tackle the water pollution problem, including:

- Building more sewage-treatment plants

- Saving more rainfall in underground tanks and cisterns

- Recycling more water (currently only 10% is recycled)

- Pumping even more water up from even deeper underground wells (up to 1000 metres deep).

### ResultsPlus
### Build Better Answers

**EXAM-STYLE QUESTION**

**Using named examples, compare the challenges facing cities in the developed and developing worlds. (8 marks)**

■ **Basic answers** (1–3 marks)
Have descriptions of urban problems but they are not linked to specific developed or developing cities. Focus on one type only or do not compare. Use some geographical terms.

● **Good answers** (4–6 marks)
Talk about a range of problems in both types of city and use examples with some detail. Tend to be two separate accounts. Good use of relevant terms.

▲ **Excellent answers** (7–8 marks)
Use detailed and appropriate examples to directly compare the challenges facing both types of city. Look at similarities and differences and make excellent use of geographical terms.

## The main advantages and disadvantages of Mexico City's water pollution measures

| Measures | Advantages | Disadvantages |
|---|---|---|
| Build more sewage-treatment plants | Will reduce water pollution. | Expensive to build. |
| Save more rainfall in tanks | Would reduce water that has to be pumped from aquifers. | People who try to store water in a tank at home find it is easily polluted by animals and insects. |
| Recycle more water | Would reduce water that has to be pumped from aquifers. | There is a limit to how much can be recycled. |
| Pump more water from deeper wells | Will help to meet demand and so reduce pollution. | Very expensive, and will only add to subsidence in the city as the ground sinks. |

Urban areas all over the world are changing, and having to face a series of challenges linked to the provision of the food, energy and raw materials that they need, as well as disposing of their waste. Cities are struggling to find ways to be more sustainable and to reduce their eco-footprints, and to develop a range of strategies to cope with the challenges they face.

## You should know...

- [ ] Why cities are growing
- [ ] Why cities in developing countries are growing very fast
- [ ] Megacities vary in terms of their population, spatial growth and economic activities
- [ ] Cities in the developed world face a series of challenges, including food, energy, transport, and waste disposal
- [ ] Cities in the developing world face challenges, including slum housing, urban pollution and the informal economy
- [ ] What an eco-footprint is
- [ ] Why some cities have a large eco-footprint
- [ ] Why eco-footprints vary in size from place to place
- [ ] How places can reduce their eco-footprint
- [ ] How we can make transport more sustainable
- [ ] How the quality of life can be improved in developing world cities
- [ ] The advantages and disadvantages of trying to develop less polluted cities
- [ ] What the main causes of environmental pollution in cities are

## Key terms

Carbon footprint
Consumption
Counterurbanisation
Eco-footprint
Fertility rate
Informal economy
Landfill
Megacities
Natural increase

Pollution
Quality of life
Re-urbanisation
Self-help schemes
Squatter settlements
Sustainable transport
Urbanisation
Waste
World cities

### Which key terms match the following definitions?

**A** Disposal of rubbish by burying it and covering it with soil

**B** Material produced by households that needs to be disposed of

**C** The presence of noise, dirt, chemicals or other substances which have harmful or poisonous effects on the environment

**D** A measure of how much land is needed to provide a place, e.g. a city, with all the energy, water and materials it needs, including how much is needed to absorb its pollution

**E** The using up of something

**F** A major urban area that has a significant role in controlling the international flows of capital and trade

**G** The degree of well-being (physical and psychological) felt by an individual or a group of people in a particular area. This can relate to their jobs, wages, food, amenities in their homes and the services they have access to such as schools, doctors and hospitals

**H** The movement of people from rural areas into towns and cities

To check your answers, look at the glossary on page 321.

**Foundation Question:** Describe how far one UK city has been successful in reducing its eco-footprint. (6 marks)

| Student answer (achieving 3 marks) | Feedback comments | Build a better answer |
|---|---|---|
| York is a city trying to cut its eco-footprint | This is an introductory sentence that does not contain enough detail to score marks. | York's eco-footprint is 5.4 ha per person – just above the UK average. |
| Energy is a lot of the eco-footprint | This is a valid point on inputs and scores 1 mark. | Energy makes up 24% of the eco-footprint of York. |
| York has a lot of waste that goes to landfill | This scores 2 marks because it is correct, but would be better with statistical detail. | York produces 85,000 tonnes of waste and recovers 70% of methane from landfill to burn to generate electricity. This is a success. |
| York has given people advice about energy saving and recycling | This is very general and lacks detail. | People are advised to buy fewer pre-packaged goods and to follow advice on reducing energy use. The success of this is hard to judge. |

Total available for spelling, punctuation and grammar = 3 marks. Marks achieved: 1.

- - - - - - - - - - - - - - - - - - - - - - - - - - - - - - - - - - - - - - - - - - - - - - - - - - -

**Higher Question:** Explain why cities such as London have a large eco-footprint. (8 marks)

| Student answer (achieving 3 marks) | Examiner comments | Build a better answer |
|---|---|---|
| London is an unsustainable city so it has a large eco-footprint. | This is an introductory sentence that does not contain enough detail to score marks. | Large eco-footprints occur when cities such as London use huge amounts of resources, recycle almost nothing and create waste and pollution. |
| It uses vast quantities of food, energy, water and building materials. | This is a valid point on inputs, with a good range of examples and scores 2 marks. | It uses vast quantities of food, energy, water and building materials. |
| The 8 million people in London have low levels of recycling. | This scores 2 marks because it is correct and includes an example. | The 8 million people in London have very low levels of recycling – around 9%. |
| In places like Freiburg in Germany and Copenhagen in Denmark they recycle around 55%, so they are more sustainable. | This part of the answer is not relevant as it is explaining why some cities have low eco-footprints, so is not relevant to the question. | Low levels of recycling means that outputs of waste going to landfill are huge and this, combined with pollution from factories, leads to very high outputs. |

Total available for spelling, punctuation and grammar = 3 marks. Marks achieved: 2.

# Chapter 16 The challenges of a rural world

260

## What are the issues facing rural areas?

### Rural areas show contrasting economic characteristics

A rural area is usually defined as one that is relatively sparsely populated and either largely given over to farming or left as wilderness. Rural areas are often described as countryside.

No two rural areas are the same. Unique combinations of relief, climate and soil produce different landscapes. These different physical combinations mean that some areas are more attractive to people and their settlements than others. Some areas are more fertile than others for agriculture.

Agriculture is often thought to be the dominant economic activity of most rural areas. However, in many developed countries, the traditional focus on farming is declining, as other activities such as tourism, leisure and recreation become more important as sources of rural employment and income. At the same time, the distinction between rural and urban is becoming less clear-cut. In short, the countryside of the developed world today is quite different to what it was, say, 200 years ago. It also contrasts with the countryside typical of much of today's developing world, which is also changing. These contrasts are reflected in the rural issues facing both types of countryside (Figure 1).

Figure1: Rural areas in the developing and developed worlds

### Rural areas in the developed world

A number of changes have resulted in the rural areas of today's developed world becoming involved in a range of economic activities – not just farming. Let us look at these changes and the ingredients of the 'new' rural economy:

- Agriculture – The most important activity in most rural areas used to be agriculture. But Figure 2 shows that it has declined as an employer in the UK. Mechanisation has cut the number of jobs. Although the amount of land being farmed has decreased, what it produces has increased. But the growth of the global economy is resulting in cheaper food being imported from overseas. For example, the UK now only grows around 60% of its food. The cutback in farming jobs has been marked in more remote rural areas. Remoteness means higher transport costs. Slightly higher costs can easily tip the balance of whether or not a farm remains profitable and stays in business. Farming has also declined close to the fringes of towns and cities as more and more of the countryside is converted from fields into building plots.

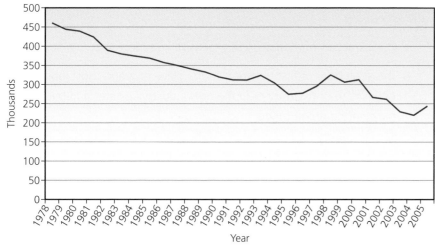

Figure 2: Employment in agriculture in the UK (1978–2005)

● Other primary activities – Mining and quarrying, forestry and fishing are long-established rural activities. They survive today but only in particular areas; they do not support large numbers of jobs.

● Recreation, leisure and tourism – One of the features of twenty-first-century living is that many people in the developed world have both spare time and **disposable income**. Both are being 'spent' on leisure, recreation and tourism. Leisure and recreational needs are being met in the rural areas that can be easily reached from towns and cities. Along the **urban fringe**, you will find sports centres, playing-fields, golf courses and country parks. Further out, the facilities may be somewhat different. In order to reach them, urban people will have to think in terms of a day-trip rather than just a short drive in the car. Maybe it will be a farm visit, a fun day at a theme park, some birdwatching around a nature reserve or an active day at a water park.

Figure 3: Recreation in the urban fringe, and in the accessible countryside

- Commuting – In the past, many people moved from the countryside into urban areas. But now there are substantial movements of people and businesses in the opposite direction. People are leaving the city and larger towns and buying homes in the countryside or smaller settlements. But many continue to work in the city or town they have just left. In short, they become commuters. This is happening in rural areas that are within easy reach of towns and cities. The two main reasons for this outward movement are:

  - the attraction of modern, less expensive and more spacious housing

  - the availability of fast transport to the place of work. Most jobs are either in the city centre or in the new industrial estates and business parks in the urban fringe.

**Activity 1**

Can you think of any others reasons why people choose to move out of towns and cities?

People feel that the time and money spent on commuting are worth the better quality of life to be found outside cities.

- Retirement – This and commuting make up what we might term a growing residential activity in rural areas. People in developed countries are living longer. The average life expectancy for women is now around 82 years and for men 78 years. Most people can expect to enjoy ten or more years of retirement. With this prospect, more and more people are moving after they have retired. They are doing this for a number of reasons:

  - it is no longer necessary to live close to what was their place of work

  - to downsize into a smaller home

  - to sell their home for something cheaper and use the difference in price as a pension

  - to move into a quieter, calmer and more attractive environment.

- Services – The presence of both commuters and retirees in rural areas creates a demand for a wide range of services, from shops to schools, medical centres to pubs.

### Rural areas in the developing world

Over most of the developing world, rural areas are important for three main reasons:

- They are inhabited by the majority of the population – levels of **urbanisation** are still quite low. In many developing countries, less than a quarter of the population lives in towns and cities.

- They provide the food necessary for the survival of both the rural inhabitants and many of the urban dwellers.

- They are an important part of the traditions and heritage of many different tribes and peoples.

*Figure 4: Growing flowers in Africa for European supermarkets*

### Agriculture

Without doubt, agriculture is the most important activity in rural areas, but whether it earns very much money is open to question. The reason for this is that much of the agriculture is **subsistence farming**. It provides work for members of a family; they consume what they produce rather than sell it. In more fertile and favourable areas, there is **commercial farming** which produces a whole range of commodities for sale – from cereals to vegetables, from fruit to beverages like tea, coffee and cocoa. Much of this output will be sold within the country where it is produced, but certain products will find their way on to the supermarket shelves in developed countries.

This export of food products is nothing new, as it was a feature of the old plantation agriculture. However, the export of fruit and vegetables (plus that of cut flowers) has been given a recent boost by the setting up of large **agribusinesses** (Figure 4). Whilst these exports may be good for the national economy, the growing of cash crops is taking over land once used by the people for growing their own food. The workers now being employed by these agribusinesses benefit from regular wages, but most of those wages now have to be spent on buying the food that they no longer grow. To make matters worse, the new agribusinesses are growing non-native crops, such as green beans and sugar snap peas, which were normally grown elsewhere but only during the summer season. Do we in the UK really need such foods to be on the supermarket shelves during the winter months? The agribusinesses are certainly making good profits, but alas those profits are not always being used to help the development of the country in which they were produced. Rather they are being moved to set up agribusinesses in other countries.

- Other primary activities – Mining occurs in very few rural locations, but fishing may be locally important along the coast, around lakes and on larger rivers.

- Tourism – As in developed countries, some rural areas of developing countries are benefiting from tourism. Involvement in this global industry is doing much to help economic development. It is giving rise to relatively new forms of work, and a job in tourism and its related services means regular wages.

## Activity 2

What sort of developing world farm products would you expect to find in your local supermarket?

*Figure 5: Big game watching – a popular form of tourism in Africa*

Thus we see that rural areas in developed countries are involved in rather more economic activities than those in developing countries. This is an important contrast. It occurs because the two types of rural areas are experiencing very different processes of change. In the developing world these are the rise of commercial farming and **rural–urban migration**, whereas in the developed world the processes at work are **suburbanisation** and **counterurbanisation**. However, there are some economic similarities. In both worlds, agriculture is the dominant land use; in both worlds, tourism is having an impact.

## Rural challenges in the developed world

It is the processes of change discussed above that are creating the challenges facing the rural areas of the world. Since the processes in the two 'worlds' are different, the challenges are different too.

We will look first at the developed world. Here, it is important to distinguish between **accessible** and **remote** rural areas – between those areas located close to major cities and those that are a long way away. They are very different scenarios.

### The survival of farming

Mention has already been made of the decline in the number of people employed in agriculture (Figure 2 on page 261). Added to that, developed countries are increasing imports of cheap food from the developing world. Thus it is that many farms, especially those in remote rural areas, are struggling to stay in business. If they are to survive, they have to look for others ways of making a living (see section on farm diversification on page 273).

### Depopulation of remote rural areas

For well over a hundred years, with the number of jobs falling, many people have been persuaded to leave the countryside. They have found work in towns and cities. This **rural–urban migration** was encouraged by:

● higher wages and more job opportunities in urban areas

● the availability of more and better services

● the perception that towns and cities offer a better **quality of life**.

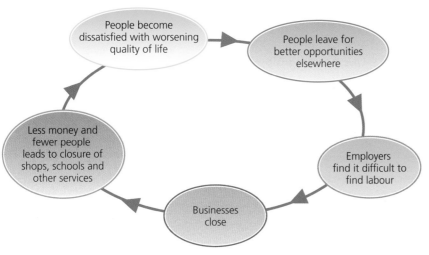

*Figure 6: The spiral of decline in rural areas*

But rural–urban migration has now largely ceased. It is only the most remote rural areas that continue to lose population, particularly young people of working age. The problem is that when such people and their families leave, the demand for local services (shops and schools) falls and they close. The loss of services makes the area even less attractive. More people are persuaded to leave. As Figure 6 shows, there is a chain reaction which leads the rural area into even faster decline. Fortunately, the downward spiral is being checked by the process of counterurbanisation. The loss of population from many remote rural areas is now being compensated by an inward movement of newcomers. Thanks to the broadband revolution, people disenchanted with town or city living and enchanted by the rural idyll are able to **telework** from home. However, whilst population is increasing in all rural areas, except the most remote, services continue to decline.

### Village decline in rural areas

A survey in 2000 found that 3000 British villages, with a combined population of 4.5 million, had no shop and no post office. One-third of British villages were dependent on neighbouring towns for even basic goods. At that time, village post offices were closing at the rate of 400 a year. The Post Office announced in 2008 the closure of a further 2500 of their branches. About half of these branches also serve as local shops.

It is now extremely difficult for more than one specialist shop, such as a butcher or baker, to survive. Shops must sell a wide range of goods if they are to compete with supermarkets and larger chain stores in neighbouring towns. The problem is made worse by villagers doing their shopping in supermarkets on the edges of nearby towns.

The closure of village shops is an important factor in persuading village residents to move home. This is especially true for households – particularly the elderly and not so well off – who do not own a car and have to rely on poor public transport. In a number of villages, when the post office has been under threat, the local residents have stepped in and taken it over themselves. In most cases, they have received grants from the Department for the Environment, Food and Rural Affairs (Defra).

*Figure 7: Protest about closure of a village post office and store*

### Quick notes (Village decline)

- The closure of village shops is happening at a fast rate. It is this that helps persuade many people to move out altogether from remote villages.

*Figure 8: The annual May Day fair at Finchingfield – a honeypot village*

## ResultsPlus
## Exam Question Report

**REAL EXAM QUESTION**

**The residents of Keswick have conflicting views about the increasing numbers of tourists. Suggest reasons why some are for and some are against the increase in tourist numbers. (4 marks)**

### How students answered

Some students only stated reasons for or against the increase in tourist numbers and failed to provide reasons for these views.

11% (0–1 marks)

Most students identified groups that would oppose and support the increase, but the reasons given were limited.

52% (2 marks)

Many students could identify the economic advantages to some groups, and the social and economic costs to others. The best answers defined their groups carefully, such as recent retirees to the area opposing development, whereas young locals needing employment opportunities welcoming it.

37% (3–4 marks)

## Tourist honeypots

The rise of tourism in the countryside is not evenly spread. Rather it is focused on what are called **honeypots** – places that offer something special that attracts large numbers of tourists. In England it might be because of:

- its immediate surroundings – e.g. Castleton in Derbyshire with its limestone caves

- the picturesqueness of the place – e.g. Finchingfield in Essex, often described as a 'chocolate-box' village (Figure 8)

- its historic associations – e.g. Tintagel in Cornwall and its legendary King Arthur's castle

- its use as the setting for a popular TV series – e.g. Holmfirth in West Yorkshire, the setting for the long-running Last of the Summer Wine series.

Becoming a honeypot village or tourist 'hotspot' is fine for those making a living out of tourism, but there is a downside – constant crowds and traffic congestion. It is not so much fun if you are a resident and place a value on peace and quiet! It is not much fun either if all the food shops have turned into tearooms and souvenir shops.

## Loss of accessible rural areas

The obvious outcome of the long-established rural–urban movement of people is the building of new homes. They are most likely to be built in two locations: around the urban fringe and in parts of the accessible countryside. In the first, housing is mainly added to the outer edge of the built-up area in the form of suburbs. In the accessible rural areas, existing towns and villages become encircled by new housing estates. In both locations, the building of new homes (suburbanisation) is taking place on what was rural space.

Up until recently, internal migration was from the countryside into urban areas. But now there are substantial volumes of people and businesses moving in the opposite direction (counterurbanisation). People are now commuting over longer distances (Figure 9). Some are even leaving their jobs in the cities and larger towns altogether and finding work and homes in the countryside or smaller settlements.

The growth of the so-called **commuter dormitories** raises two issues:

- the loss of rural land as it is built over to provide homes – can a country afford to lose its countryside?

- the growth of 'part-time' settlements – would it not be better if residents also worked there?

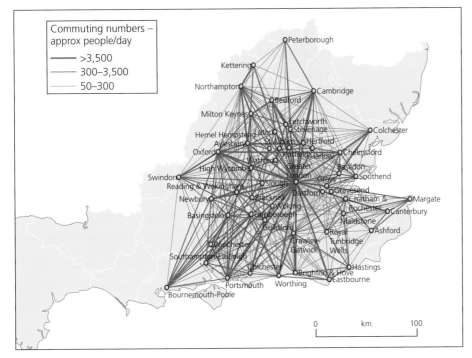

*Figure 9: Commuting in Southeast England*

## Rural challenges in the developing world

The rural areas of the developing world now find that they are being increasingly changed and challenged by a number of physical and human processes:

| | |
|---|---|
| **Environmental degradation** | This results from **desertification**, and **deforestation**. Clearance of vegetation, over-use of the soil, soil erosion and an increasing scarcity of water are causing many rural areas to dry out completely. Climate change is making these problems worse and, as all this happens, so the amount of food that can be grown steadily declines. |
| **Population change** | The high rates of natural increase add to the pressure on dwindling amounts of food, resulting in malnutrition and even starvation. This situation encourages more and more people to migrate from the countryside into towns and cities. |
| **Urbanisation** | Rural people are attracted to the towns and cities by jobs and wages that will allow them to buy their food, and by better education and healthcare. This rural–urban migration is resulting in a mainly elderly and female population being left in the rural areas to do the farming and produce the food. |
| **Human hazards** | Epidemics of diseases, such as cholera, malaria, TB and HIV/Aids – together with wars – have a major impact on the rural areas. They raise the death rates, leaving fewer people to attend to the important task of food production. |
| **Globalisation** | Some rural areas are discovering that they contain valuable resources that are of interest to the growing global economy. These might be minerals, or they might be resources, such as wildlife and spectacular scenery that attract foreign tourists. Most recently, areas of fertile soil and tropical climate are being exploited by agribusinesses growing exotic or out-of-season crops for the shelves of developed world supermarkets. Land that should be growing food for local people is now feeding people thousands of miles away. |

## Rural poverty

The outcome of these processes is rural poverty. Kenya provides a good example. Over the past 30 years, poverty has been on the rise in Kenya. More than half the country's 31 million people are poor – with three-quarters of them living in rural areas. Ironically, most of these rural poor people live in areas that have a high agricultural potential. Population densities in such areas are more than six times the national average of 55 persons per km².

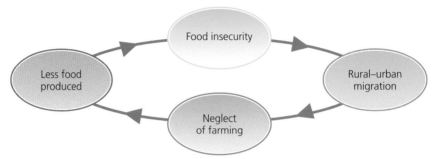

*Figure 10: The spiral of decline in rural areas of the developing world*

### Quick notes (Rural poverty)

- In Kenya, rural–urban migration and HIV/Aids deaths have done little or nothing to reduce the pressure of population on agricultural land. Rural poverty and food insecurity continue to rise.
- The challenge facing rural areas in the developing world appear to be due in part to the external processes of urbanisation and globalisation. Developing countries are not powerful players and are therefore vulnerable to exploitation. But there are also significant internal factors at work, such as population pressure, environmental degradation and human hazards.

*Figure 11: Drought can turn farmland into useless soil and sand*

The trebling of Kenya's population over the past 30 years is one of the basic causes of the rising rural poverty. But rural poverty in turn drives more population growth, to create a sort of vicious circle. Other causes of the spiral of decline (Figure 10) include:

- low agricultural productivity, made worse by environmental degradation (Figure 11), insecure land tenure and poor farming methods

- shortage of work outside subsistence farming

- difficulty in finding money to finance farm improvements

- HIV/Aids – this is particularly prevalent among the population of working age

- rural–urban migration – mainly of young males, leaving women behind and responsible for growing crops and herding livestock

- some of the best remaining farmland has been taken over by commercial farming and is now producing food for export and for foreign tourists drawn here by Kenya's wildlife.

## Activity 3

What would you suggest as good indicators of rural poverty?

# How might these rural issues be resolved?

## Improving livelihoods and opportunities

The rural issues discussed above can only be tackled by the concerted efforts of individuals, organisations and governments. This is as true for the developed world as it is for the developing world. Better livelihoods and opportunities do not just happen. They require investment, new ideas and technology. They also need the commitment of key players. The table below shows the key players in the UK, who are typical of the developed world.

| Local government | Its responsibility is to identify the issues and to come up with some possible solutions. It might respond to local community pressures. |
|---|---|
| County planners | Have a wider view of issues and their causes. Main responsibility is to protect rural areas from damaging developments. |
| National government agencies | e.g. Defra (Department for Environment, Farming and Rural Affairs), Natural England, Highlands and Islands Enterprise. They can offer new initiatives, help, guidance and funding. |
| European Union | Over the years, the EU has been helping rural areas in a variety of ways, e.g. the Common Agricultural Policy (CAP) and through its Regional Development Fund. |
| Non-government organisations | e.g. Campaign for Protecting Rural England (CPRE) and the Royal Society for the Protection of Birds (RSPB). Their role is mainly one of ensuring that rural areas are protected and not damaged by new developments. |
| Private enterprises | It is these that are likely to create new jobs and opportunities. They need to be assured that any new investments they might make will prove profitable. |

With key players, there are two different approaches to dealing with rural issues:

- **bottom-up** – where local groups and communities press for action to be taken on a particular issue. Their pressure gradually works its way up through different levels of government – local, county, regional and possibly up to national level.

- **top-down** – where a policy or programme is launched at a national level and gradually filters down for action to be taken at a local level.

In developing countries, action on rural issues is more often top-down. But this is not to say that a bottom-up approach is never followed (see pages 270–272). The main players in developing countries tend to be:

| National governments | Often distributing various forms of aid received from foreign countries. |
|---|---|
| Inter-governmental organisations (IGOs) | Examples include the United Nations Development Programme (UNDP) and the International Fund for Agricultural Development (IFAD). |
| Non-governmental organisations (NGOs) | International charities such as WaterAid, Oxfam and Médecins Sans Frontières (MSF) which target the rural poor, especially food production and health. |

Improving livelihoods and opportunities in the rural areas of developing countries lies mainly in taking actions in two main directions:

- raising food production, but in a way that does not cause more environmental degradation. Farming needs to be made both more productive and more sustainable, and food security needs to be improved

- reducing rural–urban migration by making rural areas more attractive places in which to live and work. Stopping the loss of young adults and enterprise. This will involve a whole range of initiatives, from providing healthcare, education and new types of work to improving housing conditions and the rural infrastructure.

### Crops and a shop in Ethiopia

Worke lives in the Afar region of Ethiopia (Figure 12) and she has five children. Ethiopia is one of the poorest and least developed countries in the world, ranking 169th out of 177 countries. Afar is hot and arid. It is a difficult place in which to live at the best of times, but when drought hits, it can be a killer. Worke used to live in fear of drought (Figure 13). She knew that if her animals died and her crops failed, food aid would be the only option for her and her children. So when FARM-Africa, a non-governmental organisation, came to her village several years ago, she jumped at the chance to become involved – and she formed an irrigation group with her neighbours. FARM-Africa works at a grassroots level with rural communities in eastern Africa, providing simple but long-term solutions to poverty. According to FARM-Africa, it is the millions of smallholder farmers of sub-Saharan Africa who hold the key to reversing the downward spiral of poverty and starvation now gripping large parts of the continent.

*Figure 12: The Afar region of Ethiopia*

### Quick notes (Crops and a shop in Ethiopia)

- Simple and cheap help at a grassroots level can do much to improve food security in rural areas. A regular income also means a better quality of life and less incentive to become a rural–urban migrant.

*Figure 13: The hot and arid landscape of the Afar region*

With FARM-Africa's help, Worke and her neighbours built a canal to channel water from the local river to the fields and they installed two pumps to draw river water into the canal. She also received a starter kit of seeds and went on to harvest 4000 kg of maize, worth almost £400. She now grows onions, tomatoes and peppers, which she sells locally or takes to market. But Worke's achievements have not stopped there. She has taken out a £130 loan from FARM-Africa and opened a small shop selling everyday items like soap, salt and razor blades. It has been so successful that she has already repaid the loan and its interest. That loan is an example of **micro-finance** at work. Under normal circumstances, Worke had no access at all to the services of a bank.

Thanks to the income from her shop and vegetables, four of Worke's children now attend high school and her eldest son has completed school and become a journalist in a local town. Best of all, Worke finally feels secure, particularly in terms of food. She knows that if the droughts strike again, her family will not have to rely on hand-outs. She stands proudly in her shop window: 'Before, we staggered along on one foot. Since FARM-Africa, we can stand on our own two feet.'

### Improving the rural infrastructure

It is widely recognised that the key to rural prosperity lies in improving the rural infrastructure. Improvements need to be made in the following things:

| | |
|---|---|
| **Transport** | Good access to towns gives farmers the chance to sell their surplus produce and use the money to buy better seeds and livestock. It also allows people to market other products. |
| **Water and sewage disposal** | Clean water and safe waste disposal reduce the incidence of disease and death. |
| **Healthcare and education** | These allow people to realise their potential and that of their rural surroundings. |
| **Energy** | Rural electrification improves the quality of life and opens up the possibility of setting up small industrial enterprises. |
| **Communications** | Mobile phones and access to the internet can do much to reduce the sense of remoteness. They can also help rural people to keep in touch with recent developments. |

In addition to these improvements, rural areas would greatly benefit if the global market was prepared to pay a 'fairer' price rather than a minimal price for cash crops, such as coffee, cocoa and tea.

**Activity 4**

Can you think of anything that might threaten Worke's food security?

271

**Decision-making skills**

Which of the five different aspects of rural infrastructure in the table do you think is most important?

You might make this the subject of a group discussion. Make sure that reasons are given to justify your choices.

*Figure 14: A motorbike nurse delivering local health care*

### Quick notes (Nurses on motorbikes)

- This is a simple and cheap solution to the challenge of providing healthcare in remote rural areas. It is also helping to get population numbers and food supply into a better balance.

### Quick notes (Millennium villages)

- Simple and cheap help at a grassroots level can do much to improve the rural infrastructure. Aid from developed countries would do well to focus on this sort of project.

### Nurses on motorbikes

Gladys Mahama is a community health nurse in northern Ghana. Driving around her 'patch' on a motorbike, she represents a health delivery experiment, which is beginning to have an impact on the delivery of healthcare in poor rural areas – not just in Ghana but throughout the developing world. She is one of 16 nurses who are key workers involved in a project that delivers health services alongside family planning (Figure 14).

Each village in the project area has built a detached hut in which the nurse lives and where she is able to see and treat people, in private, on a one-to-one basis. Thanks to the motorbike, she is able to visit outlying compounds and the sick who cannot move. Much of the driving is along tracks and rough ground. One immediate impact of the project is that the number of women using contraception has greatly increased. As a consequence, the birth rate has fallen, meaning that there is now less pressure on limited food supplies.

At present, the project is financed by donor money. Those who are able to make a token payment are expected to do so. The question now is whether Ghana's government can provide the funds for many more such schemes across the country.

### Millennium villages

Millennium Promise is a non-governmental organisation whose aim is to bring an end to the rural hunger, disease and poverty that are so widespread in sub-Saharan Africa. It believes that the only way to lift these areas out of the poverty trap is to use simple, practical and proven technologies.

The organisation now has eighty village projects running in ten countries. Its flagship village in Ethiopia is Koraro – a cluster of eleven small villages. The achievements so far include:

- The building of 30 safe water points and three micro-dams, which have been used to irrigate crops in this semi-arid area.

- Refurbishing classrooms in the local schools and introducing a school feeding project, which has helped to improve student performance and attendance. The education of girls is being encouraged by giving them school supplies (books, paper and pens).

- The distribution of Insecticide-treated bed nets in order to tackle the high incidence of malaria. Additional health personnel have been hired for the one clinic, in particular to tackle the growing incidence of HIV/Aids.

- Villagers have been trained in the use of improved planting techniques, green fertilisers and improved seed. Maize crop yields have increased by 16% and teff yields increased 64% on average.

Clearly there is much more to be done in Koraro, particularly making it more accessible and more in touch with the outside world. But even these actions cost money. The hope is that this and the other millennium villages will persuade developed countries to support this sort of grassroots project rather than provide food aid or hand over large sums of money to often corrupt governments.

## Making the farming economy more productive and sustainable

The developed and developing worlds face two different challenges. In the developed world, the challenge is to find ways of raising farm income, other than by raising the output of food. In the developing world, the challenge is simply to raise food output.

### Farm diversification in the developed world

Farming today in many developed countries can no longer support the large number of families it once did. Everyone wants the cost of food to be kept down but supermarkets are now paying UK farmers very low prices – and many can scarcely make a profit. The problem is that even cheaper food is being imported from overseas. So, if farmers want to stay where they are they have no choice but to diversify – by doing one of two things:

● Find other ways of making money out of the farm – while continuing to farm.

● Turn their farms into completely different businesses.

Figure 15: Examples of farm diversification – pick your own (PYO) and converting barns into housing

### Farm diversification: methods and examples

| New products | New outlets | Tourism | Leisure & recreation | Development | Energy |
|---|---|---|---|---|---|
| Organic crops | Pick Your Own (PYO) | B&B | Shooting | Converting barns into housing | Wind turbines |
| Herbs, cheese, bottled water | Farm shop | Caravan or camping site | Off-road driving | Industrial units | |
| Different animals, e.g. bees, goats, ducks, ostriches | Farmers' market | Café or restaurant | Mountain biking | Telecentres | |

**Quick notes (Farm diversification)**
- Many farms in the UK are surviving only by taking on new activities. This is necessary because of low product prices and competition from imported food.

### Activity 5

Investigate a farm near you. Has it diversified? If so, in what ways?

## Making farming greener in the developed world

Aspects of modern agriculture can have adverse impacts on the environment – the use of agrochemicals, the seepage of slurry into rivers and streams, the removal of hedgerows, and so on. The debate about the safety of growing genetically modified (GM) crops goes on. Awareness of this damage should persuade us to look for more environmentally friendly – 'greener' – ways of producing our food.

There is a range of different actions that can be taken to make farming greener, including:

| | |
|---|---|
| **Arable rotation** | Rotating vegetables with legumes (like peas and beans) that fix nitrogen helps reduce the amount of fertiliser needed. It also helps to break disease and insect pest cycles. |
| **Organic farming** | This relies on crop rotation, green manure, compost and biological pest control to maintain soil productivity and control pests. It does not use chemical fertilisers, herbicides or pesticides, livestock feed additives or genetically modified organisms. It is environmentally friendly, but organic yields are on average 20% smaller than those from 'normal' agriculture. |
| **Drip irrigation** | This is the best way to get water, fertiliser and pesticides to the roots of a crop. Computerised control systems deliver the right amount of water, fertiliser and pesticide at the right time by means of a buried tape or pipe. Waste is minimised, but the system is costly to install. |
| **Hedgerows** | Hedgerows play an important role on farms (Figure 16). They help to prevent soil erosion and water run-off, provide shelter, control livestock and protect crops from the wind. They also provide an important habitat for wildlife. During the 1960s and 1970s, many miles of hedgerow were ripped out in order to create larger fields. Fortunately, some hedgerows are now being restored. |

### Activity 6

Find out more about one of the ways in the table. What are its advantages? What are its disadvantages?

*Figure 16: Hedgerows play an important role on farms*

### Raising farm output in the developing world

There is much that can be done to raise farm output in the developing world. For example:

- Fairer trade – Freeing up global trade would allow agricultural produce from developing countries easier access to the profitable markets of the developed world. For example, the tariffs that the EU levies on imported goods means that agricultural products cost more in the shops and are not so competitively priced. But there are two problems here: (1) more commercial farming reduces the amount of land for subsistence farming; and (2) flying produce to these markets involves considerable 'food miles' (the consumption of aviation fuel and resulting carbon dioxide emissions).

- Appropriate aid – Much of the aid given to developing countries, such as loans and food during disaster situations, does not really help to improve agricultural output. What is needed is what is referred to as **intermediate technology** – seeds, stock, tools, instruction about the techniques of sustainable farming, and help with setting up local cooperatives and market groups.

- Sustainable farming techniques – These are needed to combat the recurring problems of water shortages and soil erosion. More and more people are looking to **permaculture** to do just this.

*Figure 17: Permaculture makes the best use of the land*

The ethos of permaculture is to make the best use of what the natural world provides in the drive to increase food output. It is a method of producing food in a closed loop that maintains a self-sufficient system. Within that loop, the animals, plants and micro-organisms work together in harmony. Permaculture uses no chemicals or pesticides. It grows a mixture of food and tree crops, and keeps a small number of livestock. Livestock (notably chickens) are allowed to forage for food after the crops have been harvested, and in return provide manure. Trees are planted around the fields and these soon produce mulch and nitrogen – natural fertilisers that are washed into the fields by rain. Native trees are preferred because part of the permaculture technique involves making organic insecticide sprays from their leaves, bark and wood. The trees provide shade. The soil is never left exposed to the sun and wind. It is heavily mulched to keep it cool and damp. As one farmer has put it, 'All plants, insects, animals and we who tend the garden live in a natural harmony'.

Experience has shown that permaculture can be adapted to work well in both humid and dry environments. In dry areas, training courses provide instruction in simple but effective water conservation techniques. In most cases, farmers can expect to increase their yields by four times or more within a few years. Besides delivering food security, the system allows farmers to reduce the amount of their land used for their immediate subsistence needs. Thus they have the opportunity to raise a cash crop or two for sale in nearby markets.

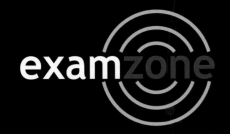

# examzone

## Know Zone
## The challenges of a rural world

Rural areas are changing. In developed countries, although population is growing in the countryside, it is no longer a place just for farming, but also for recreation and tourism. In developing countries, rural areas are areas of hardship and poverty, so many people migrate to towns.

## You should know...

- [ ] What the characteristics of a rural area are
- [ ] How the different types of rural area in developed countries – both accessible and remote – are experiencing many challenges
- [ ] How accessible rural areas are becoming 'suburbanised' as they are taken over by commuters
- [ ] How, in general, the rural population is growing from counterurbanisation, but services are declining
- [ ] What changes are taking place on farms in the developed world, with many farms having to diversify
- [ ] How there is extreme poverty and hardship in the rural areas of developing countries
- [ ] Why the move to commercial agriculture in developing countries has its costs
- [ ] How sustainable grassroots schemes run by NGOs are improving the lives of rural people in developing countries
- [ ] The future of rural areas lies in the hands of a small number of key players
- [ ] How agriculture throughout the world needs to become more sustainable and environmentally friendly

## Key terms

Agribusiness
Bottom-up approach
Cash crop
Commercial farming
Commuter dormitory
Commuting
Counterurbanisation
Depopulation
Disposable income
Diversification
Globalisation

Grassroots
Intermediate technology
Key player
Micro-finance
Permaculture
Poverty
Rural–urban migration
Subsistence farming
Suburbanisation
Telework
Top-down approach
Urban fringe
Urbanisation

### Which key terms match the following definitions?

**A** Originating within a local community rather than being imposed from above

**B** The decline of a population, both by natural processes and, occasionally, by government policy

**C** The countryside adjacent to or surrounding an urban area

**D** Commercial agriculture that is owned and managed by large corporations

**E** The amount of money which a person has available to spend on non-essential items, after they have paid for their food, clothing and household running costs

**F** The movement of people and employment from major cities to smaller settlements and rural areas located just beyond the city, or to more distant smaller cities and towns

**G** Work in which telecommunications replace work-related travel (commuting

To check your answers, look at the glossary on page 321.

**Foundation Question:** Outline the differences between farming for export and subsistence farming in developing countries. (6 marks)

| Student answer (achieving 3 marks) | Feedback comments | Build a better answer (achieving 6 marks) |
|---|---|---|
| Some crops are grown to sell for cash, but poor people grow crops to feed themselves. | *Some crops...* A correct comparison, although a context is not given. (1 mark). | In Ghana cocoa is grown to sell for money to rich countries, but some families grow a food crop (e.g. maize) to keep themselves alive. |
| The farms that export crops are often very large, while the subsistence farms are usually small. | *The farms...* A correct comparison but the point is not developed enough, perhaps needing better terminology. (1 mark). | Export crops are grown on a large scale, perhaps in plantations, while the subsistence farms are much smaller. |
| Farming for export makes loads of money but subsistence farms make no money. | *Farming for....* A correct comparison, but again it is an undeveloped point. (1 mark). | Cash-crop farms wish to sell crops to make a profit, and may operate as an agribusiness. But subsistence farms will only make some money when there is a crop surplus. |

**Overall comment:** In this question it is important to include a range of ideas, with extensions and detail, linked together. The candidate offers a good description and does well to compare, as the question instructs. However, further detail and extensions were needed. Spelling and grammar are good.

**Higher Question:** Using examples, explain how and why farms in a developed country may diversify. (8 marks)

| Student answer (achieving 4 marks) | Feedback comments | Build a better answer (achieving 8 marks) |
|---|---|---|
| Farms have great difficulty making a profit in years with poor weather so they need to look for other ways of making money. | *Farms have great . . .* identifies the fact the farmers want to make money, and a reason for diversifying – the weather conditions. (2 marks) | In England most farms are commercial, selling their produce for profit. But in a year of poor weather yields may be low and other ways of making money must be found. |
| Many farms have developed tourism, perhaps through bed and breakfast, or operating campsites. | *Many farms...* gives an example of the type of diversification that farms have used. (1 mark) | One way of making money is to use farmhouses or cottages for bed and breakfast, or providing campsites. |
| Other forms of tourism include nature trails or showing rare breeds, or providing green lanes for off-road driving. | *Other forms...* gives further examples of diversification using tourism. These could have been developed perhaps. (1 mark) | Tourism and recreation on farms may also include nature trails, shops, cafes, rare breeds, and green lanes for off-road driving. These attract a variety of people to spend money. |
| If farmers don't make a profit they will have to close down. | *If farmers...* is a vague statement and shows that the candidate has run out of ideas. | Competition with cheaper imported foods makes it difficult for English farmers to make enough profit to continue. So they have tried growing different crops such as bio-fuels. |

**Overall comment:** One explanation is provided but others needed to be given in this extended answer. Examples of diversification are given but they are all linked to one theme (tourism). There is a clear structure and SPG is good, but perhaps more geographical terms could have been used.

# Unit 3 Making geographical decisions

## Your course

Unit 3 is about making well thought-out and evidenced decisions – on topics which affect our planet.

As you will have found out when studying Units 1 and 2, the increasing demands of a rising population and economic development are putting pressure on the environment and the planet's scarce resources such as energy, food and water. These pressures may lead to conflicts between nations, between rival developers, or between those who wish to exploit the resources and those who wish to conserve the planet.

Unit 3 is 'synoptic' – it links and interrelates topics from across the specification – so it requires you to have a very thorough understanding of the core topics in Unit 1 and Unit 2 and the links between them. This will allow you to analyse a geographical problem and then draw together your knowledge and understanding to consider, select and justify your choice of proposed solutions or ways of tackling an issue. Figure 1 summarises how synopticity works.

## Your assessment

You will sit a 1-hour 30-minute written exam worth a total of 53 marks. Up to 3 marks are available for spelling, punctuation and grammar.

Your exam will relate to a Resource booklet, which will have no more than 700 words of text with maps, diagrams and photographs. You will not have seen the booklet before, so you will first need to spend around 20 minutes reading and analysing these resources very carefully before you start to answer the questions.

The questions in the exam paper will be based on the booklet but also assume some background knowledge of the topics (but not the place), as well as requiring you to use a range of skills. The paper consists of short structured questions (and, on the Foundation paper, multiple-choice questions) that cover some of the background to the issues. The final question will always require you to make a decision – to choose one of the options given – and justify that choice.

You must answer all the questions. The resources in the booklet will provide you with building blocks about the particular issue, so they will guide you through the decision-making process.

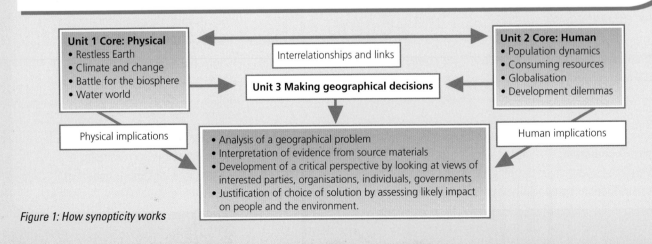

*Figure 1: How synopticity works*

# Chapter 17 Making geographical decisions

## Planning for making geographical decisions

### Long-term preparation

Throughout this book you will find suggestions as to how you can build up your skills, knowledge and understanding of geographical decision making (see Figure 2). In this book's core chapters, 1–4 and 9–12, there are decision-making skills boxes which set mini activities aimed at developing your skills throughout the course. Because the units you are studying cover a variety of scales from local to global you will be able to practise your skills using resources that focus on all different sorts of environments and places.

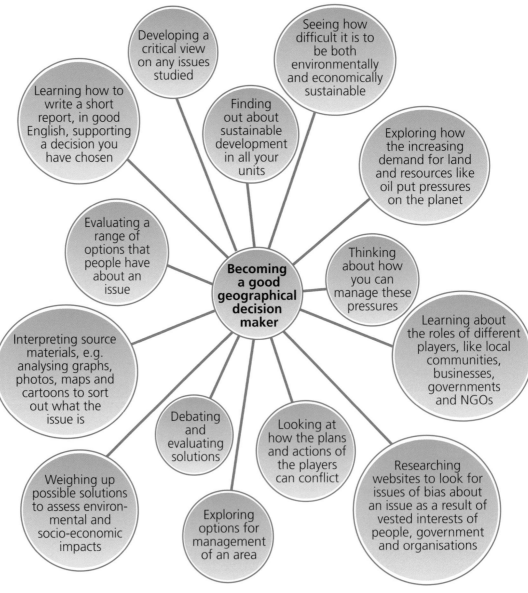

*Figure 2: Becoming a good geographical decision maker*

In your exam you must be able to demonstrate certain skills, so it is very worthwhile developing them over the course, so that when you are faced with resources about an issue or a place which you may not have studied, you feel confident about exploring the geographical problem. You must be able to demonstrate the following skills:

1. Reference skills – ability to accurately use a variety of sources for obtaining information.
2. Communication skills – ability to present information in a clear and appropriate way through written communication, maps and diagrams.
3. Interpretative skills – ability to give meaning to data and diagrams, and make sense of photographs.
4. Evaluative skills – ability to critically analyse and interpret the full range of viewpoints, evidence and options as well as weigh up and justify decisions.
5. Problem-solving skills – ability to enquire and to think and communicate clearly, critically and constructively about solutions to a problem.
6. Synoptic skills – ability to link and interrelate topics from across the specification when focusing on a geographical problem.

Your resources will focus on one or more of the key ideas shown in Figure 3.

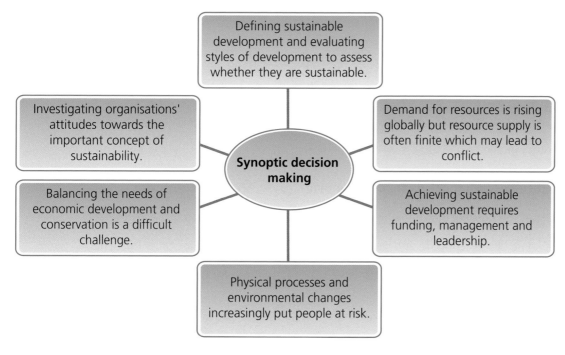

*Figure 3: The key ideas*

## Sample resource booklet

# Development dilemmas for Bolivia

### Some guidance on your reading time

- Spend around 30 minutes reading the booklet first. This leaves you an hour to answer the questions.

- Begin by reading through the materials and underlining key words and identifying links with Unit 1 and Unit 2.

- Make short notes on any resources, such as graphs and photographs, to identify any key trends and features.

- Read the summary about the problem and pick out the key words (eg. 'top-down', 'bottom-up').

- Read the views about the problem and try to identify any differences or conflicts.

### Section 1: Profile of Bolivia

- This section provides statistics about Bolivia's current level of development. It is one of the poorest countries in South America.

- Look at the various measures of development.

- Think about Rostow's model of development and what stage the evidence suggests Bolivia is at.

- How would Bolivia's profile be a hindrance to development?

### Section 2: The difficulties facing Bolivia for development

- This section looks at the possible environmental problems which could occur when Bolivia develops.

- It also considers some of the economic and social factors which might make development difficult.

- It will also give you a chance to think about what would be the best way for Bolivia to develop.

### Section 3: Development schemes for Bolivia

- This section looks at three possible schemes for Bolivian development.

- You will need to evaluate these schemes and then choose the one which you think will be best for Bolivia, environmentally, socio-economically and politically. Think about both the Bolivian environment and its people. [See prompts in the text.]

- You should also think about *why* you might not want some schemes, because to *justify,* you have to say why you say 'yes' and why you say 'no'.

- The next section will tell you about the problems and will summarise the key issues.

## Information on the problem: What is the best way for Bolivia to develop?

◉ Some people believe it is best for Bolivia to develop economically by exploiting her natural resources. These large-scale developments would need to be top-down.

◉ Other people believe that Bolivia should concentrate on small-scale, bottom-up developments to improve the health, education and quality of life of the people living below the poverty line.

## Four views on how Bolivia should develop

◉ Highlight the key points in the views, as shown below.

◉ Look at who wrote them – are they an official organisation? What standpoint are they coming from, and how could this lead to bias?

◉ Can the views be linked to any one of the decisions?

### View A

All development of resources such as lithium, gas and the rainforest should be controlled by the government, not the TNCs who take wealth out of the country and destroy the environment. We need the money to improve health and education.

**President Evo Morales**, a native Indian elected until 2015

*Suggests might be left-wing, as against TNCs and favours state ownership of resources*

*Likely to support poor people*

### View B

A third of our country is above 3000 metres, often with very steep slopes. The country is landlocked. Our ability to export goods is terrible. Our geography is against us as it adds huge costs to resource development.

**Ferdinand Molina**, a famous Bolivian journalist

*Major problems of geography*

*Very difficult for export of resources*

*An independent viewpoint, perhaps?*

### View C

With so many people living below the poverty line, mainly in rural areas, we need sustainable development of agriculture, with bottom-up development of rural tourism and craft industries, supported by renewable energy development to empower the people.

**Comibol** (a Bolivian NGO)

*Favours bottom-up support for poor people*

## View D

We have many concerns about Bolivia. These include:
- Chopping down the rainforest for large-scale agriculture
- Destroying fragile ecosystems in the salt flats or rainforest to extract resources.
- Loss of water supplies from glacier melt in the High Andes

**Green activist** from a large international NGO

*Very focused on environmental issues. Likely to be against resource development?*

*This does not suggest that this person is from Bolivia*

# Resources for Section 1: Profile of Bolivia

## Bolivia – one of the poorest countries in South America?

*How low is this compared to very poor countries in Africa?*

**Table 1: Some facts about Bolivia**

*Comparison*

| | Bolivia | USA |
|---|---|---|
| GDP per capita ($US) | 4,500 | 47,500 |
| % in primary employment | 40%+ | Less than 1% |
| % living below the poverty line | 51% (Up to 80% in rural Andes) | 4% |
| Human Development Index (Rank) | 0.74 (113/192) | 0.96 (13/192) |
| Life expectancy | 66.9 years | 78.1 years |

*Primary = agriculture and mining*

*In the middle for ranking*

*How does this compare to the poorest countries in Africa?*

- When you analyse data, try to think of ways of describing it, e.g. *Americans are over 10 times wealthier than Bolivians.*

- Think back to other countries you have studied. How poor is Bolivia in comparison?

# The regions of Bolivia – a divided country

**Photo A: The Altiplano** – *the poorest part of Bolivia, where nearly 70% of Bolivians (mainly native Indians) live*

What is the landscape like? What type of farming is it?

Why is the Altiplano such a harsh environment to live in?

**Photo B: The Amazon rainforest** – *a vast, mostly unexplored, area of natural rainforest*

What 'condition' is the rainforest in?

What could the rainforest be used for?

Can you see any ways of exporting wood?

**Photo C: The Salar de Uyuni** – *the world's largest reservoir of lithium*

Is this an attractive area?

What damage would 'digging' for lithium do?

**Photo D: The Eastern Lowlands** – *the richest area of Bolivia*

Compare the farming here to the farming in the Altiplano.

● Try to write as detailed a description of the landscape and activities shown as you can.

## Resources for Section 2: The difficulties facing Bolivia for development

### Environmental problems

The major environmental problem is shortage of water in the Altiplano and Andes foothills, especially in the La Paz area and its rapidly growing suburbs of El Alto, the most densely populated area of Bolivia.

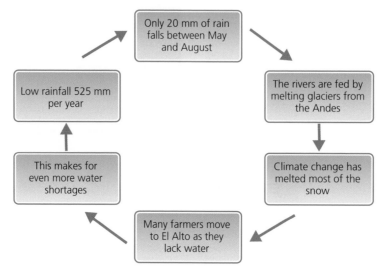

*Figure 4: The water shortage problem*

### Economic problems

The major economic problem facing Bolivia is poverty.

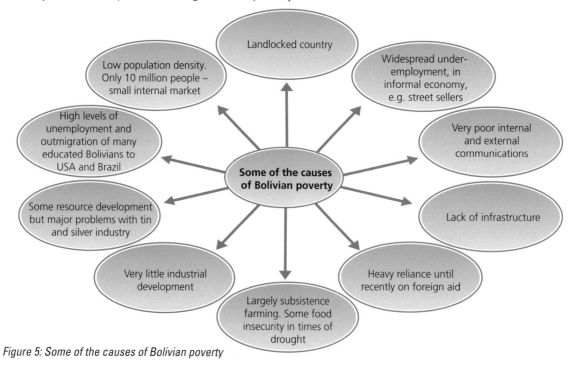

*Figure 5: Some of the causes of Bolivian poverty*

● Work your way round the circles and, for each, explain how it could contribute to Bolivian poverty.

## Social problems

The major social problems in Bolivia are related to the low standard of living.

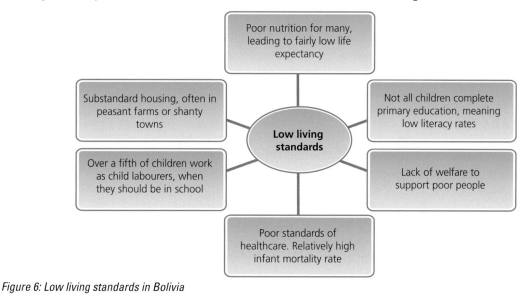

Figure 6: Low living standards in Bolivia

# Resources for Section 3: Development schemes for Bolivia

## Option A: Developing the resource base of Bolivia

- Bolivia was recently found to contain up to 50% of the world's resources of lithium.

- Lithium is one of the most important substances for twenty-first-century living, as it is used in non-rechargeable batteries for electric vehicles and in mobile phones and laptops.

- There are very high costs in extracting lithium from the salt lakes (see Photo C, page 285).

- Financing the development is a major challenge for Bolivia, because the political situation means that the State must be in charge, not transnational corporations.

    What are the problems caused by the political situation?

- Bolivia has already taken over control from TNCs of its natural gas production, with very mixed success.

- If the lithium resources were to be developed, Bolivia could become the Saudi Arabia of the electric car era.

- Combined with natural gas and timber, Bolivia could become a really wealthy country based on exporting resources, but it would destroy the landscape.

    What are the advantages and disadvantages of resource development/ exploitation?

## Option B: Improving the quality of life of the Bolivian people

- NGOs would work with the government, communities and agricultural cooperatives to develop policies and finance to improve access to energy and services.

- In 2010 only 68% of the Bolivian population had access to electricity – in rural areas the rate is only 30%.

- Micro hydro-power schemes and solar schemes would provide electricity for cooking, lighting and simple irrigation pumps which farmers need to overcome drought.

*Renewable energy*

- This would limit the need to cut down trees for fuel wood.

- The electricity could be used by the cooperatives to carry out knitting and weaving and also to start ecotourism businesses, as well as to power basic equipment.

*Why are these activities so important for development?*

- The government is financing health services (such as local clinics) and primary schools for all rural areas and the urban shanty towns such as El Alto.

- The main focus of all these schemes is to help both the rural poor in the Altiplano and the urban poor out of the poverty trap.

## Option C: Developing Bolivia's agriculture

◉ Around 40% of Bolivia's people are working in agriculture.

*Why is agriculture so important?*

◉ In the Eastern Lowlands there are huge farms of 5000 hectares, growing soya beans, rice, maize and coffee. The rich landowners, who are often not from Bolivia, sometimes use forced labour to make money from exporting crops.

◉ Flooding can be a major problem in the Eastern Lowlands.

*Why does it need developing?*

◉ Many people would like to develop the Amazon rainforest – as Brazil has done – for logging, cattle ranching, rice growing and sugar cane. This would mean the destruction of valuable rainforest, but many see it as the best way to increase food supplies and improve the well-being of the local population.

◉ The Eastern part of Bolivia is very rich, in contrast to the High Andes. It is threatening to declare independence because it does not agree with current government policies which favour the Indian people in the Altiplano.

◉ The Indian people who live in the Altiplano and Andean valleys farm at just above subsistence level, growing potatoes and beans and rearing sheep, llamas and alpacas. Soil erosion and drought are major problems.

◉ Many people argue that any farming developments should be in this poor Altiplano region, rather than in the commercial farms of the East, or by destroying the rainforest to grow crops.

◉ Bolivia can feed its people except when there are major disasters such as droughts, but it could also export many valuable crops too.

*Why would this be good for its economy?*

## Sample exam questions for the Foundation paper

1. (a) Study Table 1: Some facts about Bolivia
     (i) Define the terms 'Gross Domestic Product per capita' and 'Human Development Index'. (2 marks)
     (ii) Outline the differences between GDP per capita and the Human Development Index as a measure of Bolivia's development. (2 marks)
     (iii) Choose **four** further facts which suggest Bolivia is a poor country. (4 marks)
   (b) Look at Photo A: The Altiplano and Photo D: The Eastern Lowlands. Use the boxes below to describe the main features of each. (6 marks)

|  | **Photo A: The Altiplano** | **Photo D: The Eastern Lowlands** |
|---|---|---|
| Landscape |  |  |
| Economic activities |  |  |
| Settlement |  |  |

(Total for Question 1: 14 marks)

2. (a) State how the following may have contributed to a water problem in the Andes:
     A. Rainfall (2 marks)
     B. Global warming. (2 marks)
   (b) Choose **two** groups of people living in the Andes and suggest how they would be affected by the water problem.
     1 .................................................................................................................................(2 marks)
     2 .................................................................................................................................(2 marks)

(Total for Question 2: 8 marks)

3. Bolivia has some advantages for development but many disadvantages.
   (a) Identify **one** advantage and suggest how it helps development. (4 marks)
   (b) Choose **four** features of the geography of Bolivia and suggest how they might hinder its development.
     (4 x 2 = 8 marks)                                     (Total for Question 3: 12 marks)

4. (a) Why are TNCs likely to be important in developing Bolivia? (3 marks)
   (b) Why does the Bolivian government not want them to take part? (3 marks)

(Total for Question 4: 6 marks)

*Remember – spelling, punctuation and grammar will be assessed in your answer to this question. (3 marks)*

5. Study the **three** options for Bolivian development – Options A, B and C.
   (a) Select **one** option you think would be best for the people and the environment of Bolivia.
     Explain its advantages. (6 marks)
   (b) Explain why you have **not** chosen **one** of the other two options. (3 marks)

(Total for Question 5: 9 + 3 = 12 marks)

## Sample exam questions for the Higher paper

Study Table 1: Some facts about Bolivia

1. (a) Define:
    (i)   GDP per capita (2 marks)
    (ii)  The Human Development Index. (2 marks)
   (b) State how the data shown suggests that Bolivia is a relatively poor country. (5 marks)
   (c) (i)   State what stage of Rostow's Development Model Bolivia has reached. (1 mark)
       (ii)  Give the reasons for your choice. (4 marks)
   (d) Using Photo A and Photo D, describe the landscapes shown, to compare the Altiplano and the Eastern Lowlands. (4 marks)

(Total for Question 1: 18 marks)

Study the section on the environmental and economic problems facing Bolivia

2. (a) (i)   Explain why most of the Altiplano of Bolivia is experiencing a water problem. (4 marks)
       (ii)  Choose **two** examples of groups of people who will be affected by the water problem and explain why. (4 marks)

(Total for Question 2: 8 marks)

3. (a) Suggest how the following may have hindered development in Bolivia:
    (i)   High altitude
    (ii)  Being a landlocked country
    (iii) Out migration
    (iv)  Poverty of its people. (4 X 3 = 12 marks)

(Total for Question 3: 12 marks)

*Remember – spelling, punctuation and grammar will be assessed in your answer to this question. (3 marks)*

Study the **three** options for Bolivian development – Options A, B and C.

4. (a) Select **one** option you think would be best for the people and environment in Bolivia.
   (b) Justify your choice.

   Use information from the resource and your own knowledge from Units 1 and 2 to support your answer.

(Total for Question 4: 12 + 3 = 15 marks)

# Unit 4 Investigating geography

## Your course

This is the controlled assessment part of your course. The controlled assessment is an investigation in which you will be asked to undertake a fieldwork task. Having completed your fieldwork, you will then be required to write a report based on your results, completed under examination conditions and supervised by your teacher.

## Your assessment

This list shows you how the marks (50 in total) are allocated between the different elements and *what is actually being assessed*.

### Purpose of the investigation (6 marks)
*Your definition of the question you will be investigating, including the location of your fieldwork.*

### Methods of collecting data (9 marks)
*Your description and explanation of the methods of data collection you will use.*

### The methods of presenting data (11 marks)
*The data-presentation techniques you will use and their quality.*

### Analysis and conclusions (9 marks)
*How the findings of your investigation are brought together and conclusions are drawn.*

### Evaluation (9 marks)
*Your evaluation of your methods of data collection and presentation, and the analysis and conclusions drawn.*

### Planning and organisation (6 marks)
*The planning and organisation of your report, including your use of geographical terminology and the quality of your writing.*

# What are the controls in the controlled assessment?

You need to be aware of the level of control which occurs at each stage in the investigation. There are two levels which affect the way you must work:

### High level of control

Certain stages of your investigation must be completed individually – by yourself without help from others, in the classroom, under the close supervision of your teachers. This work cannot be taken home – it has to be handed in at the end of each lesson, so that you cannot continue working on it in your own time.

### Limited level of control

For some stages of your investigation, you will be able to work in your own time – at home, in the library or elsewhere. You will also be able to work in groups.

# Chapter 18 Your fieldwork investigation

## Objectives

- Identify geographical questions suitable for small-scale fieldwork.

- Develop clear methods to gain data in response to the question asked.

- Be able to present geographical data.

- Understand the geographical processes responsible for patterns seen in the data.

- Be able to analyse and explain geographical processes.

- Be able to evaluate the strengths and weaknesses of the investigation when it has been completed.

## How you will be marked

The purpose of your investigation (6 marks)

| Mark | Reasons for marks given |
|------|--------------------------|
| 0 | No location or issue is identified. |
| 1–2 | Issue or question is weakly identified. Location is mentioned but unclear. |
| 3–4 | A clear statement identifies the issue or question. Location is mentioned. |
| 5–6 | A focused statement identifies and evaluates the issue or question. Location is focused on the place of the investigation. |

An important element of geography is its use in the real world. It is a subject which allows us to investigate issues and problems with the intention of making a situation better, or developing a better understanding of a chosen issue. Most geographical investigations cannot be achieved by the use of a laboratory or computer simulation alone – they require us to collect information through the use of fieldwork. This 'controlled assessment' is an opportunity to develop the skills which are important in becoming a geographer, skills which are also transferable to other situations.

## Purpose of your investigation

All investigations require a purpose. Geographical enquiry is not simply a case of walking around a place and seeing what we can see. In this assessment, you will be asked to react to a question which is deliberately written to allow you to develop a more focused purpose of your own. For example, a typical task question might read: *How effective is the coastal management at your chosen location?*

This is obviously a very wide-ranging question – which would need a great deal of research and a huge amount of writing to answer properly. You therefore need to choose a smaller focus from such a question – a specific element which you can investigate. This is what we call a 'focus concept' – a specific issue, given in the form of a question. In the case of the coastal management question above, you might ask a specific question about the environmental damage caused by the management of the coast at a location, or about whether its benefits outweigh its costs.

One of the first things that you must do in developing your work, is to identify a focus which is clearly linked to the general question given to you, and then give a clear explanation as to why your focus is important and helps to answer the general question.

### Locating your investigation

To make your investigation informative for the person reading it, you must first locate it by describing where you carried out your fieldwork. This description will normally include various pieces of information (Figure 1) and maps to show the location graphically.

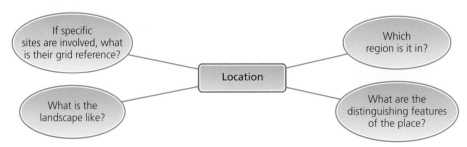

*Figure 1: Questions to consider when writing a location description*

You need to think carefully about the level of detail given in locating your investigation. If your work is focused on the question 'How has service provision changed in your chosen rural area?', for example, you might simply be locating and describing one or two villages, and the areas within them which were investigated. However, if you were focused on the question 'How do channel characteristics vary along your chosen river?', you would need to identify and describe specific points along a river, using a map and perhaps some grid references to pinpoint the location of all the individual sites used. This shows that it is important to consider the scale at which you locate your investigation. The most important factor is that your description needs to be clear to others so that they have a clear idea of where your investigation took place.

The location of your investigation is an ideal opportunity to use websites such as Google Earth or Multimap. Figure 2 shows a possible pair of images for showing the location of two villages used to investigate service provision. Used with a short written description, this would lead to a clear impression for the reader. Figure 3 does the same thing for an investigation of a river. Note the difference in the scale of the images used, and how in both cases, the images have been annotated so that they have some clear information regarding the locations of the investigation.

**Top Tip**

Make sure you annotate any maps you include, giving details about the locations which are important, and perhaps the data collection methods you have used at particular locations.

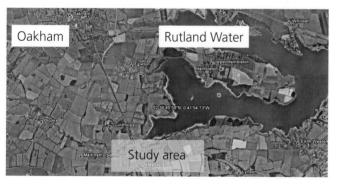

© 2008 Europa Technologies   © 2007 Google
© 2008 Tele Atlas
© 2008 Infoterra Ltd & Bluesky

Figure 2: Images taken from Google Earth used to locate study villages

© 2008 Tele Atlas   © 2007 Google
© 2008 Infoterra Ltd & Bluesky

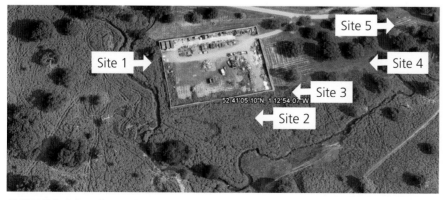

© 2008 Tele Atlas   © 2007 Google
©2008 Infoterra Ltd & Bluesky

Figure 3: An image used to locate sites for a river investigation (note the difference in scale from those images used in Figure 2)

**Control**

You will be able to research some information about your location in groups, but when writing up this section as part of your report you will need to do this individually.

## How you will be marked

The methods of collecting data (9 marks)

| Mark | Reasons for marks given |
|---|---|
| 0 | There is no evidence of data collected or method(s) of collection. |
| 1–3 | There is limited evidence of primary and secondary data collected. There is little explanation of why the methods were used to collect primary and secondary data. The contribution of the student to the primary data collection is briefly described. Limited evidence of risk assessment. No obvious evidence of the use of GIS. |
| 4–6 | The primary and secondary data has been collected by the student and is appropriate for the investigation. There is some explanation of why the methods were used to collect data. The contribution of the student to the primary data collection is clearly described. Clear evidence of risk assessment having been undertaken. Some limited use of GIS. |
| 7–9 | The primary and secondary data has been accurately collected by the student and is appropriate. There is detailed explanation of why the methods were used to collect primary and secondary data. The contribution of the student to the primary data collection is described in detail. Clear reference to risk assessment, explicitly linked to the investigation. Use of GIS is clear and appropriate. |

## Methods of collecting data

When you have explained the purpose, and described the locations to be used, you will then need to decide which methods you are going to use for collecting data. This requires careful planning, and includes a consideration of the types of data that you will use – primary or secondary data, or both.

| | |
|---|---|
| **Primary data** | This is data which is collected first hand. In the case of a school-based investigation this is the data which is collected by a group of students whilst undertaking an exercise of fieldwork. |
| **Secondary data** | This is data which has already been collected by others for a particular purpose, and then 'published' on websites, in books, official reports, etc. It can be included to support primary data in an investigation. A good example is the UK census, which is taken every ten years and provides a large amount of data about the population for academic studies. |

### Sampling

As well as deciding on the types of data which are to be collected, the sampling method must be considered. Sampling is nearly always necessary because it is just not possible to measure every item or interview every person. A carefully chosen sample will be much easier and quicker to investigate and will still give fairly accurate results. Choosing a representative sample – the *when*, *what*, *which* and *where* of the data collection – is an important part of the process. If you were using questionnaires, for example, you would need to make sure that the right people were targeted. There is little point conducting a survey to find out which services young people in a town would like to see developed if you only go to a skate park for views, as only a single interest group is likely to be asked. This would give you a very untypical data set and would call into question any eventual analysis of that data. You need to decide on the timing of your data collection as well as being sure of the groups you wish to measure to make sure that the results gained are not biased. There are a number of ways in which sampling can be carried out, and you should decide which is most appropriate for your investigation.

### Sampling methods

There are a number of ways in which samples can be taken, so you should be careful to plan – and identify – how you have carried out your sampling.

**Random sampling** – the locations for data collection are chosen by chance. An example might be the use of a quadrat to count the number of plant species at a number of sites. As you finish at one site, you would throw the quadrat and count the next site as being where it landed.

**Systematic sampling** – the locations of the sites used are found at equal intervals from each other. This might be each fourth shop within a shopping area, or at points on a grid if sampling the size of pebbles on a beach.

**Line-intercept sampling** (also known as a transect) – sites are sampled along a line. This might be used to measure the environmental quality across a city centre, or a river at points along its course.

Having decided on the sampling strategy of the investigation, and having perhaps gained some secondary data which can be used as background information to support your work, you need to decide on the primary data collection methods you wish to use. In geography there are a whole range of techniques which can be used to collect data, and you should look beyond the few introduced below for further ideas.

### Sketching/photos

It is often useful to include a clear picture of your main sites, or close-up details of the ideas which you discuss, e.g. photos of graffiti (see Figure 4). Sketches should be annotated to show features, processes, etc., giving a clear description and explanation of important features (see Figure 5).

### Control

When you collect your data, there is only limited control. This means you can work as part of a group if appropriate.

### Top Tip

Remember to explain why you have used a particular sampling method.

*Figure 4: A typical beach*

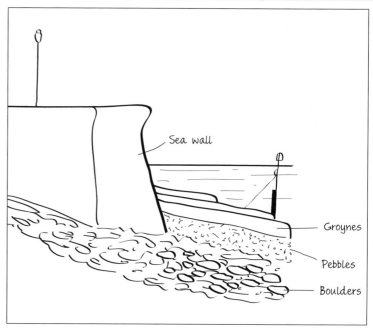

*Figure 5: An annotated sketch of Figure 4, emphasising its important features*

## Mapping

Geography uses mapping a great deal. There are many different types of mapping. You might be looking at parking provision, and need to shade in areas of car parking on a street map. You might be looking at the pattern of plant growth on some sand dunes, and use a map of the dunes to shade in where different plant species are found. A frequently used method is that of land use mapping, where different types of shop or other building use are identified by differently shaded colours. You could look up different types of map in books to see what might be possible, and then consider the type of map you need. At this stage, you may well need to ask for help in finding the correct map. You might use:

- an Ordnance Survey map

- a town street plan

- a sketch map that you have created yourself.

## Questionnaires

Questionnaires are a very useful way in which to collect data – but you need to consider carefully what you are trying to find out. There are a number of choices you need to make:

- Which questions will allow you to collect the information you need, without making the questionnaire too long?

- Will the questions be *open* (allowing respondents to offer opinions or supply information) or *closed* (requiring respondents to choose from a set of fixed alternatives, such as Yes/No, 1/2/3/more than 3)?

- Can some questions be answered without being asked, such as the sex of the person you are interviewing, and possibly estimating their age rather than asking for it?

- Will you fill in the questionnaire yourself, from someone's answers, or will they fill it in by themselves?

- How will you introduce yourself when you first ask someone to respond to your questionnaire?

Once you have thought about these issues, you should draft a questionnaire and then ask a friend to fill it in, to make sure that it works as you want it to.

## Measuring physical features

Measurement is central to the study of physical geography. It includes activities such as:

- measuring the width of a river

- measuring the density of species found in a woodland

- measuring wind speed.

In each case it is important to make sure that the measurements are taken accurately. By using appropriate sampling strategies, and measuring particular features, a good understanding can be gained as to the processes occurring in a physical environment (see Example 1).

## Example 1: Basic measurements used to investigate river characteristics

When investigating the changing characteristics of a river as it flows downstream, a number of different measurements can be taken:

◉ Channel width, by using a tape measure

◉ Channel depth, taken in several equally spaced places across the stream so that a cross-section can be drawn. This can allow you to calculate the cross-sectional area.

◉ Water velocity, using a flow meter or float and stop watch

◉ Sediment size, measuring the size of sediment on the river bed. This might be done using a ruler (where the sediment is large) or sieves (where the sediment is small).

© 2008 Tele Atlas   © 2007 Google
©2008 Infoterra Ltd & Bluesky
*Figure 6: Location of Site 3*

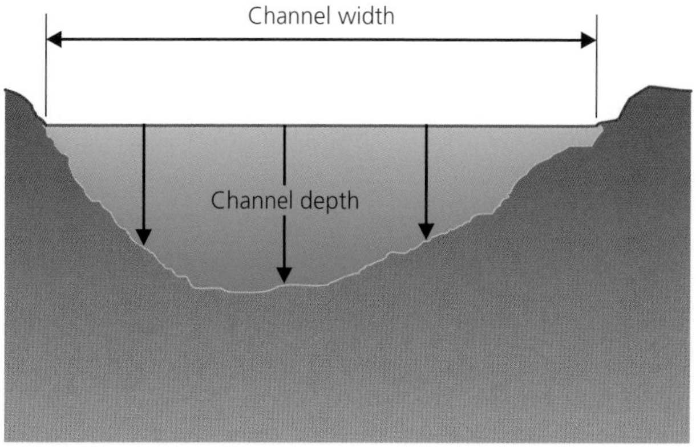

*Figure 7: Cross-section of the river channel at Site 3*

## Data sheets

You should be confident about developing a clear and efficient data sheet from your work on your local area. Remember that you should leave plenty of space to ensure that results can be clearly accommodated. If this means you need to use more than one sheet of paper, then do so.

## Risk assessment

Before you go on your fieldwork trip, it is important that you carry out a risk assessment so that you are prepared to tackle any safety issues if they should arise. You also need to include your risk assessment in your final report.

In order to complete a risk assessment, you need to consider the following:

◉ The risks – the things that have the potential to do harm;

◉ The severity of each risk – how bad a potential injury from them might be;

◉ The management of the risks – the plans in place or guidance to reduce the risks and the potential injury.

It may be useful to present this information in the form of a table.

## Methods of presenting data

Once you have collected your data, you then need to decide how to present it, so that those reading your investigation can fully understand it. As with data collection, there are a number of ways in which you can present your data.

### Tables and graphs

The simplest way of representing data is by using tables with the numeric results in them. Graphs and charts are more visually interesting and they can make a description and analysis of data very clear, but you need to be careful that you use the correct type for the data.

Some people simply work their way along a spreadsheet toolbar to provide some variety in their presentation, but this shows a lack of understanding of the use of graphs. You should use the type of graph that does the job most accurately and efficiently. Some basic types are explained below.

### Line graphs

Line graphs can be used to plot continuous data, such as a population size increase over a number of years, or the temperature of a particular location over a number of hours (Figure 8). The variable should be plotted on the y-axis, whilst the time should be plotted on the x-axis. It is also possible to plot more than one set of a particular data type on one graph – for example, the temperature change at three sites – allowing for comparison.

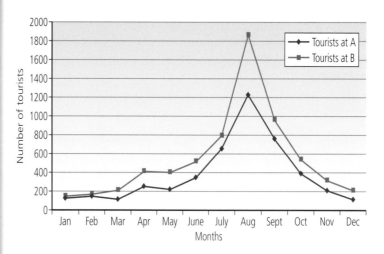

Figure 8: A line graph showing the distribution of monthly tourist numbers at two sites

### Pie charts

This graph type is used to present group values such as the number of different transport types observed at a road junction over the course of a day (Figure 9). The values are first converted into percentages, and then into degrees to allow the plotting of the data. If you use a graphics package on a computer, this will normally be done for you automatically. Pie charts tend to be overused, and are best used when there are several categories of data in the group.

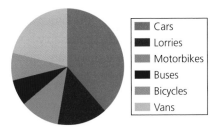

Figure 9: A pie-chart showing traffic composition at an urban location

## Bar charts

As with pie-charts, bar charts are a common form of graph used in investigations. The y-axis is used for a numeric scale, such as the frequency of an event (Figure 10) while the x-axis is used to identify the categories of the data.

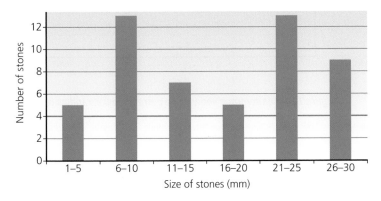

Figure 10: A bar chart showing the number of stones of different sizes found on a stream bed

## Scattergraphs

Scattergraphs are more complex than the other types of graphs described here, as they do more than just present data visually. Scattergraphs are used to plot two sets of data to find out if there is a link between them. For example, you might count the number of services or amenities in twelve settlements, and find out what the population of each is. You would then plot the results as a number of points (Figure 11). These show a pattern from bottom left to top right. This is what is called a *positive correlation*, because as one variable (population size) gets larger, so does the other (number of services).

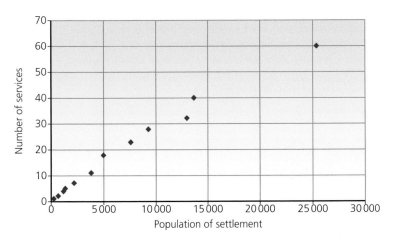

Figure 11: A scattergraph showing the relationship between settlement population and the number of services. (In this case, a positive correlation is shown.)

If a scattergraph shows a clear pattern of points from top left to bottom right (Figure 12), this shows a *negative correlation* – as one variable gets larger the other gets smaller. Both these patterns would show that there is probably some link between the two variables being plotted. But if the points are random and show no pattern (Figure 13), there is no apparent correlation, so the two variables are probably not linked.

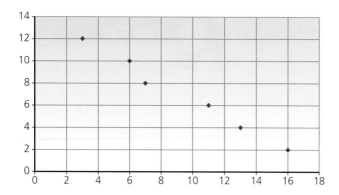

Figure 12: A scattergraph showing a negative correlation

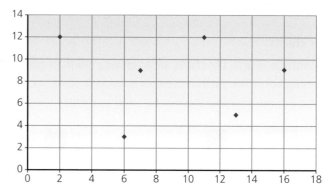

Figure 13: A scattergraph showing no correlation (i.e. random results)

## Simple statistics

You may also want to calculate some simple statistics using your data. The most frequently used statistics are the *mean*, *mode*, and *median*. These can help in describing either the most frequently occurring or largest group of data in a data set, and can therefore help quantify and make clearer the patterns in the data you have collected.

| | |
|---|---|
| **Mean** | The mean is the statistical average of a set of numbers. This is found by adding the values together and dividing the result by the number of values present. For example, the average of the four values 3, 4, 6 and 8 is 3+4+6+8 divided by 4 = 21 divided by 4 = 5.25. |
| **Median** | The median is the middle value in a set of numbers. It is found by arranging the values in order, and identifying the middle value. If the number of values is even, the two middle values are added together and divided by two. For example, with values 3, 6, 2, 7, 9 and 4, the median is found by first ordering the values: 2, 3, 4, 6, 7, 9, and then – because there are an even number of values – adding the middle two together (4+6) and dividing the result by two. Hence the median value is 10 divided by 2 = 5. |
| **Mode** | The mode is the most frequently appearing value in a group of numbers. For example, if a set of values is 5, 3, 4, 8, 5, 7, 5, 2, 3, 5 and 1, then the mode value is 5, because it appears more frequently than any other value. |

## Maps

Maps are another valuable way of presenting your data. You should try to make them clear and colourful. As with graphs, there are a number of different types of map which can be used.

Choropleth maps (Figure 14) use shading to show patterns in data, with shading normally becoming darker with larger numbers in the data. This type of map is used to compare areas in terms of grouped values, such as infant mortality rates, or house prices. Data must be sorted into groups, with clear boundaries, such as county boundaries, or regional boundaries.

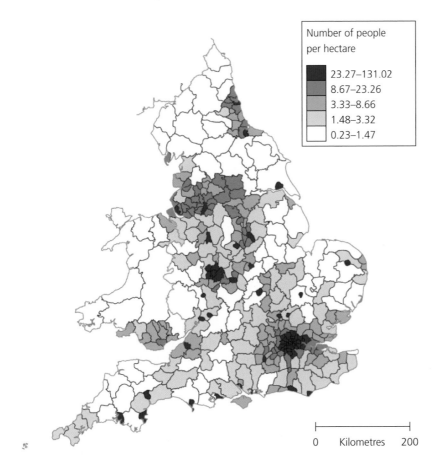

Number of people per hectare

| | |
|---|---|
| ■ | 23.27–131.02 |
| ■ | 8.67–23.26 |
| ■ | 3.33–8.66 |
| ■ | 1.48–3.32 |
| □ | 0.23–1.47 |

0    Kilometres    200

*Figure 14: A choropleth map of population density in England and Wales*

Flow lines (Figure 15) are used to show data relating to movement. Examples might include the number of pedestrians who walk down a certain street or the amount of traffic on a road. The direction in which the flow is moving is indicated by an arrow head at one end, and the width of the flow line is dependent on the volume of the flow. The scale used to determine the width of the lines must be chosen carefully, to allow the lowest and highest value to be shown clearly on the map.

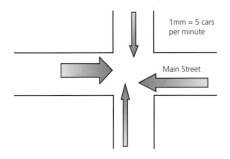

1mm = 5 cars per minute

Main Street

*Figure 15: A flow line diagram showing traffic numbers at a crossroads*

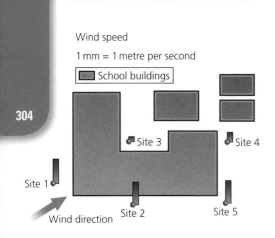

Wind speed

1 mm = 1 metre per second

☐ School buildings

Site 3    Site 4

Site 1

Wind direction    Site 2    Site 5

*Figure 16: A map of a school's buildings, with wind speeds plotted at five sites*

## How you will be marked

### Analysis and conclusions (9 marks)

| Mark | Reasons for marks given |
|------|-------------------------|
| 0 | There is no analysis or conclusion. |
| 1–3 | Data has been extracted and described.<br>Some basic conclusions have been drawn which vaguely relate to the question or issue investigated. |
| 4–6 | Data is described in some detail with analytical comments.<br>Plausible conclusions are reached using the evidence which is presented in the investigation report. |
| 7–9 | There is analysis, which draw together the student's findings.<br>The conclusions are accurate and substantiated and refer to the correct theory where appropriate. |

## Control

You must complete the write-up of your analysis, conclusions and evaluation individually and in class time under exam-style conditions.

Where data cannot be easily grouped and shown on a choropleth map, you may want to plot separate graphs or charts on to a map (Figure 16). This is a way of showing simple data distributions relative to each other in space, such as environmental quality values at given sites in an area. Again, you need to consider scale carefully so that the highest and lowest values can be clearly shown on the same map.

If you are using maps to present data, this might be a good opportunity to use GIS applications. These packages are designed to allow for the presentation of spatial data, and will also make decisions on issues such as shading and scale far easier to handle.

## Developing your analysis and reaching your conclusions

Once you have presented your results, you will need to consider how to make the best use of them in explaining the patterns you see. For students, the analysis of data can be the most challenging part in the process of completing a fieldwork-based project – often because they find it hard to distinguish between *describing* the data and *explaining* the data. Analysis always includes explaining what you have found.

### Making sense of your data

It is useful to start an analysis by laying out all of the results which you gained during the data collection and data presentation phases. Make a list of the data collected. Next to each set of data, describe what you think it shows (using any graphs, tables or maps you have produced to help). Finally, try to explain the data, highlighting *why* you think the results appear as they do. Remember that you will need to *describe* your results – *what* do they show – before *explaining* your results – *why* are the results as they are? The description of results should be completed fairly quickly, merely outlining what a group of data shows.

If you look back at Figure 14 on page 303, for example, you might describe the population density data given by highlighting the general pattern, i.e. that in general terms the population density is greater in and immediately surrounding large conurbations such as London, Birmingham and Liverpool, while rural areas have a much lower population density. This should then be supported by some specific examples such as Greater London having the highest population density of between 23.27 and 131.02 people per hectare, while North Devon has a low density of 0.23 to 1.47 people per hectare. Therefore, any description of the data should give a general impression of the patterns and include specific examples such as numbers, or quotes from interviews.

304

# Explaining the patterns in your data

Your description of the data should be brief, while a greater focus should be given to explaining their patterns. This section of your project – where you give reasons for the patterns in your data – is crucial. You have presented and described your data, but can you now suggest why the results and associated patterns look like they do?

Look again at a question given at the start of this section – *'How do channel characteristics vary along your chosen river?'*. The results for the size of the sediment in the channel from several sites might appear like this.

| Site | Average size in millimetres | Sediment shape |
|------|-----------------------------|----------------|
| 1 (upstream) | 12 | Angular |
| 2 | 8 | Sub-angular |
| 3 | 7 | Sub-angular |
| 4 | 5 | Sub-rounded |
| 5 (downstream) | 2 | Rounded |

*Size of sediment found at five different sites along a stretch of river*

We can *describe* the pattern as one showing decreasing sediment size as we move downstream, from an average of 12 mm at Site 1 to 2 mm at Site 5. We then need to explain this. There might be a number of explanations, but we could argue that the decrease in size is due to erosive processes in the channel, causing pebbles and gravel to constantly hit each other, leading to attrition of the sediment. It is at this point that we might then draw together some of our other results to make our argument stronger. If we have collected data on the shape of sediment at the five sites and can show that the sediment is becoming increasingly rounded, this might give extra evidence that attrition is occurring and wearing down the pebbles/gravel. Your explanations therefore need to give reasons for the data presented and, at the same time, they should also attempt to make links between different elements of your data, rather than simply reading it like a detailed list.

**Top Tip**

Remember that it is important to explain your results, not just describe them. Make sure you understand the difference between the two.

## Concluding your study

Having explained your results, you should write a conclusion. This should summarise your main findings, and include the following two elements:

1. You should summarise your main findings in relation to how far they answer the question you posed at the start. Having posed that question, have you been able to answer it to some extent, and what is your answer?
2. You should refer back to any theory which is related to your study. For example, if studying service provision in rural areas, theory would suggest that smaller settlements will have fewer services. Do your results agree with theory, or are they different? If they are different can you explain why they might be different?

Therefore, having presented your findings in graphs, maps, etc., you then need to use these results to describe and explain what you have found, before concluding your study by relating what you have found to your original question, and to relevant geographical theories.

## Evaluating your study

The evaluation of a piece of work is another area which students often find very difficult. This is the part of a study which aims to reflect on the process of collecting data, and how that process might have impacted on the quality of the results gained. An evaluation should consider the collection and analysis of data and how these might impact on the conclusions made at the end of the study. It is very important that you accept that no study is perfect – even those carried out by university academics – and it is therefore perfectly reasonable to highlight where you think the shortcomings of your work are.

## What is an evaluation?

An evaluation should be based on three basic questions:

### How you will be marked

**Evaluation (9 marks)**

| Mark | Reasons for marks given |
|---|---|
| 0 | There is no evaluation. |
| 1–3 | There is limited evaluation of the investigation. Either all aspects of the investigation have been evaluated in limited detail or some aspects of the investigation have been evaluated in more detail. |
| 4–6 | There is evaluation of the investigation which varies in completeness between the aspects. Some of the limitations of the evidence collected have been recognised. |
| 7–9 | There is detailed evaluation of the investigation which reflects on the limitations of the evidence collected. |

## What problems did you encounter when collecting your data?

When collecting data, you need to be aware of the possible problems involved. If you were measuring the impact of tourism on a local environment, collecting your data on a bank holiday Monday might have given unusual pedestrian flow counts. While this is useful in showing the extremes of use that the environment sees, it might not be representative. If you were focusing on traffic volumes at locations around a CBD, you would not have been able to be in all the locations at once. This means that some of the differences in traffic might be due to time differences. For example, you might have visited two locations during the rush hour but, by the time you reached the third, rush hour might have been over.

Whatever your focus, you should consider the shortcomings of your data collection. Remember that this evaluation does not make your results incorrect – it demonstrates that you have a clear understanding of the difficulties involved in any data collection exercise.

## How might the problems you had affect your results and therefore your analysis?

If you have identified any problems experienced in collecting your data, you next need to suggest how they might have affected your results. If we take the example of traffic flows around a CBD, having described the problems in collecting the data, you might then go on to say that, given that the third location was visited after the end of rush hour, its results might have been lower because of the time difference. Hence, your analysis that the third location is much quieter than the other two may be correct, but you have to accept that it may in part be inaccurate and it is possible that the site might be much busier during rush hour, much like the other two sites. Having explained this, you should finally explain how this might impact on your conclusions. Again, you should remember that this should not be seen as showing that you have done a poor piece of work, but that you are aware of the impact of any problems on your results.

## How would you change your approach if you were to do the investigation again?

Finally, having identified the problems you had when collecting your data, you should now suggest how you would try to alter your collection methods if you did the study again. Therefore, if you identified the problems with collecting traffic flow data within a CBD, you might suggest that you would alter the locations used to ensure that all of them could be covered in the rush hour, or that you would ask a friend or parent to collect data in one or more locations, so that the timings were as close to each other as possible. This section, therefore, is focused on developing solutions to the problems identified in the first part of the evaluation.

## Planning and organising your investigation

As you complete your fieldwork investigation, you will begin to gain a lot of paper, data, photos, and other evidence. During the process it is important that you organise your work so that you do not lose information, and when it comes to writing up your study, it is very important that you organise your work and thoughts into a well-planned and coherent end product.

### Putting a study together

When an investigation is planned, it needs to be clear what the parts of the finished product will be and how they will fit together. It is possible to complete the fieldwork investigation using more than one medium but, if that is your plan, you should be clear before you start about how the different elements will come together to make sense.

The elements of a completed fieldwork investigation are shown in the bullet points below and should act as your basic framework when planning and organising your work. Remember that the word limit for the controlled assessment is 2,000 words. This means you should carefully consider the amount you write for each element of the report, and in each case make sure your writing is focused.

- Report outline
- Purpose of investigation
- Methods used to collect data
- Presentation of data
- Analysis and conclusions
- Evaluation.

You must remember that it will only be possible for each element to be worked on for a limited period of time. So you should plan your time carefully, to ensure you finish each element – you do not want to run out of time, leaving some elements unfinished. You should always allocate time to check each element – making sure that you have presented your information well, and that the spelling and grammar is accurate (see the section on Spelling, punctuation and geographical terminology) as well as ensuring that everything flows properly from one section to the next.

## How you will be marked

Planning and organisation (6 marks)

| Mark | Reasons for marks given |
|---|---|
| 0 | The investigation report lacks any planning or organisation. Geographical terminology is absent. Spelling, punctuation and the rules of grammar errors are frequent. |
| 1–2 | The work may be incomplete and not organised into a clear sequence. Geographical terminology may not be used accurately or appropriately. |
| 3–4 | There is a sequence of enquiry in the investigation report. Content is clear, for example page numbers are all present. Spelling, punctuation and the rules of grammar are accurate. Geographical terminology is used appropriately in the investigation report. |
| 5–6 | The sequence of the enquiry in the investigation report is clear. Diagrams are integrated into the text with appropriate sub-headings. Spelling, punctuation and the rules of grammar are accurate. There is accurate and appropriate use of geographical terminology. |

## Including illustrations such as graphs, photos, tables, etc.

A well-presented project will have a series of illustrations – graphs, photos, and perhaps maps. You should ensure, however, that you only include illustrations which show useful and relevant information. And it is important that they are properly integrated into the written element of the work – you should not simply paste them in and assume that the reader will understand how they relate to the text. You should give each one an appropriate title, starting with 'Photo', 'Figure' or 'Table' and numbered – Photo 1, Photo 2, Figure 1, Figure 2, etc. The title should then describe what the illustration is showing, in a way that helps the reader understand it. In your text, when you describe or explain results, you should link the writing to the relevant diagram by referring to it in brackets, as shown in Example 2 below.

### Example 2: Labelling a photo and referring to it in the text

The village had a small number of shops, such as a newsagent (Photo 1), and a chip shop, but few other services.

There are a number of reasons for the lack of services in the village . . .

*Photo 1: Local services in village A*

## Spelling, punctuation and geographical terminology

Finally, you need to be careful with your spelling and punctuation. If you are using a word processor, the software may pick up on spelling, grammatical and punctuation 'errors', but do not assume that the computer is always right. You must always read through what you have written and check it carefully. Remember that marks are obtained for good use of language, and that poor spelling and punctuation can lead to a loss of marks.

You should also use geographical terminology where possible. For example, rather than writing 'amount of water flowing through the river', use 'discharge'. You will gain credit for using geographical terminology – if it is used correctly – because it shows a higher level of understanding on your part, and can often lead to briefer, but better explanations.

## Variety of report formats

It is important that you carefully consider how you intend to present your fieldwork investigation. You have the opportunity to present your work in a number of different formats – just as professionals in a commercial setting might use different ways to present their information and ideas.

The most obvious format to use is a written report of approximately 2,000 words. Where this is chosen, it is simply a case of using a word processor to write up the various sections of your report. However, alternative formats can be used:

- DVDs – perhaps used where interviews are part of the data collection, or filming yourself explaining some of your results, perhaps carried out whilst collecting data on a stretch of river, or explaining changes in a transect across a city centre.

- PowerPoint presentations – which might be written in a similar fashion to a word processed piece, but completed as a presentation (Figure 17).

- Personalised GIS maps – these could be used in data presentation to show results, and possibly in annotating satellite images to develop issues or ideas in a more graphic form.

- Web pages – these can be developed to create a website format rather than a simple word processed project. One advantage of this format might be the ability to link different parts of the report together so that the reader can move backwards and forwards between elements. You might also decide to include hyperlinks to other websites which might help give background information or provide sources of secondary data (Figure 18).

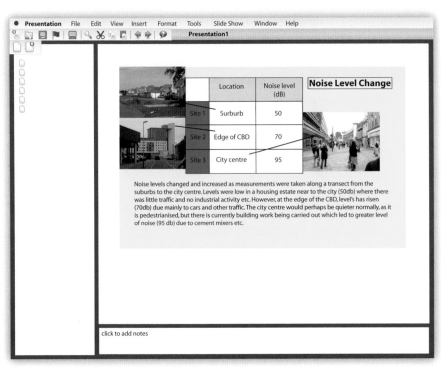

*Figure 17: Example of a PowerPoint slide*

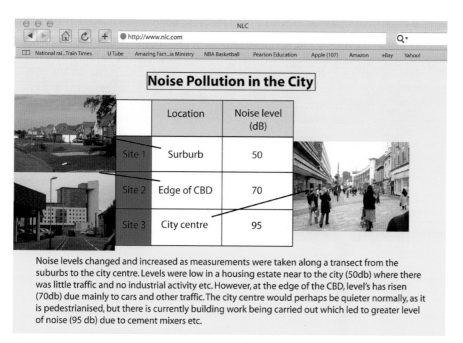

Figure 18: *Example of a website used to discuss noise level change*

You can use more than one format, but your teacher may decide to use particular formats having considered both the available technology within your school and/or any other organisational restrictions which might apply.

If you are given a choice of formats, you should consider the following points in planning and developing your ideas:

◉ You need to understand how the elements and formats of your report will fit together. Will the reader understand how to 'read' your work? Are you trying to use too many formats?

◉ Given that you have limited time, you should be confident that you can use the format you have chosen efficiently. A website might sound like an exciting idea, but if you do not know how to write one, you will not have enough time to teach yourself.

◉ Remember that the geography is what is important. Students can often get carried away in spending time on the design aspect of a PowerPoint presentation or website, and forget that it is the geographical content that they are gaining credit for.

You should decide on a format, if given a choice, at the initial planning stage and discuss it with your teacher so that you are both happy with your decision.

Welcome to ExamZone! Revising for your exams can be a daunting prospect. In this section of the book we'll take you through the best way of revising for your exams, step-by-step, to ensure you get the best results that you can achieve.

# Zone In!

Have you ever become so absorbed in a task that it suddenly feels entirely natural? This is a feeling familiar to many athletes and performers: it's a feeling of being 'in the zone' that helps you focus and achieve your best. Here are our top tips for getting in the zone with your revision.

## UNDERSTAND THE PROCESS

Understand the exam process and what revision you need to do. This will give you confidence but also help you to put things into proportion. These pages are a good place to find some starting pointers for performing well in exams.

## BUILD YOUR CONFIDENCE

Use your revision time, not just to revise the information you need to know, but also to practise the skills you need for the examination. Try answering questions in timed conditions so that you're more prepared for writing answers in the exam. The more prepared you are, the more confident you will feel on exam day.

## DEAL WITH DISTRACTIONS

Think about the issues in your life that may interfere with revision. Write them all down. Think about how you can deal with each so they don't affect your revision. For example, revise in a room without a television, but plan breaks in your revision so that you can watch your favourite programmes. Be really honest with yourself about this – lots of students confuse time spent in their room with time revising. It's not at all the same thing if you've taken a look at Facebook every few minutes or taken mini-breaks to send that vital text message.

## FRIENDS AND FAMILY

Make sure that they know when you want to revise and even share your revision plan with them. Help them to understand that you must not get distracted. Set aside quality time with them, when you aren't revising or worrying about what you should be doing.

## GET ORGANISED

If your notes, papers and books are in a mess you will find it difficult to start your revision. It is well worth spending a day organising your file notes with section dividers and ensuring that everything is in the right place. When you have a neat set of papers, turn your attention to organising your revision location. If this is your bedroom, make sure that you have a clean and organised area to revise in.

## KEEP HEALTHY

During revision and exam time, make sure you eat well and exercise, and get enough sleep. If your body is not in the right state, your mind won't be either – and staying up late to cram the night before the exam is likely to leave you too tired to do your best.

# Planning Zone

The key to success in exams and revision often lies in the right planning. Knowing what you need to do and when you need to do it is your best path to a stress-free experience. Here are some top tips in creating a great personal revision plan:

# JUNE

### 1. Know when your exam is
Find out your exam dates. Go to www.edexcel.com/i-am-a/student/timetables/pages/home.aspx to find all final exam dates, and check with your teacher. This will enable you to start planning your revision with the end date in mind.

### 2. Know your strengths and weaknesses
At the end of the chapter that you are studying, complete the 'You should know' checklist. Highlight the areas that you feel less confident on and allocate extra time to spend revising them.

### 3. Personalise your revision
This will help you to plan your personal revision effectively by putting a little more time into your weaker areas. Use your mock examination results and/or any further tests that are available to you as a check on your self-assessment.

### 4. Set your goals
Once you know your areas of strength and weakness you will be ready to set your daily and weekly goals.

### 5. Divide up your time and plan ahead
Draw up a calendar, or list all the dates, from when you can start your revision through to your exams.

### 6. Know what you're doing
Break your revision down into smaller sections. This will make it more manageable and less daunting. You might do this by referring to the Edexcel GCSE Geography B specification, or by the chapter objectives, or by headings within the chapters.

### 7. Link it together
Also make time for considering how topics interrelate. For example, when you are revising your case studies it would be sensible to cross-reference them to other parts of your work. It would be very useful to make a list of examples and case-studies that you have studied and think laterally about how else you might use them in an examination. You know more than you think!

### 8. Break it up
Revise one small section at a time, but ensure you give more time to topics that you have identified weaknesses in.

### 9. Be realistic
Be realistic in how much time you can devote to your revision, but also make sure you put in enough time. Give yourself regular breaks or different activities to give your life some variety. Revision need not be a prison sentence!

### 10. Check your progress
Make sure you allow time for assessing progress against your initial self-assessment. Measuring progress will allow you to see and celebrate your improvement, and these little victories will build your confidence for the final exam.

### Finally –
stick to your plan!

# Know Zone

Remember that different people learn in different ways – some remember visually and therefore might want to think about using diagrams and other drawings for their revision, whereas others remember better through sound or through writing things out. Think about what works best for you by trying out some of the techniques below.

## REVISION TECHNIQUES

**Highlighting:** work through your notes and highlight the important terms, ideas and explanations so that you start to filter out what you need to revise.

**Key terms:** look at the key terms highlighted in bold in each chapter. Try to write down a concise definition for this term. Now check your definition against the glossary definition on page 321.

**Summaries:** writing a summary of the information in a chapter can be a useful way of making sure you've understood it. But don't just copy it all out. Try to reduce each paragraph to a couple of sentences. Then try to reduce the couple of sentences to a few words!

**Concept maps:** if you're a visual learner, you may find it easier to take in information by representing it visually. Draw concept maps or other diagrams. These are particularly good at showing links. For example, you could create a concept map which shows how to learn about sustainability.

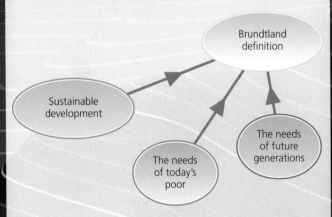

**Mnemonics:** this is when you take the first letter of a series of words you want to remember and then make a new word or sentence. An example of this is SNAP. This stands for Singapore Needs Additional People.

**Index cards:** Write important events, definitions and processes on index cards and then test yourself.

**Quizzes:** Learning facts can be dull. Why not make a quiz out of it? Set a friend 20 questions to answer. Make up multiple-choice questions. You might even make up your own exam questions and see if your friend can answer them!

And then when you are ready:

**Practice questions:** go back through all the ResultsPlus features with questions to see if you can answer them (without cheating!). Try writing out some of your answers in timed conditions so that you're used to the amount of time you'll have to answer each type of question in the exam. Then, check the guidance for each one and try to mark your answer.

Use the list below to find all the ResultsPlus questions.

# Don't Panic Zone

Once you have completed your revision in your plan, you'll be coming closer and closer to the big day. Many students find this the most stressful time and tend to go into panic-mode, either working long hours without really giving their brain a chance to absorb information, or giving up and staring blankly at the wall. Follow these tips to ensure that you don't panic at the last minute.

## TOP TIPS

1. Test yourself by relating your knowledge to geography issues that arise in the news – can you explain what is happening in these issues and why?

2. Look over past exam papers and their mark schemes. Look carefully at what the mark schemes are expecting of candidates in relation to the question.

3. Do as many practice questions as you can to improve your technique, help manage your time and build confidence in dealing with different questions.

4. Write down a handful of the most difficult bits of information for each chapter that you have studied. At the last minute focus on learning these.

5. Relax the night before your exam – last-minute revision for several hours rarely has much additional benefit. Your brain needs to be rested and relaxed to perform at its best.

6. Remember the purpose of the exam – it's for you to show the examiner what you have learnt.

## LAST MINUTE LEARNING TIPS FOR GEOGRAPHY

● Remember that an intelligent guess is better than nothing – if you can't think of an example of an LIC city then take a guess – you cannot lose marks.

● Know your categories – don't go into the examination unclear about basic definitions: developed and developing, urban and rural, erosion and weathering. Check out the glossary.

● Many examination questions ask you to interpret resources. Make sure that you revise the skills that help you do this effectively.

Here is some guidance on what to expect in the exam itself: what the questions will be like and what the paper will look like.

## ASSESSMENT OBJECTIVES

The questions that you will be asked are designed to examine the following aspects of your geography. These are known as Assessment objectives (AO). There are three AOs.

| AO1 | Recall, select and communicate knowledge and understanding of places, environments and concepts. |
| AO2 | Apply knowledge and understanding in familiar and unfamiliar contexts. |
| AO3 | Select and use a variety of skills, techniques and technologies to investigate, analyse and evaluate questions and issues. |

## THE TYPES OF QUESTION THAT YOU CAN EXPECT IN YOUR EXAM

The examination papers are designed so that the opening part of each question is the easiest part. The difficulty becomes progressively harder as you move through the question. The level of difficulty is controlled by the command word and content required in your answer. The Foundation tier papers have questions which have more 'scaffolding' (helping you to structure and develop your answer) to make these papers more accessible.

There are four different types of question:

| TYPE |
| --- |
| **Short** – single word answers or responses involving a simple phrase or statement. |
| **MCQ** – Multiple Choice Question. |
| **Open** – free-response questions that involve a limited amount of continuous prose. |
| **Long** – free-response questions where candidates have the opportunity for extended writing and allow opportunities for assessing the quality of your written communication. |

## UNDERSTANDING THE LANGUAGE OF THE EXAM PAPER

It is vital that you know what 'command' words ask you to do. Common errors are:

1. Confusing *describe* with *explain*.

2. Adding *explanation* when you are only asked to *describe*.

| Identify... | Name a process or a location |
| --- | --- |
| Complete | Finish off a task that has already been partly done |
| Name | Like 'identify' |
| Describe... | Give the main characteristics of a topic or issue |
| Explain... | Give reasons why something is as it is |
| Examine... | Describe something with some detail |
| Outline... | Give the main features of something |
| Define... | Say what something means |
| Suggest reasons... | Say why something might have happened or occurred |
| Give the reasons... | Say why something happened or occurred |
| Comment on... | Give some reasons why or how something is as it is |
| State... | Like 'name' or 'identify' |

# Meet the exam paper

This section shows you what the exam paper looks like. Check that you understand each part. Now is a good opportunity to ask your teacher about anything that you are not sure of here.

Print your surname here, and your other names in the next box. This is an additional safeguard to ensure that the exam board awards the marks to the right candidate.

Here you fill in your personal exam number. Take care when writing it down because the number is important to the exam board when writing your score.

Ensure that you understand exactly how long the examination will last, and plan your time accordingly.

Here you fill in your school's centre number. You will be given this by your teacher on the day of your exam.

Ensure that you read the instructions carefully and that you understand exactly which questions from which sections you should attempt.

Note that the quality of your written communication will also be marked. Take particular care to present your thoughts and work at the highest standard you can for maximum marks.

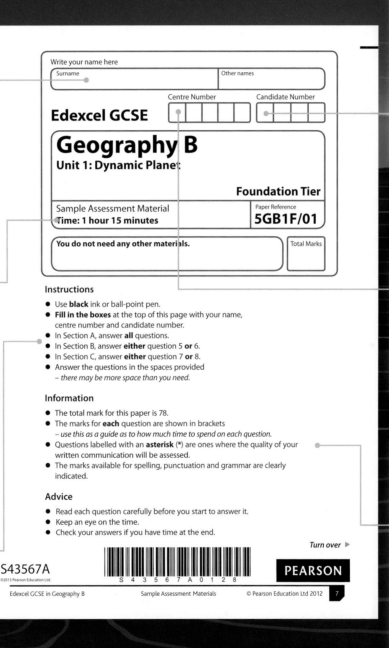

Write your name here

Surname

Other names

**Edexcel GCSE**

Centre Number

Candidate Number

# Geography B
## Unit 1: Dynamic Planet

**Foundation Tier**

Sample Assessment Material
Time: 1 hour 15 minutes

Paper Reference
**5GB1F/01**

You do not need any other materials.

Total Marks

**Instructions**

- Use **black** ink or ball-point pen.
- **Fill in the boxes** at the top of this page with your name, centre number and candidate number.
- In Section A, answer **all** questions.
- In Section B, answer **either** question 5 **or** 6.
- In Section C, answer **either** question 7 **or** 8.
- Answer the questions in the spaces provided
  – there may be more space than you need.

**Information**

- The total mark for this paper is 78.
- The marks for **each** question are shown in brackets
  – use this as a guide as to how much time to spend on each question.
- Questions labelled with an **asterisk** (*) are ones where the quality of your written communication will be assessed.
- The marks available for spelling, punctuation and grammar are clearly indicated.

**Advice**

- Read each question carefully before you start to answer it.
- Keep an eye on the time.
- Check your answers if you have time at the end.

*Turn over* ▶

S43567A
©2013 Pearson Education Ltd.

Edexcel GCSE in Geography B         Sample Assessment Materials         © Pearson Education Ltd 2012   7

**PEARSON**

If the Changing Economy of the UK is the topic that you have studied in class and you wish to answer the question on it, remember to indicate this where you are asked on the paper.

It is not always one examiner who will mark your entire paper. Sometimes, one examiner will mark one question and another will mark a different question. So, you must indicate which question you have answered so that your paper is sent to the correct examiner!

Pay attention to any text highlighted in bold. It is highlighted to alert you to important information, so be sure to read it and take note!

Read the instructions each time – they are there to provide guidance.

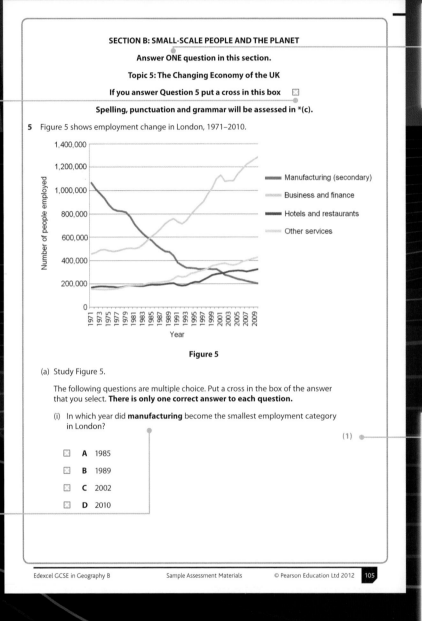

The marks for each question are shown on the right-hand side of the page. Make sure that you note how many marks a question is worth as this will give you an idea of how long to spend on that question.

---

**SECTION B: SMALL-SCALE PEOPLE AND THE PLANET**

**Answer ONE question in this section.**

**Topic 5: The Changing Economy of the UK**

**If you answer Question 5 put a cross in this box** ☒

**Spelling, punctuation and grammar will be assessed in \*(c).**

5   Figure 5 shows employment change in London, 1971–2010.

Legend:
- Manufacturing (secondary)
- Business and finance
- Hotels and restaurants
- Other services

Y-axis: Number of people employed (0 to 1,400,000)
X-axis: Year (1971 to 2009)

**Figure 5**

(a)  Study Figure 5.

The following questions are multiple choice. Put a cross in the box of the answer that you select. **There is only one correct answer to each question.**

(i)   In which year did **manufacturing** become the smallest employment category in London?

(1)

☒   **A**   1985

☒   **B**   1989

☒   **C**   2002

☒   **D**   2010

This is the 'stem' of a question – it often includes important information that you need to think about in your answers.

This is the resource – be careful, it may not be exactly like resources that you have seen before.

These are the command words that tell you what to do.

Words in bold are highlighted to catch your attention so be sure to note them.

---

**SECTION C: LARGE-SCALE DYNAMIC PLANET**

**Answer ONE question in this section**

**Topic 7: Oceans on the Edge**

**If you answer Question 7 put a cross in this box** ☒

**Spelling, punctuation and grammar will be assessed in *(c) (ii).**

**7**   Figure 7 shows a marine food web.

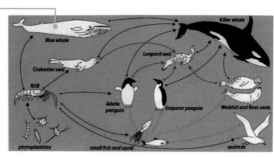

**Figure 7**

(a)  Study Figure 7.

The following questions are multiple choice. Put a cross in the box of the answer that you select. **There is only one correct answer to each question.**

(i)   Which of the following would be the **most** likely impact of a decline in the population of phytoplankton?

(1)

☒   **A**   An increase in the number of krill.

☒   **B**   An increase in the number of all other species.

☒   **C**   A decrease in the number of krill but no other impact.

☒   **D**   A decrease in the number of all other species.

Well done, you have finished your exam. So, what now? This section provides answers to the most common questions students have about what happens after they complete their exams.

## About your grades

Whether you've done better than, worse than or just as you expected, your grades are the final measure of your performance on your course and in the exams.

### When will my results be published?

Results for summer examinations are issued in August, with GCE first and GCSE second.

### Can I get my results online?

Visit www.resultsplusdirect.co.uk, where you will find detailed student results information including the 'Edexcel Gradeometer' which demonstrates how close you were to the nearest grade boundary. Students can only gain their results online if their centre gives them permission to do so.

### I haven't done as well as I expected. What can I do now?

First of all, talk to your subject teacher. After all the teaching that you have had, tests and internal examinations, he/she is the person who best knows what grade you are capable of achieving. Take your results slip to your subject teacher, and go through the information on it in detail. If you both think that there is something wrong with the result, the school or college can apply to see your completed examination paper and then, if necessary, ask for a re-mark immediately. The original mark can be confirmed or lowered, as well as raised, as a result of a re-mark.

### What can I do with a GCSE in Geography?

Geography is well known as a subject that links to all other subjects of the curriculum, so a GCSE in Geography is a stepping stone to a whole range of opportunities. A good grade will help you to move on to AS, Applied A Level or BTEC course. You may want to continue your study of Geography or take a course such as a BTEC National in Travel and Tourism which has a more work-related approach.

The skills that you develop can lead you to employment opportunities in journalism, media, engineering, ICT, travel and tourism, environmental management, marketing, business management and teaching. Geographers are everywhere!

# Glossary
## A-C

**Abrasion**: (in rivers) erosion caused by the river picking up stones and rubbing them against the bed and banks of the channel in the flow.

**Accessible countryside/rural areas**: countryside within easy reach of urban areas.

**Adaptation**: changes that take place to react to a situation or condition. (They may or may not be successful.)

**Ageing population**: a population with a rising average age.

**Agribusiness**: commercial agriculture that is owned and managed by large corporations.

**Air masses**: large bodies of air (many thousands of km² in area) that form over polar or tropical source regions such as north Africa.

**Alternative energy**: energy sources that provide an alternative to fossil fuels.

**Altitude**: the height of the land.

**Alternative fuels**: fuel sources that provide an alternative to fossil fuels.

**Amenities**: things like restaurants, health clubs, shops and cinemas that people want access to.

**Anti-natalist**: policies that seek to limit population growth by birth control.

**Appropriate technology**: equipment that the local community is able to use relatively easily and without much cost.

**Aquaculture**: commercial fish farming, e.g. rearing fish or prawns in ponds or submerged cages.

**Aquifer**: an underground store of water, formed when water-bearing (permeable) rocks lie on top of impermeable rocks.

**Asteroid collision**: a large extra-terrestrial object – such as a meteor – passes intact through the atmosphere and impacts with the Earth's surface.

**Asthenosphere**: the upper part of the Earth's mantle, where the rocks are more fluid.

**Atlantic depressions**: weather systems that bring stormy conditions and frontal rainfall to the western coastlines of Europe. (Depressions form when polar and continental air masses meet over the Atlantic ocean.)

**Attrition**: (in rivers) gradual wearing down of particles by erosion as they collide with each other, making them smaller and rounder.

**Automation**: the use of machinery, rather than people, in manufacturing and data processing.

**Backwash**: water from a breaking wave which flows under gravity down a beach and returns to the sea.

**Bay**: a feature produced when erosion creates an indent in the coastline, often located between two headlands.

**Biodiversity**: the number and variety of living species found in a specific area.

**Biofuels**: fuel sources derived from agricultural crops.

**Biome**: a plant and animal community covering a large area of the Earth's surface.

**Biosphere**: the living part – plants and animals – of the Earth.

**Birth rate**: the number of births per 1,000 people in a year.

**Bleaching**: degradation of coral reefs under conditions of increased acidity in sea water.

**Boserupian theory**: the view that when population grows it stimulates technological changes that produce increases in output, ensuring that living standards can be maintained for the growing population.

**Bottom-up strategy**: development projects that originate in local communities rather than in central government or external agencies.

**Branch plant**: a subsidiary business (usually a factory) set up by a company to meet an increasing demand or need. They are often located away from the parent company, but near some new market or cheap resource.

**Brownfield site**: a piece of land that has been used and abandoned, and is now awaiting some new use.

**Bus lane**: a marked lane in a road in which only public transport vehicles such as buses and taxis are permitted.

**Call centre**: an office equipped to handle a large volume of telephone calls (especially for taking orders or serving customers).

**Carbon footprint**: a measurement of all the greenhouse gases we individually produce, through burning fossil fuels for electricity, transport, etc., expressed as tonnes (or kg) of carbon-dioxide equivalent.

**Carrying capacity**: the maximum number of people that can be supported by the resources and technology of a given area.

**Chocolate box village**: a rural settlement that appears to match the picturesque, pretty image sometimes used on boxes of chocolates, etc.

**CITES**: Convention on International Trade in Endangered Species of Wild Fauna and Flora – an international agreement.

**Clark-Fisher model**: a generalised description of how societies' employment structures change as they develop.

**Climate change**: long-term changes in temperature and precipitation.

**Coastal flooding**: the inundation of low-lying areas in coastal areas and regions.

**Coastal management**: the processes and plans applied to coastal areas by local authorities and agencies.

**Collision plate boundary**: a tectonic margin at which two continental plates come together.

**Commercial farming**: a type of agriculture producing crops and livestock for sale and processing.

**Commodity/production chains**: the linkages between a product and the sources of its basic materials and/or its components.

**Commune**: a group of people with common interests, living as a community and sharing benefits equally (often based on farming).

**Commuter belt**: a residential area within relatively easy reach of (and often surrounding) a city, where many of the residents travel to and from the city daily.

# C-D

**Commuter dormitories:** largely residential settlements lying within the commuting catchments of towns or cities.

**Commuters:** people who travel from their home to their place of work, the distance being such that the journey most often involves some form of transport.

**Concordant coast:** a coastline created when alternating hard and soft rocks occur parallel to the coast, and are eroded at different rates.

**Congestion charging:** a system of traffic control that charges drivers who enter the congested central area of a city.

**Conservation:** managing the environment in order to preserve, protect or restore it.

**Conservative plate boundary:** where two tectonic plates slide past each other.

**Constructive plate boundary:** tectonic plate margin where rising magma adds new material to the diverging plates.

**Constructive waves:** small, weak waves with a low frequency that tend to add sand and other sediment to the coastline because they do not break with much force.

**Consumer industries:** industries that produce goods for people to use/consume.

**Consumption:** the using up of something.

**Continental crust:** the part of the crust dominated by less dense granitic rocks.

**Continental shelf:** the submerged edge of a continental land-mass.

**Convection currents:** (in tectonics) circulating movements of magma in the mantle caused by heat from the core.

**Coral reef:** a hard stony ridge, just above or below the surface of the sea, formed by the external skeletons of millions of tiny creatures called polyps.

**Core:** (in tectonics) the central part of the Earth, consisting of a solid inner core and a more fluid outer core, and mostly composed of iron and nickel.

**Core region:** the most important social, political and economic area of a country or global region – the centre of power.

**Corrosion:** chemical erosion caused by the dissolving of rocks and minerals by water.

**Corruption Perception Index:** a measure, produced by Transparency International, to show how corrupt the public sector of a country is, based on judgements by experts.

**Counterurbanisation:** the movement of people and employment from major cities to smaller settlements and rural areas located just beyond the city, or to more distant smaller cities and towns.

**Cultural background:** the origins of an individual's or group's belief system.

**Cultural dilution:** where a particular culture is changed and weakened, usually by exposure to other competing cultures.

**Death rate:** the number of deaths per 1,000 people in a year.

**Deforestation:** the chopping down and removal of trees to clear an area of forest.

**Degradation:** the social, economic and environmental decline of an area, often through deindustrialisation.

**Deindustrialisation:** the decline in industrial activity in a region or an economy.

**Dependence:** a condition in which something (e.g. a country) is only able to survive by relying on outside support (e.g. from another country).

**Dependency theory:** a theory that suggests that the poorer countries of the world supply resources, and also wealth, to the richer countries through an economic system, involving finance and trade, that favours the developed countries. Colonialism was a stage of this and, today, free trade, loans, and the role of big corporations continue this relationship, so that the poor countries are dependent on the rich countries.

**Depopulation:** the decline of a population, both by natural processes and, occasionally, by government policy.

**Deposition:** the dropping of sediment that was being carried by a moving force.

**Deprived area:** an area in which there is a damaging lack of the material benefits that are considered to be basic necessities – employment, housing, etc.

**Deregulation:** removing state regulations. Often applied to the financial institutions in England after 1986 when many regulations about banks, building societies and other financial intuitions were relaxed or abolished.

**Derelict land:** land on which factories or houses have been demolished.

**Desertification:** the process by which land becomes drier and degraded, as a result of climate change or human activities, or both.

**Destructive plate boundary:** tectonic plate margins where oceanic plate is subducted.

**Destructive waves:** large, powerful waves with a high frequency that tend to take sediment away from the beach, because their backwash is greater than their swash.

**Developed countries:** countries at a late stage of development. They are generally quite rich, with a high proportion of people working in secondary and, especially, tertiary occupations. Also known as More Economically Developed Countries (MEDCs).

**Developing countries:** countries at an early stage of development. They are generally poor, with a high proportion of people working in primary occupations. Also known as Less Economically Developed Countries (LEDCs).

**Development:** economic and social progress that leads to an improvement in the quality of life for an increasing proportion of the population.

**Discordant coast:** a coastline created when alternating hard and soft rocks occur at right angles to the coast, and are eroded at different rates.

**Disparity:** a great difference – e.g. between parts of a country in terms of wealth.

**Disposable income:** the amount of money which a person has available to spend on non-essential items, after they have paid for their food, clothing and household running costs.

# E-G

'Do nothing': (in coastal management) an approach that allows natural processes to take their course without any intervention.

Drainage basin: the area of land drained by a river and its tributaries.

Eco footprint: a measure of how much land is needed to provide a place (e.g. a city) with all the energy, water and materials it needs, including how much is needed to absorb its pollution and waste.

Economic development: the progress made by a country or area in creating wealth through businesses, industry and trade.

Economic recession: a time of decline in business and industry, usually marked by a decrease in wealth, an increase in unemployment, and closure of businesses.

Economic migrant: a person who moves in order to find employment.

Ecosystem: a community of plants and animals that interact with each other and their physical environment.

Emerging countries: countries that have begun to experience high rates of economic growth, as for example Brazil, Russia, India, China and South Africa (the so-called BRICS countries); also known as the recently industrialising countries (RICs).

Emigrant: a person leaving a country or region to live somewhere else (for at least a year).

Employment structure: the proportions of people who work in primary, secondary, tertiary or quaternary jobs.

Enhanced greenhouse effect: the increased greenhouse effect resulting from human action (emission of greenhouse gases) and leading to global warming.

Enterprise Zones: areas designated by the government to promote economic growth, by being able to offer financial benefits such as grants for buildings and machinery and a relaxation of planning regulations.

Environment Impact Assessment (EIA): a method of evaluating the effect of plans and policies on the environment.

Environmental degradation: negative impacts on the natural environment, generally through human action.

Environmental Performance Index: a measure, produced by Yale University, that looks at the environmental conditions that people live in and the health of ecosystems.

Environmental pollution: the degradation of the environment through the emission of toxic waste material.

Erosion: the wearing away and removal of material by a moving force, such as a breaking wave.

Estuary: a river mouth that is wide and experiences tidal conditions.

Eutrophication: the loss of oxygen in water after too much nutrient enrichment has taken place.

Evacuation: the removal of people from an area, generally in an attempt to avoid a threatened disaster (or escape from an actual one).

Exploitation: making full use of something (often implying that the use is unfair and has a negative impact).

Extinction: the permanent loss of something, generally used with reference to species of plants or animals, when there are no living examples left.

Extreme climate: a climate that is unusually challenging, usually in terms of its temperature conditions or type and extent of precipitation.

Farmers' market: a set of stalls run by farmers and food growers from the local area.

Fauna: animals.

Fetch: the distance of sea over which winds blow and waves move towards the coastline.

Flood plain: the relatively flat area forming the valley floor on either side of a river channel, which is sometimes flooded.

Flood risk: the predicted frequency of inundation (floods) in an area.

Flora: plants.

Flows: the movement of objects, people and ideas between places.

Food chains: the interconnections between different organisms (plants and animals) that rely upon one another as their source of food.

Food miles: the distance covered supplying food to consumers.

Food web: an illustration of the grouping of animals and plants found in an ecosystem, showing the sources of food for each organism.

Foreign direct investment (FDI): when a business from one country invests money in a company in another country or builds its own factory or office in another country.

Formal sector: (of the economy) work where the people are formally employed, with permanent jobs and regular pay (and they pay their taxes).

Fragile: (environment) easily disturbed and difficult to restore and therefore lacking in natural resilience. Plant communities in fragile areas have evolved in highly specialised ways to deal with challenging conditions. As a result, they cannot tolerate environmental changes.

Frontal rainfall: precipitation formed when a warm (tropical) air mass rises above a denser, colder (polar) air mass. As the tropical air cools, condensation and precipitation occur.

GDP per capita: Gross Domestic Product per person, is the total wealth created within a country, divided by its population.

GECF: Gas Exporting Countries Forum.

Gender Inequality Index: the part of the UN Development Programme reporting system that considers the disadvantages (e.g. health, education) facing females in all countries.

Gene pool: the genetic material contained by a specific population.

Geological climate events: climate changes that result from major geological events such as volcanic eruptions.

Geological structure: the way that rocks are arranged, both vertically and horizontally.

Geology: the science and study of the Earth's crust and its components.

Glacial region: an area that is covered by ice (either a valley glacier or much larger ice sheets).

Global economy: the evolving economic system that increasingly links the countries of the world; it involves the exploitation of resources and the production and marketing of goods and services.

Global city: a major urban area that has a significant role in controlling the international flows of capital and trade.

Global shift: the movement of manufacturing from developed countries to cheaper production locations in developing countries.

Global warming: a trend whereby global temperatures rise over time, linked in modern times with the human production of greenhouse gases.

# G-M

**Globalisation:** the process, led by transnational companies, whereby the world's countries are all becoming part of one vast global economy.

**GNI per capita:** a measurement of economic activity that is calculated by dividing the gross (total) national income by the size of the population. GNI takes into account not just the value of goods and services, but also all of the income earned from investments overseas.

**Goods:** produced items and materials.

**Gradient:** the slope of the land.

**Grassroots scheme:** a scheme that originates within a local community rather than being imposed from above.

**Green belt:** an area around a city composed mostly of farmland and parkland in which development is strictly controlled. Its purpose is to stop the outward spread of the city.

**Green sector:** that part of economic activity that pays attention to environmental issues.

**Greenfield site:** a piece of land that has not been built on before, but is now being considered for development.

**Greenhouse gases:** those gases in the atmosphere that absorb outgoing radiation, hence increasing the temperature of the atmosphere.

**Groundwater:** water contained beneath the surface, as a reserve.

**Gulf Stream:** a warm ocean current in the North Atlantic that flows from the coast of Florida (USA) towards northern Europe.

**Habitat:** an animal or plant's natural home.

**Happy Planet Index:** a measure, produced by the New Economics Foundation, that measures sustainable progress towards the well-being of people.

**Hard engineering:** using solid structures to resist forces of erosion.

**Hard rock coast:** a coastal region composed of resistant materials.

**Headland:** a part of the coastline that protrudes into the sea.

**Holistic approach:** an approach to environmental management that treats the whole area as an interrelated system.

**Honeypot:** a place of special interest or appeal that attracts large numbers of visitors and tends to become overcrowded at peak times.

**Hot arid regions:** parts of the world that have high average temperatures and very low precipitation.

**Human Development Index:** a measure of development that uses four economic and social indicators to produce an index figure that allows comparison between countries.

**Hydro-electric power (HEP):** the use of fast flowing water to turn turbines which produce electricity.

**Hydrogen economy:** a proposed system based on the delivery of energy that is derived from hydrogen and so avoids the negative aspects of using fossil fuels.

**Hydrograph:** a graph which shows the discharge of a river, related to rainfall, over a period of time.

**Hydrological cycle:** the global stores of water and linking processes that connect them.

**Hydropower:** electricity generated by turbines that are driven by moving water.

**Ice age:** a period in the Earth's past when the polar ice caps were much larger than today.

**Immigrant:** a person arriving in a country or region to live (for at least a year).

**Impermeable:** not allowing water to pass through.

**Industrial stage:** the economic stage when manufacturing industry develops.

**Industrialisation:** the process whereby industrial activity (particularly manufacturing) assumes a greater importance in the economy of a country or region.

**Infant mortality rate:** the number of deaths of children (under the age of one) per thousand live births a year.

**Infiltration:** the process whereby water soaks into the soil and rock.

**Informal sector:** (of the economy) forms of employment that are not officially recognised, e.g. people working for themselves on the streets of developing cities.

**Infrastructure:** the basic physical and organisational structures that are required to support the development of businesses and industry (e.g. roads, power supplies).

**Integrated Coastal Zone Management (ICZM):** the system of dividing the UK coastline into zones that can be managed holistically.

**Integrated river management:** a holistic system of managing rivers that takes an overview of the whole river basin and the relationship between its different parts.

**Interlocking spurs:** areas of high land which stick out into a steep-sided valleys.

**Intermediate technology:** a technology that the local community is able to use relatively easily and without much cost.

**Joints:** lines of weakness in a rock that water can pass along.

**Land degradation:** the declining quality and quantity of land, generally because of human action.

**Landfill:** disposal of rubbish by burying it and covering it over with soil.

**Latitude:** the position of a place north or south of the Equator, expressed in degrees.

**Levees:** natural embankments of sediment along the banks of a river.

**Life expectancy:** the average number of years a person might be expected to live.

**Little Ice Age:** a period of slight global cooling that lasted from around the mid-fifteenth century to the mid-nineteenth century.

**Long profile:** the gradient of a river, from its source to its mouth.

**Long-term planning:** planning that looks beyond immediate costs and benefits by exploring impacts in the future.

**Longshore drift:** the movement of material along a coast by breaking waves.

**Lower course:** that part of a river system that is close to the mouth of the river.

**Magnitude:** the size of something.

**Malthusian theory:** the view that population growth is the main reason why a society would collapse.

**Mangrove swamp:** a tidal swamp dominated by mangrove trees and shrubs that can survive in the salty and muddy conditions found along tropical coastlines.

# M-P

**Marine ecosystem:** the web of organisms that live in the ocean or a part of an ocean.

**Maritime:** a coastal environment or climate that lacks extremes of temperature, and experiences higher rainfall, when compared with land-locked areas at a similar latitude (distance from the Equator).

**Market economy:** a system for businesses, industry and people based on free trade which is influenced by supply and demand (with little or no government intervention).

**Mass movement:** the downslope movement, by gravity, of soil and/or rock by the processes of slumping, falling, sliding and flowing.

**Meanders:** the bends formed in a river as it winds across the landscape.

**Megafauna:** very large mammals, such as those that lived during the last ice age.

**Micro-finance:** the provision of financial help (mainly capital) to small businesses and private enterprises which do not have access to banking services.

**Micro-hydro schemes:** small-scale HEP systems that generate electricity locally.

**Mid-course:** the central section of a river's course.

**'Middle-Income':** a World Bank category for countries that are not very rich ('High-Income') nor very poor ('Low-Income').

**Migration:** the process of people changing their place of residence, either within or between countries.

**Millennium Development Goals (MDGs):** the development goals agreed by world governments at the UN summit in September 2000.

**Natural causes:** those processes and forces that are not controlled by humans.

**Natural change:** the change (an increase or a decrease) in population numbers resulting from the difference between the birth and death rates over one year.

**Natural increase:** the difference between birth rate and death rate.

**Natural resources:** those materials found in the natural world that are useful to man, and that we have the technology and willingness to use.

**Net in-migration:** the increase in a country's population as a result of more people arriving than leaving.

**Network:** a system of linkages between objects, places or individuals.

**Newly industrialised countries (NICs):** countries which experienced rapid economic growth during the second half of the twentieth century as a result of industrialisation, as for example Malaysia, Hong Kong (now part of China), Taiwan and Singapore.

**New economy:** the emergence of new types of economic activity and employment in the last few decades.

**Nomadic pastoralism:** a type of farming where farmers have no permanent land and migrate with their cattle, etc. from one place to another.

**Non-renewable resource:** those resources – like coal or oil – that cannot be 'remade', because it would take millions of years for them to form again.

**Nutrient cycle:** a set of processes whereby organisms extract minerals necessary for growth from soil or water, before passing them on through the food chain – and ultimately back to the soil and water.

**Oceanic crust:** the part of the crust dominated by denser basaltic rocks.

**OPEC:** Organisation of the Petroleum Exporting Countries.

**Orbital changes:** changes in the pathway of the Earth around the Sun and in its axial geometry.

**Organic agriculture:** farming systems that use no artificial chemicals.

**Orographic rainfall:** precipitation formed when an air mass rises above a relief obstacle (mountains) leading to cooling and the condensation of water vapour.

**Outsourcing:** a process in which a company subcontracts part of its business to another company.

**Over-abstraction:** when water is being used more quickly than it is being replaced.

**Overfishing:** taking too many fish (or other organisms) from the water before they have had time to reproduce and replenish stocks for the next generation.

**Overpopulation:** a situation where the population of an area cannot be fully supported by the available resources. The symptoms include a low (even declining) standard of living, overcrowding and high unemployment.

**Ox-bow lake:** an arc-shaped lake which has been cut off from a meandering river.

**Periphery:** the outer limits or edge of an area, often remote or isolated from its core.

**Permaculture:** an intensive form of food production that is both high-yielding and sustainable because it is based on natural ecological processes.

**Permafrost:** permanently frozen ground, found in polar (glacial and tundra) regions.

**Permeable:** allowing water to pass through.

**Plate margin:** the boundary between two tectonic plates.

**Players:** individuals and groups who are interested in and affected by a decision-making process.

**Polar:** relating to the North or South Pole. In polar regions the land is covered with ice (glacial) or frozen (tundra).

**Polar continental:** an air mass whose source region is an area of land in cold, northern latitudes (Siberia) and which may move westwards, bringing cold, dry conditions to Europe in winter.

**Political freedom (index):** a measure calculated by several organisations (e.g. Freedom House) that looks at the level of democracy in each country.

**Pollution:** the presence of chemicals, noise, dirt or other substances which have harmful or poisonous effects on an environment.

**Population pyramid:** a diagrammatic way of showing the age and sex structure of a population.

**Population structure:** the composition of a population, usually in terms of its age and gender.

**Pores:** small air spaces found in a rock or other material that can also be filled with water.

**Post-industrial stage:** that period in the development of a society when manufacturing industry declines in importance, and is replaced by other forms of employment.

# P-S

**Poverty:** a state of shortage of money and goods, usually measured in terms of average wealth and income in a society.

**Poverty cycle:** a set of processes that maintain a group or society in poverty.

**Pre-industrial stage:** that period in the development of a society when manufacturing industry has yet to develop.

**Precipitation:** when moisture falls from the atmosphere – as rain, hail, sleet or snow.

**Prediction:** forecasting future changes.

**Preparation:** the process of getting ready for an event.

**Preserve:** maintain (something) in its existing state.

**Primary employment:** working in the primary sector – extracting and exploiting raw materials.

**Primary sector:** the economic activities that involve the working of natural resources – agriculture, fishing, forestry, mining and quarrying.

**Production chain:** the sequence of activities needed to turn raw materials into a finished product.

**Pro-natalist:** policies that encourage human reproduction and population growth.

**Pull factor:** something that attracts people to a location.

**Push factor:** something that make people wish to leave a location.

**Quality of life:** the degree of well-being (physical and psychological) felt by an individual or a group of people in a particular area. This can relate to their jobs, wages, food, amenities in their homes, and the services they have access to, such as schools, doctors and hospitals.

**Quaternary Period:** the most recent major geological period of Earth's history, consisting of the Pleistocene and the Holocene.

**Quaternary sector:** the economic activities that provide intellectual services – information gathering and processing, universities, and research and development.

**Ramsar:** The Ramsar Convention on Wetlands is an intergovernmental treaty for the conservation and wise use of wetlands.

**Rebranding:** the creation of a new name, term, or design, for an existing organisation or place.

**Redevelopment:** development that aims to stimulate growth in areas that have experienced decline.

**Regeneration:** growth in areas that have experienced decline in the past.

**Regulated flow:** the steady movement of water through a drainage basin that will not bring flash flooding.

**Renewable resource:** resources, such as forests, that can be maintained by management.

**Remote countryside/rural area:** rural areas that are distant from and thus little affected by urban areas and their populations.

**Response:** the way in which people react to a situation.

**Retirement communities:** settlements where most of the residents are retired.

**River cliff:** steep outer edge of a meander where erosion is at its maximum.

**River pollution:** the emission of harmful or poisonous substances into river water (or their presence in the river).

**Run-off:** water that flows directly over the land towards rivers or the sea after heavy rainfall.

**Rural depopulation:** the decline of population in rural areas and regions.

**Rural idyll:** the common perception that rural areas are quiet and attractive – and therefore good places to live in.

**Rural–urban migration:** the movement of people from the countryside into towns and cities.

**Sea-level rise:** the increase in the level of the sea, relative to the land.

**Seasonality:** marked differences in temperatures and/or precipitation occurring during different seasons of the year.

**Secondary sector:** the economic activities that involve making things, either by manufacturing (TV, car, etc.) or construction (a house, road, etc.). The sector also includes public utilities, such as producing electricity and gas.

**Sediment:** usually sand, mud or pebbles deposited by a river.

**Services:** those things that are provided, bought and sold that are not tangible.

**Short-term emergency relief:** help and aid provided to an area to prevent immediate loss of life because of shortages of basics, such as water, food and shelter.

**Siltation:** the deposition of silt (sediment) in rivers and harbours.

**Slip-off slope:** inner gentle slope of a meander where deposition takes place.

**Socialist:** describes a political approach where a government takes control of businesses, industries, and infrastructure for the benefit of the community as a whole.

**Socio-political development:** the progress made by a country in terms of improving the lives of people, and also establishing ways in which people can participate in decision-making.

**Soft rock coast:** a coastal area made up of easily eroded materials.

**Soils:** the weathered remains of rock (sand, silt and clay) to which decayed organic matter (such as the remains of leaves) has been added.

**Solar output:** the energy emitted by the Sun.

**Solifluction:** the movement downhill of soggy soil when the ground layer beneath is frozen. It often occurs in tundra regions.

**Spit:** material deposited by the sea which grows across a bay or the mouth of a river.

**Stack:** a detached column of rock located just off-shore.

**Stakeholder:** a person, group or organisation that has a direct or indirect interest in the outcomes of a particular development or decision. Stakeholders can either influence the outcomes or be affected by them.

**Strategic realignment:** the reorganisation of coastal defences that is often part of managed retreat.

**Stump:** a stack that has collapsed, leaving a small area of rock above sea-level.

**Sub-aerial processes:** weathering and mass movement.

**Subsistence farming:** a type of agriculture producing food and materials for the benefit only of the farmer and his family.

**Suburbanisation:** the outward spread of the built-up area, often at lower densities compared with the older parts of a town or city.

# S-Z

**Superpower countries:** the world's most powerful and influential nations – the USA and, increasingly, China and India.

**Supply chain:** a sequence of steps or stages involved in moving a product or service from the supplier to the customer. In the case of a product, the chain may involve the processing of raw materials into components, and the assembly of components into finished products. Such a sequence is often referred to as a 'production chain'.

**Sustainability:** the ability to keep something (such as the quality of life) going at the same rate or level. From this stems the idea that the current generation of people should not damage the environment in ways that will threaten future generations' environment (or quality of life).

**Sustainable development:** development that meets the needs of the present without compromising (limiting) the ability of future generations to meet their own needs.

**Sustainable resources:** resources – such as wood – that can be renewed if we act to replace them as we use them.

**Swash:** the forward movement of water up a beach after a wave has broken.

**Sweatshop:** a place of work where very poorly paid employees work long hours in unsatisfactory and often unsafe conditions.

**Tectonic hazards:** threats posed by earthquakes, volcanoes and other events triggered by crustal processes.

**Telecottaging:** working from a home in the country, using computer communication.

**Teleworking:** any form of work in which telecommunications replace work-related travel (commuting).

**Temperate climate:** a climate that is not extreme (in terms of heat, cold, dryness or wetness).

**Tertiary sector:** the economic activities that provide various services – commercial (shops and banks), professional (solicitors and dentists), social (schools and hospitals), entertainment (restaurawnts and cinemas) and personal (hairdressers and fitness trainers).

**Thermal growing season:** the number of months of the year when it is warm enough for crops to grow (6 °C or above).

**Throughflow:** water that flows slowly through the soil until it reaches a river.

**Tipping point:** the point at which the momentum of a change becomes unstoppable.

**Top-down projects:** projects set up and organised by governments, often with little consultation with local communities.

**Transnational company/corporation (TNC):** a large company operating in several countries.

**Tropical continental:** an air mass whose source region is an area of land in the tropics (north Africa) and which may move northwards, bringing hot, dry conditions to Europe in summer.

**Tundra:** the flat, treeless Arctic regions of Europe, Asia and North America, where the ground is permanently frozen.

**Underpopulation:** a situation where the resources of an area could support a larger population without any lowering of the standard of living or where a population is too small to develop its resources effectively.

**Unsustainable:** unable to be kept going at the same rate or level.

**Upper course:** the source area of a river, often in an upland or mountainous region.

**Urban development corporations:** organisations set up by central government to coordinate rapid improvements in derelict urban areas. Their aims were to improve the environment, to give cash grants to attract firms, to renovate buildings and to give advice to firms thinking of moving in.

**Urban fringe:** the countryside adjacent to or surrounding an urban area.

**Urban sprawl:** urban growth, usually weakly controlled, into surrounding rural and semi-rural areas.

**Urbanisation:** the development and growth of towns or cities.

**Volcanic activity:** the escape of molten rock, ash and gases from an opening in the Earth's surface (or when there is evidence that it is imminent).

**Water flow:** movement processes of the Earth's water, including evaporation, precipitation and overland flow.

**Water harvesting:** storing rainwater or used water ('grey water') for use in periods of drought.

**Water insecurity:** when safe water availability is insufficient to ensure the population of an area enjoys good health, livelihood and earnings. The condition can be caused by water insufficiency or poor water quality.

**Water insufficiency:** a lack of adequate water supplies needed to meet a society's economic and social needs.

**Water management schemes:** programmes to control rivers, generally organised by local or central government.

**Water store:** a build-up of water that has collected on or below the ground, or in the atmosphere.

**Water table:** the level in the soil or bedrock below which water is usually present.

**Water transfers:** movements of water and water vapour through the biosphere, lithosphere and atmosphere.

**Waterfall:** sudden descent of a river or stream over a vertical or very steep slope in its bed.

**Water unreliability:** when rainfall and/or river flows vary from season to season, sometimes unpredictably, resulting in periods of water scarcity.

**Weathering:** the breakdown and decay of rock by natural processes, without the involvement of any moving forces.

**Wilderness:** uncultivated, uninhabited and inhospitable regions.

**World cities:** the leading cities of the world, such as London, New York and Tokyo; major centres in the economic networks being produced by globalisation. They are major centres of finance, business and political influence, and are home to the headquarters of many TNCs.

**Youthful population:** a population in which there is a high percentage of people under the age of 16 (or sometimes 18).

**Zero population growth:** when natural change and migration change cancel each other out, and there is no change in the total population.

# Index
## A-C

# C–E

# E-H

# H-N

# N-R

# R-U

## V-Z

Acknowledgements continued from page 2.

**We are grateful to the following for permission to reproduce copyright material:**

**Figures**

Figures 1.1 from "This dynamic earth"; Figure 1.2 'The convectional currents in the mantle'; Figure 1.6 from 'Figure 10i-6: Collision of a oceanic plate with a continental plate'; and Figure 1.8 from 'Figure 10i-7: Collision of two continental plates', Source: U.S. Geological Survey; Figure 1.11 from http://earth.rice.edu/MTPE/geo/geosphere/hot/volcanoes/volcanoes_map.gif, copyright © Smithsonian Institution, Global Volcanism Program, 2012; Figures 1.12 from "The Hawaiian chain of islands"; and Figure 1.14 from "A cut-away view of a composite volcano", Source: U.S. Geological Survey; Figure 1.15, "Volcanic Explosivity Index volume graph", 2005, Wikipedia, http://commons.wikimedia.org/wiki/File:VEIfigure.jpg, granted under the GNU Free Documentation License (GFDL); Figure 1.19 'The San Francisco seismic net on-line', Source: U.S. Geological Survey; Figure 2.1 from "Changes in the Earth's average temperature during the last million years", copyright © 2012 Schlumberger Excellence in Educational Development, Inc. All rights reserved. For more information, visit our Web site, at www.planetseed.com; Figure 2.2 from Forestry Commission, http://www.forestry.gov.uk/images/emissionspiechart.jpg/$File/emissionspiechart.jpg, Crown Copyright material is reproduced with permission under the terms of the Click-Use License; Figure 2.6 'A map showing the size of countries in proportion to how much CO2 they emit' from http://www.worldmapper.org/, copyright © SASI Group (University of Sheffield) and Mark Newman (University of Michigan); Figure 2.8 adapted from UKCIP 09: 'The climate of the UK and recent trends', http://www.ukcip.org.uk, copyright © Met Office 2007; Figure 3.1 from *Elements of Ecology*, 6th ed. (Smith, T.M. and Smith, R.L.) Figures 23.4, 23.12, 23.19, pp.500, 506, 511, copyright © 2006. Printed and Electronically reproduced by permission of Pearson Education, Inc., Upper Saddle River, New Jersey; Figure 3.1 from *Ecology of World Vegetation* by O.W. Archibold, copyright © 1995 Chapman Hall. Used by kind permission from Springer Science+Business Media B.V.; Figure 3.8 adapted from National Geographic, http://ngm.nationalgeographic.com/2007/01/amazon-rain-forest/amazon-map-interactive, copyright © National Geographic Image Collection; Figure 4.9 'The Colorado River regime before and after construction of the Glen Canyon Dam', U.S. Geological Survey; Figure 5.4 adapted from "Landforms produced by the erosion of a headland", http://www.georesources.co.uk/sea6.gif, copyright © David Rayner (Georesources); Figure 5.5 'The Swanage coast' copyright © www.collinsbartholomew.com Ltd. Reproduced with kind permission of HarperCollins Publishers; Figures 5.7 'Destructive Waves' and Figure 5.8 'Constructive Waves', adapted from http://cgz.e2bn.net/e2bn/leas/c99/schools/cgz/accounts/staff/rchambers/GeoBytes%20GCSE%20Blog%20Resources/Images/Coasts/Constructive_Waves.jpg, copyright © Rob Chambers; Figure 5.10 from "The process of Longshore drift", http://geographyfieldwork.com/LongshoreDrift.htm. Reproduced by permission of Barcelona Field Studies Centre http://geographyfieldwork.com; Figure 5.12 from http://www.le.ac.uk/bl/gat/virtualfc/217/images/ley2.jpg, copyright © Ted Gaten; Figure 5.18 adapted from *Geography in Action Book 2*, Heinemann (ed Andy Owen, 1995) p18, copyright © Pearson Education Limited;

Figure 6.2 from *Geography and Change*, Hodder Arnold (Flint, D., Flint, C. and Punnett, N. 1996); Figure 6.6 'Formation of an ox-bow lake' adapted from http://cgz.e2bn.net/e2bn/leas/c99/schools/cgz/accounts/staff/rchambers/GeoBytes%20GCSE%20Blog%20Resources/Images/Rivers/ox-bow_lake.gif, copyright © Rob Chambers; Figure 6.9 from 'River Severn Hydrograph July 2007' Bewdley Case Study, www.geography.org.uk, copyright © Geographical Association and the Environment Agency; Figure 6.15 from www.swenvo.org.uk, copyright © Environment Agency, copyright © Environment Agency; Figure 7.4 adapted from "Aquatic food chain" *Encyclopaedia Britannica*, copyright © 2006 by Encyclopaedia Britannica, Inc. Reprinted with permission; Figure 7.5 from "De-oxygenated "dead spots": areas of water where eutrophication has occurred on a large scale" UNEP, United Nations Environment Programme, www.unep.org, copyright © UNEP; Figure 7.14 from The Pacific Garbage Patch – areas of "rubbish soup Whale Harvests in the Antarctic Ocean from the 1904 to 1981" http://www.env.go.jp/en/wpaper/1994/eae230000000035.html, Source: The International Whaling Commission, 2008 compiled by Greg Donovan; Figure 7.15 "Out of sight, out of mind" by John Papastan and John Bradley, copyright © Greenpeace; Figure 8.10 adapted from "Native Peoples and Languages of Alaska Map" by Krauss, Michael E. 1974. Revised 1982, http://www.uaf.edu/anlc/, Fairbanks, Alaska. Reproduced by permission of Alaska Native Language Center; Figure 9.1 from US Census Bureau, International Data Base, http://www.census.gov/ipc/prod/wp02/wp-02003.pdf, Source: US Census Bureau; Figures 9.4 adapted from *Population and Migration*, Philip Allan Updates (Witherick, M., 2004) Figure 10; and Figure 9.5 adapted from *AQA AS Geography*, Philip Allan Updates (Barker, A. et al, 2008) Figure 5.5. Reproduced with permission of Philip Allan Updates; Figure 9.6 from 'Population pyramid for China, 2005' Source: US Census Bureau, International Data Base; Figure 9.7 from 'Population pyramids for Indonesia, Mexico and the UK, 2005', Source: US Census Bureau, International Data Base; Figure 9.8 from *Population and Migration* Philip Allan Updates (Witherick, M., 2004) Figure 32. Reproduced by permission of Philip Allan Updates; Figure 9.13 from 'UK residents born abroad (1951 – 2001)', Office for National Statistics, Crown copyright © 2002, Crown Copyright material is reproduced with permission under the terms of the Click-Use License; Figure 9.16 adapted from "Economic migrants from Eastern Europe", (Witherick, M.) Figure 2, *Geo Factsheet*, 184 (September 2005), copyright © Curriculum Press; Figure 9.18 from *Edexcel AS Geography*, Figure 12.4, Philip Allan Updates (Warn, S. et al, 2008). Reproduced by permission of Philip Allan Updates;

## Acknowledgements

Figure 10.1 from "New Reserves: As Prices Surge, Oil Giants Turn Sludge Into Gold - Total Leads Push in Canada To Process Tar-Like Sand; Toxic Lakes and More CO2 - Digging It Up, Steaming It Out", *The Wall Street Journal*, 27 March 2006 (Gold, R.), copyright © 2006, Dow Jones and Company, Inc; Figure 10.3 from 'The World's leading diamond producing nations', http://www.pangeadiamondfields.com/diamond-sector.htm, copyright © WWW International Diamond Consultants; Figure 10.4 'World wealth distribution in 2002 by country' from www.worldmapper.org/ copyright © SASI Group (University of Sheffield) and Mark Newman (University of Michigan; Figures 10.6, 10.7 from BP Statistical Review of World Energy 2007 www.bp.com, copyright © BP plc; Figure 10.11 from *Beyond the Limits: Confronting Global Collapse, Envisioning a Sustainable Future* (Meadows, et al 1992), copyright © Donella Meadows, with permission of Chelsea Green Publishing (www.chelseagreen.com); Figure 11.11 from *Global Challenge: A2 Level Geography for Edexcel B*, Longman (McNaught, A. and Witherick, M., 2001) Figure 4.21, copyright © Pearson Education Limited; Figure 11.12 from *Edexcel AS Geography*, Philip Allan Updates (Warn, S. et al, 2008). Reproduced by permission of Philip Allan Updates; Figure 12.5 'Worldmapper map showing Human Development Index increases from 1975 to 2002' from http://www.worldmapper.org/, copyright © SASI Group (University of Sheffield) and Mark Newman (University of Michigan); Figure 12.7 adapted from "Tax Regions Map", http://www.tra.go.tz/regions.htm, copyright © Tanzania Revenue Authority; Figures 14.9 from "Complimentary Set Changing Environments for as" from *Changing Environments*, Longman (Warn & Naish, 2000) Figure 1.8; and Figures 15.6 and 15.7 adapted from *Understanding GCSE Geography AQA*, Heinemann (Ann Bowen and John Pallister, 2009) pp.156, 157 copyright © Pearson Education Limited; Figure 15.12 adapted from "The eco-footprints of different cities and countries", from *Living Planet Report 2006* http://assets.panda.org/downloads/living_planet_report.pdf copyright © WWF; Figure 15.13 adapted from "How York's eco-footprint is made up" The Eco Footprint of York, York lifestyles and their environmental impact, www.york.ac.uk, copyright © University of York; and Figure 16.9 adapted from "The Polycentric Metropolis: Learning from Mega-City Regions in Europe" by Professor Peter Hall and Dr Kathy Pain, (2006) London: Earthscan, www.lboro.ac.uk/gawc/rb/images/rb225f3.jpg, granted with permission.

### Text
Extract on page 96 from http://www.environment-agency.gov.uk/, 2009 copyright © Environment Agency;

### Tables
Table on page 193 'Regions of Brazil: GDP per capita in R$ (Brazilian reals) 2005', Source: IBGE, 2007; Table on page 206 "Changing industrial structure in the UK, 1964–2011" (% of total employment), ONS (2012) Labour Market Statistics, September UK National Accounts, Source: Office for National Statistics licensed under the Open Government Licence v.1.0; Table on page 217 "Foreign migrants in UK industries, 2011 (workforce share, recent migrants), Labour Force Survey 2011, Source: Office for National Statistics licensed under the Open Government Licence v.1.0.

In some instances we have been unable to trace the owners of copyright material, and we would appreciate any information that would enable us to do so.